EVOLUTIONARY THEORY: PATHS INTO THE FUTURE

EVOLUTIONARY THEORY: PATHS INTO THE FUTURE

Edited by

JEFFREY W. POLLARD

MRC Group in Human Genetic Diseases, Department of Biochemistry, Queen Elizabeth College, University of London

A Wiley–Interscience Publication

JOHN WILEY & SONS

Chichester · New York · Brisbane · Toronto · Singapore

Library of Congress Cataloging in Publication Data:

Main entry under title:

Evolutionary theory.

'A Wiley–Interscience publication.'
Includes bibliographies and indexes.
1. Evolution. I. Pollard, Jeffrey W.
QH366.2.E8726 1984 575.01 83-21606

ISBN 0 471 90026 5 (U.S.)

British Library Cataloguing in Publication Data:

Pollard, Jeffrey W.
Evolutionary theory.
1. Evolution
I. Title
575.01 QH366.2

ISBN 0 471 90026 5

Printed in Great Britain by Galliard (Printers) Ltd, Great Yarmouth

List of contributors

Daniel R. Brooks,
Department of Zoology,
University of British Columbia,
Vancouver,
British Columbia V6T 2A9,
Canada

Christopher A. Cullis,
John Innes Institute,
Colney Lane,
Norwich,
Norfolk NR4 7UH,
UK

William Engels,
Department of Genetics,
University of Wisconsin,
Madison,
Wisconsin 53706,
USA

Brian C. Goodwin,
School of Biological Sciences,
University of Sussex,
Falmer,
Brighton,
Sussex
BN1 9QG,
UK

Reginald M. Gorczynski,
Ontario Cancer Institute,
500 Sherbourne Street,
Toronto,
Ontario M4X 1K9,
Canada

Mae-Wan Ho,
Biology Discipline,
Open University,
Milton Keynes,
Buckinghamshire MK7 6AA,
UK

Philippe Janvier,
Centre National de la Recherche Scientifique,
ERA 963,
Laboratoire de Paléontologie des Vertébrés,
Université Paris VI,
4 place Jussieu,
75230 Paris Cedex 05,
France

David Penny,
Department of Botany and Zoology,
Massey University,
Palmerston North,
New Zealand

Jeffrey W. Pollard,
MRC Group in Human Genetic Diseases,
Department of Biochemistry,
Queen Elizabeth College,
University of London,
Campden Hill,
London W8 7AH,
UK

Karl R. Popper,
Emeritus Professor of Logic and Scientific Method,
London School of Economics and Political Science,
University of London,
Houghton Street,
London WC2A 2AE,
UK

Anna Riddiford,
Department of Botany and Zoology,
Massey University,
Palmerston North,
New Zealand

Donn E. Rosen,
Department of Ichthyology,
The American Museum of Natural History,
Central Park West at 79th Street,
New York,
New York 10024,
USA

Peter T. Saunders,
Department of Mathematics,
Queen Elizabeth College,
University of London,
Campden Hill Road,
London W8 7AH,
UK

Edward J. Steele,
Department of Immunology,
The John Curtin School of Medical Research,
Australian National University,
Canberra,
Australia 2601

Howard M. Temin,
Department of Oncology,
McArdle Laboratory,
University of Wisconsin,
Madison,
Wisconsin 53706,
USA

Ed O. Wiley,
Museum of Natural History and Department of Systematics and Ecology,
University of Kansas,
Lawrence,
Kansas 66045,
USA

Contents

Preface

The exact moment that my interest in evolution was re-awakened can be identified with some precision. My initial education had been as a zoologist and I had come to accept the conventional neo-Darwinian theory as a complete explanation for evolution. I had, however, read Popper and consequently I was somewhat disquieted that this theory was being accepted *a priori* and therefore no longer being subjected to critical tests to attempt to falsify it, and as such in a Popperian sense, it was ceasing to be a scientific theory. It was during a conversation when I was expressing these ideas that I found out that Ted Steele, who was also a post-doctoral fellow at the Ontario Cancer Institute in Toronto, held similar views. Later in that evening in the fall of 1978 he described to me his somatic selection hypothesis and it was this conversation that re-awakened my interest in evolution. During that winter, this and following discussions led me to re-examine my whole understanding of the evolutionary process and eventually to break through the conceptual barrier provided by the synthetic theory and to demand a new and better theory of evolution. Consequently, it was with some pleasure, and not a little trepidation, that I accepted an invitation from the Life Sciences Editor at John Wiley in 1982 to edit a book that would look to the future and to try to delineate some of the paths that evolutionary thought might take during the next decade. The resultant volume is, because of the enormous scope of the subject and the freewill of the authors, but a drop in this ocean. However, I hope the reader will discover in this book the need to break through the framework of hypotheses that formulate the synthetic theory and find some of the considerations necessary for a new and roomier framework of a more adequate theory of evolution.

Prefaces are also the place for thanks and there are many. In particular, it was the richness of Popper's ideas that were the intellectual starting point for this book and it has given me considerable pleasure that he agreed to contribute a chapter to this volume. I would like to single out Dr Ted Steele for thanks for the many stimulating discussions over many years, not only about evolution. My gratitude also goes to Professor Garth Chapman, Dr Brian Gardiner, and Francis Hitchin for encouragement in the early stages of this project. I would also like to thank Liz Moor for patience, under a heavy work load, in typing all the innumerable letters. Finally, and not least, to my research group for putting up with me while I should have been talking about protein synthesis but instead would go on about cladistics and things.

JEFFREY W. POLLARD

Introduction: Paths into the future

The neo-Darwinian or synthetic theory of evolution states that the random accumulation of small genetic changes in the germline genes guided by natural selection is both necessary and sufficient to explain evolution. And further, that trans-specific evolution is nothing but an extrapolation and magnification of such events that take place within populations of a species (Mayr, 1963). This theory as a complete description of evolution although often enjoying the status of *a priori* knowledge, has recently come under intense criticism from a variety of sources. The extent of this criticism is such that Gould (1980) has stated:

> 'That theory (the synthetic theory) as a general proposition, is effectively dead, despite its persistence as text book orthodoxy' (Gould, 1980, p. 120).

In the same article Gould questions if a new and more general theory of evolution is emerging. My view is that one is necessary but that we have a considerable distance to travel both conceptually and experimentally before one will be forthcoming. The purpose of this volume, therefore is not to continuously attack neo-Darwinism, this has been done in many places before, but to indicate some of the considerations and constraints that a new theory must have.

The synthetic theory is an amalgamation of several distinguishable theories. Riddiford and Penny (Chapter 1) have divided it into three sub-hypotheses:

1. The theory of descent—that is that complex organisms have been derived from simpler ones; or that evolution has occurred.
2. Microevolution—that is, that the population produces more descendents than will survive and that some of the variability can alter the probability of survival of individuals.
3. That the mechanisms acting in (2) are sufficient to account for (1).

This sub-division clears up the confusion caused by mixing the concepts of evolution (hypothesis 1) and the mechanism which causes evolution. This confusion is often generated on purpose by such groups as the creationists who seek to use the logically flawed device of 'disproving' evolution by attempting to show that our view of the mechanism of evolution is incorrect. This confusion has also been increased by mis-interpretation of Popper's view (1976) that evolutionary theories were historical statements and therefore outside the bounds of science and were at best metaphysical research programmes. This has been discussed by Riddiford and Penny who point out that with the development of newer analytical techniques at least most evolutionary predictions are, in fact, testable. In particular, they point to the data that show the congruence of evolutionary trees derived from different protein sequences and conclude that because of the closeness of trees the theory of the existence of an evolutionary tree has been rigously tested in a Popperian sense. This is so because the concept of hierarchy would be sorely tested if there was no congruence between data derived from morphological observations and DNA and protein sequence studies. Thus, it is not the existence of the hierarchy that needs to be debated but its pattern and more importantly how this pattern is brought about. Consequently it is in this area that we need to develop new theories.

The synthetic theory's explanation of the mechanism of evolutionary change again has a number of readily distinguishable and often confused sub-hypotheses. The main ones are:

1. Heredity—that is that the offspring reproduce the parents genotype with a high but not complete degree of fidelity.
2. Variation—small accidental and hereditary mutations are present prior to selection.
3. Natural selection—the whole of the hereditary material is controlled by elimination of what is not adaptive.
4. Weismann's doctrine—change only occurs in the germlines genes which are inviolate from somatic or environmental influences.

We need not argue here about hereditary since one of the achievements of molecular genetics is the description of DNA, its mechanism of replication, and the demonstration that gene sequences exhibit variation within a species. Nor do we need to argue that natural selection may occur. We know this from Kettlewell's moths, mutant reconstruction experiments, and the Luria–Delbruck experiments. But what must be argued about is the central contention of the synthetic theory that small variations in the germline genes gradually selected by natural selection (or accumulated by chance if we include the neutralists view) is *all* and *sufficient* to explain *all* the variety of organisms on this planet. In other words, that the accumulations of micromutations in populations, whose behaviour we can describe with some precision, is sufficient to produce the entire observable hierarchy by a combination of selection (or chance fixation) and reproductive isolation. It is this contention that is under fire and that lead Gould to state that the synthetic theory is effectively dead. In fact as long ago as 1943 Goldschmidt pointed out the logical flaw in equating population genetics with the transmutation of species. Thus it is the mechanism of producing the hierarchy that the remainder of the volume seeks to investigate.

Initially, however, the pattern that needs to be explained must be correctly described. As this of necessity encompasses palaeontology this will also tell us something about the tempo of change and since the synthetic theory demands a gradual rate of change in the fossil record any deviation from this will tend to invalidate this theory. In fact, Popper (1976) has pointed out that gradualism is one of the major predictions of the synthetic theory. Many palaeontologists, however, have described a pattern of punctuated equilibrium in the fossil record; that is, a period of rapid change followed by long periods of stasis (Eldredge and Gould, 1972; Stanley, 1979). Consequently these workers have been unable to observe the gradual change from one species to another demanded by the synthetic theory (Gould, 1980). This view has of course, been challenged from traditional sources with much of the argument revolving around a debate over the methods of classification. To this debate cladistics is centre stage. Janvier (Chapter 2) has written a lucid explanation of the principles of cladistic analysis for the non-specialist and he has demarcated the limits of its reliability as a method of ascertaining hierarchies. He also sets the record straight on the position of the so-called transformed cladists with respect to evolution. Nevertheless Janvier has stated:

> 'The processes inferred from fossils must not be taken for more than what they are, that is models applied to occasional cases where the fossils seem to be more complete than usual. I fear that such hypotheses about process (evolutionary process) still remain beyond the limits of reliability of palaeontology—the prospect in experimental biology looks more promising in the search for evolutionary theories' (Janvier, Chapter 2, p. 65).

Similar conclusions are outlined in Rosen's contribution (Chapter 3). He argues for the inclusion of ontogeny in the construction of evolutionary hierarchies by suggesting that von Baer's law of ontogeny recapitulating phylogeny is correct, at least to the extent that it shows the ancestral character. The notion that understanding ontogenic transformations will enrich our understanding of evolution is discussed in more detail by Goodwin (Chapter 4). He suggests that evolution proceeds by a series of generative transformations which specify the potentiality for a basic form upon which innumerable variations may arise. To illustrate this concept he concentrates on the pentadactyl limb demonstrating that each limb form is a different solution to the same field equation. The concept of a common ancestor for each limb type is therefore artificial and consequently transitional forms would not be expected in the fossil record. Thus different morphologies will be discretely separated from one another and species will be distinct forms *ab initio*. This quantal nature of species formation implies, in contrast to the synthetic theory, that natural selection has no role in the creation of new species but only has a secondary role in eliminating forms which do not correspond to a particular environment. Natural selection can therefore be considered as a filter through which new species pass. Ontogenic transformations are not arbitrary, however, but they are constrained and

Goodwin argues that it is these constraints that we should be interested in if we are to understand evolution.

In a similar vein Saunders and Ho (Chapter 5) have discussed the increase in complexity that is perceived in evolution. They enunciate the principle of the minimum increase in complexity. That is, that changes occur in patterns which are constrained to those that are most easily brought about. Similar biological solutions therefore, should be expected in widely differing groups. This is clearly the case for segmentation and may perhaps be an explanation for the phenomenon of parallel evolution. Such developmental constraints to the possibilities of change are best understood within the concept of bauplans of the type discussed by Goodwin.

Emergence of new forms is the essence of speciation but rather in the terms described above, than those postulated by the synthetic theory. Clearly one would expect allopatric speciation to be the most common mode (as discussed by Gould (1982) and Bush (1982)) and I would suggest that there are two forms of change: macroevolution, generating completely new potentialities, and micro-evolution, perhaps even of the kind described by the synthetic theory, honing up the fit between environment and form. Relatively therefore, it is the macro-evolutionary event that is important in evolution because it will produce rapid, saltatory speciation. Brooks and Wiley (Chapter 6) have discussed models of speciation taking into account developmental constraints and have developed a non-equilibrium theory of evolution in terms of entropic phenomena. They claim to have developed a new evolutionary theory which predicts why organisms are related in a hierarchial manner and why allopatric speciation is expected as the most common mode. They have also attempted to approach the problem posed by Popper of how to explain singular historical events within the content of more general laws by suggesting that these laws are those of non-equilibrium thermodynamics. They also propose that the theory shows not only that characters should be good predictors of history, as was discussed by Janvier, but also that developmental biology should provide data bearing on the validity of the theory and provides a rationale for ontogeny recapitulating phylogeny. Thus as these workers state:

> 'If ontogeny and phylogeny are linked through non-equilibrium evolution we should expect that phylogenetic analysis applied to various cell types in a specialized lineage would provide a summary fate map indicating the various pathways of development' (Brooks and Wiley, Chapter 6, p. 164).

In many of the papers, therefore, the view is developed that in order to understand evolution we must understand development mechanics. Unfortunately our molecular understanding of development is scanty and certainly we are far from Rosen's demand of a theory of process that is deterministically tied to the evolutionary pattern it seeks to explain. But our knowledge of how the genome may change to produce new developmental patterns is rapidly advancing. This has led to the overthrow of the concept of a fixed, rigid, particulate genome and replaced it with the view of DNA as a metabolic molecule constantly in flux, moving, enlarging, and turning over. This concept of course, fits nicely with the idea of evolution produced by macromutational events giving rise to punctual speciation.

The fluid genome is best characterized by the movable genetic elements which are capable of transposing from one site in the genome to another. These elements reviewed by Temin and Engels (Chapter 7) are thought to be represented throughout the plant and animal kingdom and may represent a major fraction of the genome (10% to 20% in *Drosophila*). The intra-genomic elements are also structurally related to the retroviruses and Temin and Engels have suggested repeated cycles of evolution between viruses and transposable elements, thus implying a close relationship between genomic evolution and the evolution of its parasites.

The function of any of the transposable elements remains a mystery. It is, however, attractive to consider that they may play a role in the regulation of differentiation. Examples of this so far described may include the mating strain switching in yeast and trypanosome antigen variation. Similarly sequences related to transposable elements have been described which are confined to particular gene families expressed in specific tissues. Sutcliffe *et al.* (1982) have suggested that these sequences are identifier sequences for gene expression. Unequivocable roles in the regulation of gene batteries, however, have yet to be demonstrated and thus enthusiasm for these molecules to be the missing regulatory molecules must be tempered for the time being (but see Davidson and Posakony, 1982). What is certain is that transposable elements cause mutations (up to 90% at certain loci) by insertion both within and without the gene, produce complex genomic rearrangements of up to 10% per chromosome per generation and also inactivate areas of the genome.

Rapid chromosomal rearrangements and changes in genome size may also be mediated by transposable elements. Such processes may result in reproductive isolation by either developmental or genomic incompatibility. Thus for example, if transposable elements mediate the re-setting of developmental pathways or the expression of latent characters by transposition to a new regulatory site then a rapid bifurcation could occur in temporarily isolated species. Similarly the acquisition or loss of a transposable element can result in genomic incompatibility as demonstrated by hybrid dysgenesis in *Drosophila melanogaster* and provides a good mechanism for producing reproductive isolation necessary for speciation (see Rose and Doolittle, 1983, for a further discussion of these points). Transposable elements may also transport pieces of the genome around allowing for example, the re-alignment of exons to create different enzyme specificities. However, we await, as Temin and Engels point out, much more experimental detail before a proper consideration of these elements in evolution can be preformed.

Rapid changes in genomic size may also be achieved by gene amplification of both coding and non-coding sequences. This gene amplification is clearly important in evolutionary biology since it is not only a means of changing coding potential but it also provides a further mechanism whereby genomes may become incompatible (Flavell, 1982). This mechanism has been somewhat overlooked in the present volume not as an underestimate of its importance but because two recent books have been aimed directly at this topic (Dover and Flavell, 1982; Schimke, 1982). Cullis (Chapter 8), however, has examined gene amplification and its interaction with the environment. His model system is the flax plant which under certain environmental conditions undergoes an enlargement in genome size that may become inherited. Cullis discusses other examples of gene amplification in the light of the flax model and proposes that under certain circumstances the environment can cause disproportionate replication of particular segments of the genome resulting in gene amplification. He also makes the evolutionary point that even in cases when specific gene sequence amplification is selected such as in the case of resistance to the anti-tumour drug methotrexate, that large areas of unrelated DNA are also amplified. Potentially, therefore, selection of one gene can result in the amplification of unselected genes. It is also very interesting that in a recent report (Coderre *et al.*, 1983) amplification of the bifunctional thymidylate synthetase-dihydrofolate reductase gene in methotrexate resistant *Leishmania tropica* is lost in non-selective conditions but is immediately recalled if the organism is returned to selective conditions allowing it to survive. This is clearly outside most evolutionary theories and it will be fascinating to see if such mechanisms exist in multicellular organisms.

The area of genome evolution is perhaps the most exciting subject in evolutionary biology at present and yet, it would seem that we have viewed only a fragment of the potential flexibility of the genome. A more radical but as yet unconfirmed view of genome flexibility is considered by Steele *et al.* (Chapter 9). They propose a genetic interchange between the soma and the germline mediated by retrovirus-like vectors. Such a mechanism would enable genetic traits selected within the lifespan of an organism to become inherited. This proposal is in contravention to Weismann's doctrine of the continuity of the germplasm which is one of the cornerstones of neo-Darwinism as described above, and incidently is one of the few rigorously testable aspects of it. Steele and co-authors point to a substantial body of evidence that suggests that there is a genetic interchange between the soma and the germline and if this is correct then this will dramatically change evolutionary concepts.

All these observations on the flexible genome point to mechanisms whereby the genome may evolve rapidly and in some cases co-ordinately. These events pose a challenge to conventional population genetics which seeks to relate the accumulation of micromutations to the generation of new species and will demand a change in emphasis from peptide variability to genomic variability. Such a process is happening already but if we are to take a cue from this book it would appear that we must concentrate more on the molecular biology of ontogenic transformations so that we may understand the material that evolutionary processes are working on. However, we need not only understand change but also the stasis that follows the evolutionary event. This stasis may only reflect the rarity of evolutionary events but it may also reflect some fundamental of evolution. For example the acquisition from outside of a new transposable element that causes rapid genetic variation and genetic incompatibility with other members of the species followed by the harmonization (or loss) of the element within the new host genome and ensuing relative genetic stability. Such behaviour for transposable elements may be found in *Drosophila* (see Chapter 7; Kugimiya *et al.*, 1983). Such speculations, however, are idle until we know more about genetic variation particularly with respect to developmental changes.

It is of course, the attempt to define man's place in nature that was one of the main spurs to evolutionary studies. Popper in his contribution (Chapter 10) asks what is the difference between an amoeba and Einstein. He argues that the difference lies in the evolution, as a consequence of natural selection, of the descriptive function of human language, this being the prerequisite for critical thinking. Therefore, unlike the amoeba, Einstein was able to formulate his theories in language, when they became objects, and in this way they could be placed outside himself. As objects theories can be critically tested and so modified accordingly, resulting in the evolution of better and better theories by a process analogous to Darwinian evolution. These hypotheses then form a store of knowledge (World 3) greater than can be held or known by a single individual. Popper provides a scenario of how the biological evolution of language may have occurred. He has also summarized his entire epistemology which is so important for experimental scientists to grasp, in order for them to ensure that their theories are correctly formulated for critical testing. I hope also, that this volume in its totality will stimulate the readers critically to reassess their framework of evolutionary theories and perhaps it will open up new concepts that will be the stimulus for further evolutionary investigations.

REFERENCES

Bush, G. L. (1982) What do we really know about speciation, in Milkman, R. (ed.) *Perspectives on Evolution*, Sinauer Associates, Massachusetts, pp. 119–128.

Coderre, J. A., Beverley, S. M., Schimke, R. T., and Santi, D. V. (1983) Overproduction of a bifunctional thymidylate synthetase-dihydrofolate reductase in methotrexate-resistant *Leishmania tropica*, *Proc. Natl. Acad. Sci. USA*, **80**, 2132–2136.

Davidson, E. H., and Posakony, J. W. (1982) Repetitive sequence transcripts in development. *Nature*, **216**, 633–635.

Dover, G. A., and Flavell, R. B. (eds.) (1982) *Genome Evolution*, Academic Press, London and New York.

Eldredge, N., and Gould, S. J. (1972) Punctuated equilibrium: An alternative to phyletic gradualism, in Schopf, T. J. M. (ed.) *Models in Paleobiology*, Freeman Cooper and Co., California, pp. 82–115.

Flavell, R. B. (1982) Sequence amplification, deletion and rearrangement: Major sources of variation during species divergence, in Dover, G. A., and Flavell, R. B. (eds.) *Genoma Evolution*, Academic Press, London and New York, pp. 301–323.

Goldschmidt, R. B. (1943) Ecotype, ecospecies and macroevolution, in Piternick, L. K. (ed.) *Richard Goldschmidt: Controversial geneticist and creative biologist: A critical review of his contributions with an introduction by Karl Von Frisch*, Birkhauser Verlag, Basel, pp. 140–153.

Gould, S. J. (1980) Is a new and general theory of evolution emerging? *Paleobiology*, **6**, 119–130.

Gould, S. J. (1982) The meaning of punctuated equilibrium and its role in validating a hierarchial approach to macroevolution, in Milkman, R. (ed.) *Perspectives on Evolution*, Sinauer Associates, Massachusetts, pp. 83–105.

Kugimiya, W., Ikenaga, H., and Saigo, K. (1983) Close relationship between the long terminal repeats of avian leukosis-sarcoma virus and copia-like moveable genetic elements of *Drosophila*. *Proc. Natl. Acad. Sci. USA*, **80**, 3193–3197.

Mayr, E. (1963) *Animal Species and Evolution*, Belknap Press of Harvard University Press, Cambridge, U.S.A.

Popper, K. R. (1976) *Unended Quest: An Intellectual Autobiography*, Fontana/Collins, U.K.

Rose, M. R., and Doolittle, W. F. (1983) Molecular biological mechanisms of speciation. *Science*, **220**, 157–162.

Schimke, R. T. (ed.) (1982) *Gene Amplification*, Cold Spring Harbor Press, Cold Spring Harbor Laboratories, New York.

Stanley, S. M. (1979) *Macroevolution: Pattern and Process*, Freeman, Cooper and Co., San Francisco.

Sutcliffe, J. G., Milner, R. J., Bloom, F. E., and Lerner, R. A. (1982) Common 82 nucleotide sequence unique to brain RNA. *Proc. Natl. Acad. Sci. USA*, **79**, 4982–4946.

Evolutionary Theory: Paths into the Future
Edited by J. W. Pollard

Chapter 1

The scientific status of modern evolutionary theory

ANNA RIDDIFORD AND DAVID PENNY
Department of Botany and Zoology,
Massey University,
Palmerston North,
New Zealand

1.1. INTRODUCTION

The impact on scientific thought of Darwin's *On the Origin of Species* has been likened to the impact of Newton's laws (Huxley, 1974; Manser, 1965; Scriven, 1959). The theory of evolution occupies a central position in biological thinking, and evolutionary explanations have wide application in disciplines as diverse as biochemistry, ecology, palaeontology, and genetics. Nevertheless, the value of such explanations has been subject to considerable debate. Modern evolutionary theory, which we also call the synthetic theory, has been severely criticized from both within and from without the scientific community and the scientific status of the theory has become a subject of contention. The purpose of this chapter is to identify the different types of criticism and to examine them in detail in order to locate any real difficulties with the theory. To achieve these ends, current thought in both evolutionary biology and philosophy of science has been examined.

In most cases the main criticisms can be placed in one of the following categories which are summarized below.

1. Modern evolutionary theory is metaphysical rather than scientific.

> 'I have come to the conclusion that Darwinism is not a testable scientific theory, but a *metaphysical research programme*—a possible framework for testable scientific theories' (Popper, 1976, p. 168).

2. It is better not to assume that evolution has occurred.

'...it has been realized that more and more of the evolutionary framework is inessential, and may be dropped. ... in all early work in cladistics, the nodes are taken to represent ancestral species. This assumption has been found to be unnecessary, even misleading, and may be dropped' (Patterson, 1980).

3. The principle of natural selection is a tautology.

'...these "theories" [of evolution] are actually tautologies and, as such, cannot make empirically testable predictions. They are not scientific theories at all' (Peters, 1976, p. 1).

4. Natural selection explains too much and lacks predictive power.

'The real problem with Darwinian selection theory, however, is that it can explain everything, and therefore, nothing. By logical necessity what survives (or what produces more offspring) is more fit than what doesn't, what is more fit is therefore better adapted, and what is better adapted is therefore selected for (or in other words, survives)' (Rosen, 1978, p. 370).

5. There are no evolutionary laws.

'Perhaps Darwin should be called "biology's Karl Marx" rather than its "Newton". In both cases there is a basic picture which seems to render a complex mass of facts comprehensible without giving either the power of control or of prediction. ... Indeed neither "explain" in the sense that they enable events to be deduced from a set of initial conditions together with universal laws, in the way that physics and chemistry explain within their respective fields' (Manser, 1965, pp. 30–31).

6. The principle of natural selection is just the application of prevailing political and social ideas to biology.

'He looked at the whole living world from the point of the political economists ... It is impossible to understand why Darwin influenced the sociological theorists so profoundly, unless we realize his whole teaching really constitutes a sociology of nature, and that Darwin merely transferred the prevailing English political ideas and applied them to nature' (Radl, 1930, p. 18).

Some of the criticisms are more serious than others. The suggestion that the synthetic theory as a whole is not a scientific theory is the most fundamental criticism. It raises the question of how science is distinguished from non-science, as well as the status of evolutionary theory itself.

In order to analyse these criticisms, we give a statement of our understanding of the synthetic theory, follow this with a discussion of scientific method, and then consider the specific criticisms. We use a framework developed from Popper's theory of knowledge; his approach has been criticized as requiring more stringent standards than most scientists practise, but we wish to test the theory of evolution against such standards. We distinguish between cases where it is difficult to make good tests of predictions, and where it is impossible to make tests. The first case is still science, the second is (for the present) metaphysics.

Our conclusion is in three parts:

1. The synthetic theory of evolution meets the criteria of a normal scientific theory. Problems in evolution raise interesting and important questions on the nature of science and on the role of falsification, but the problems are not specific to evolutionary theory.
2. The theory is incomplete. Additional sub-hypotheses are required before the theory can be applied to some problems; more information is required, such as on the nature and amount of variation in regulatory genes; and improved mathematical methods are needed to give quantitative, rather than just qualitative, predictions.
3. We expect that future studies, particularly in molecular biology will greatly improve the precision of evolutionary theory.

1.2. STATEMENT OF EVOLUTIONARY THEORY

It is necessary to distinguish modern evolutionary theory from that first presented by Darwin in *On the Origin of Species*. He had no knowledge of how the characteristics of a species were preserved and passed from one generation to another. Consequently the source and maintenance of variation was a great mystery. Mendel's laws of inheritance and the idea of genotype and phenotype were not yet known, nor was it understood that the hereditary material of an individual is not altered by any phenotypic features that the individual acquires during its lifetime.

The modern theory of evolution which is known as neo-Darwinism, the modern synthesis, or the synthetic theory, retains Darwin's principle of natural selection, combining it with the discoveries of population genetics, paleontology, systematics and molecular biology.

> 'The new "synthetic theory" of evolution amplified Darwin's theory in the light of the chromosome theory of heredity, population genetics, the biological concept of a species and many other concepts of biology and paleontology. The new synthesis is characterized by the complete rejection of the inheritance of acquired characteristics, an emphasis on the gradualness of evolution, the realization that evolutionary phenomena are population phenomena and a reaffirmation of the overwhelming importance of natural selection' (Mayr, 1978, p. 52).

The principle of natural selection, once formulated as 'survival of the fittest', is now more acceptably stated as 'differential reproduction' resulting in the survival of some genotypes over others.

'The classics of evolutionism described natural selection as the survival of the fittest. We prefer to describe it as differential perpetuation of genotypes or of genetic systems' (Dobzhansky, 1968, p. 115).

It is traditional to divide the study of evolution into two areas which can loosely be called macroevolution and microrevolution. (Note that we are using macroevolution to mean the evolution of all living groups, not evolution by macromutation; Løvtrup, 1981). The first considers the question of whether evolution has occurred and by what pathways. It was called the theory of descent by Darwin and the fact of evolution by Julian Huxley. It is also called historical reconstruction, or simply evolution. The other area is the study of the mechanism of evolution and this is usually called microevolution. Darwin suggested that natural selection was the main mechanism that could explain the non-random aspects of evolution.

These two aspects of description and explanation (mechanism) are usually distinguished in science, and in most areas for historical reasons they are clearly identified. For instance, the theory of continental drift was proposed long before plate tectonics was suggested as the mechanism for continental drift; Kepler's laws described the motion of the planets but again it was many years before Newton was able to show that his proposed gravitational force would account for these observations. Although it is important to make this distinction, in practice

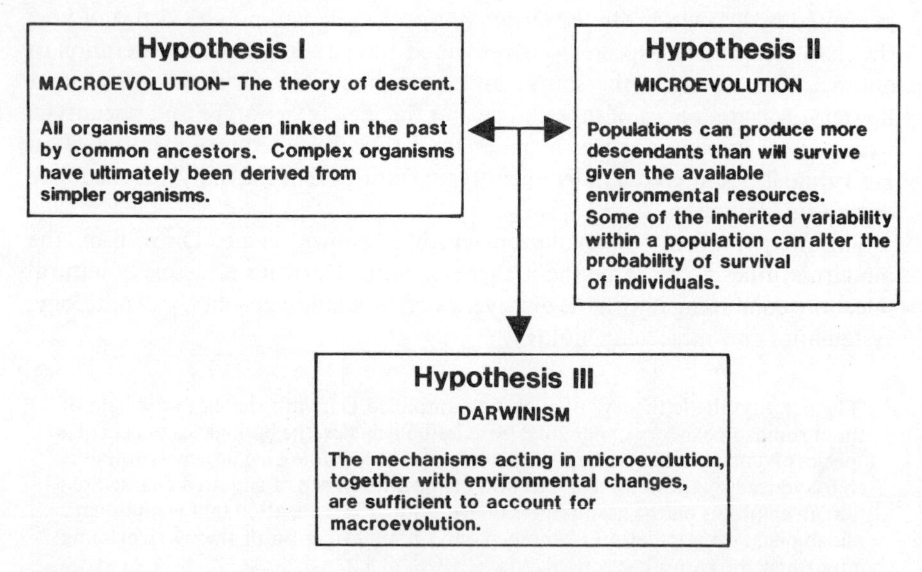

Figure 1. The three main hypotheses of the synthetic theory. They are shown with arrows in each direction because they are mutually supporting in that, for example, the theory of descent is supported by the existence of a mechanism that could lead to species modification and divergence; but the theory of descent has also led to a search for a mechanism that could result in descent with modification

there is a continual interaction between the hypotheses about descriptions and explanation.

Not everyone has used this division into description and mechanism (Løvtrup, 1982) and we have found it particularly helpful to change from this two-way division and to introduce a third category that formally states that microevolutionary processes are sufficient to account for the historical process of evolution. This trinity (or troika) is shown in Figure 1 with the three main divisions being:

Hypothesis I. The theory of descent (macroevolution, the fact of evolution, evolution, historical reconstruction of evolution).

Hypothesis II. Microevolution (natural selection and chance). This includes an ecological and a genetic aspect and is based on both field and laboratory studies. A more detailed statement is given in Figure 2.

Hypothesis III. Darwinism. Microevolution is sufficient to account for macroevolution.

Each hypothesis is dependent upon a number of interrelated auxiliary or sub-hypotheses and for microevolution the relationship of these is shown in Figure 2. The separation of the hypotheses of Darwinism and microevolution allows a

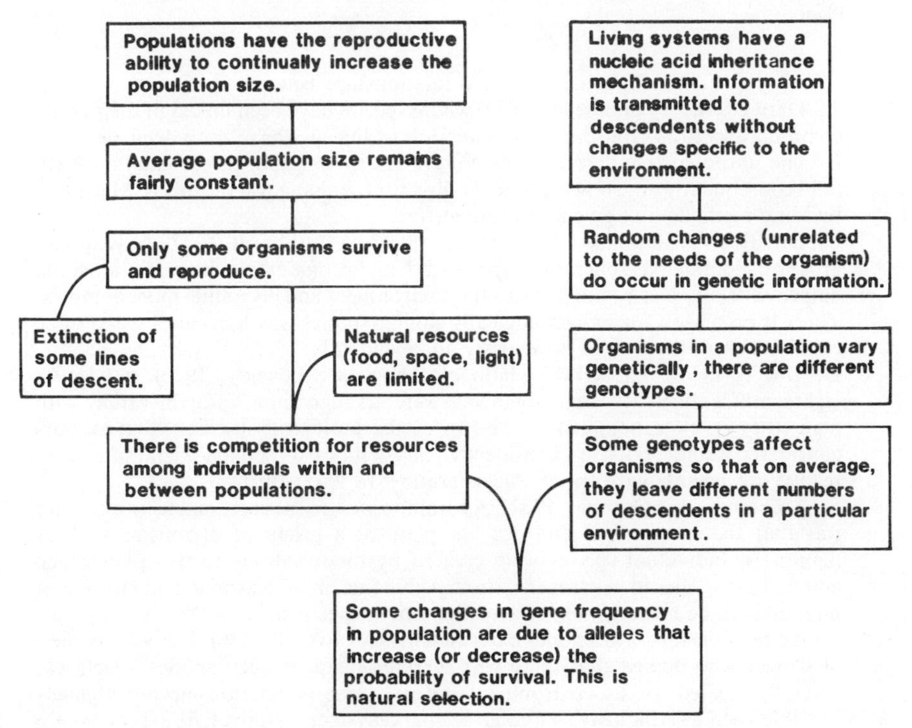

Figure 2. Sub-hypotheses of Hypothesis II (Microevolution)

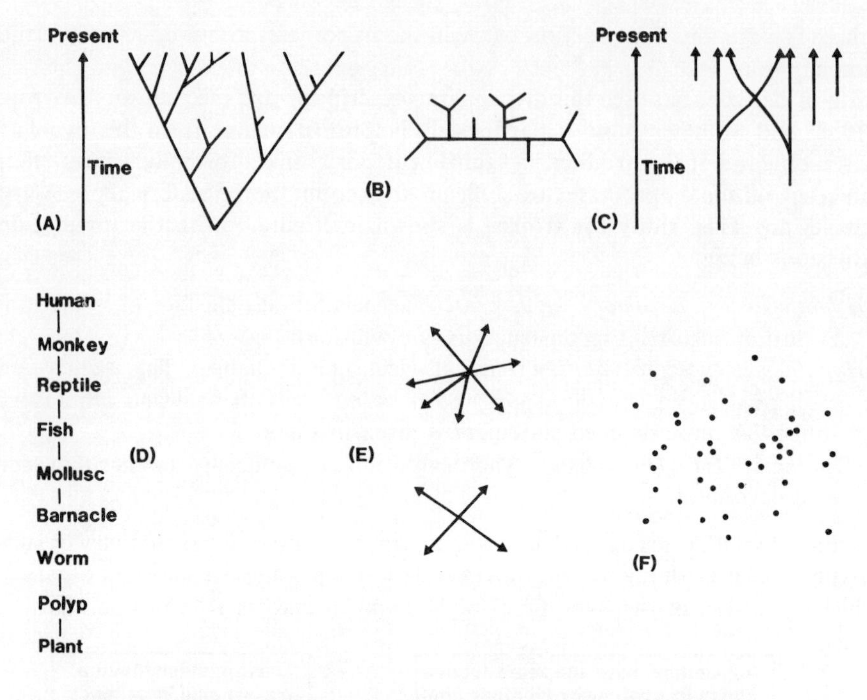

Figure 3. Some suggested relationships between species.
A. Darwin's theory of descent (1837) where species have been linked in the past by common ancestors; splitting and extinction of lineages have both occurred.
B. The unrooted tree derived from Figure 3A by omitting the original ancestor (and consequently not showing time, or direction of change on the links). Many tree building methods first give an unrooted tree.
C. Lamarck's evolutionary theory (1802, 1809; see Burkhardt, 1977). New species arose by spontaneous generation; species did not become extinct because they could 'improve' themselves by adapting to the environment and becoming more complex. There is no reason for morphologically similar species (say lions and tigers) to be closely related, they could have arisen independently.
D. One form of 'The Great Chain of Being' (see Lovejoy, 1936), a popular eighteenth century view that organisms were arranged in a linear hierarchy with man (that is, male humans) at the top of the biological species. Most authors considered it static, but a few considered it an evolutionary sequence, and some even read it downwards with species 'degenerating' to lower forms.
E. Archetypes (see Ospovat, 1981). A common view in the early nineteenth century was that the creator had an 'idea' or 'plan' of a group of organisms such as mammals. Individual species were created by modifications to this plan which adapted the species to its place in nature. Again there is no reason to find any pair of mammals to be more closely related than any other pair.
F. No relationships—all species independent (see Lovejoy, 1936). This was the view of Cuvier who denied that there was a relationship between species—each was specially created for its environment. Other forms of relationship are logically possible such as MacLeay's circular 'quinary' system (Ospovat, 1981), or even a two-dimensional rectangular system analogous to the periodic table of chemistry

more precise discussion of the criticisms. We have found that the difficulties with the synthetic theory lie principally with testing the components of Hypothesis III. Other formulations of the components of evolutionary theory can be found in Williams (1970) and Manier (1980).

Darwin made important contributions to all three hypotheses. His role in developing Hypotheses II and III is well known but it is not generally realized that his formulation of Hypothesis I (the theory of descent) was novel in 1837 when he first developed it (see Ospovat, 1981). The theory of descent is represented in Figure 3A and some alternatives are shown in the rest of Figure 3. Darwin appears to have first been convinced by uniformitarianism, an idea similar to Hypothesis III. This idea is that geological features can be explained if present day processes operated over millions of years in the past. By 1837 Darwin had accepted a limited form of the theory of descent and was searching for a mechanism that would account for this type of evolution. A year later he realized that natural selection would account for the fit between the structure of organisms and their environment. It was apparently not until the 1850s that he fully accepted Hypothesis III (Ospovat, 1981).

1.3. CRITICISM 1. MODERN EVOLUTIONARY THEORY: METAPHYSICS OR SCIENCE?

'Scientists armed with half-baked philosophical ideas—usually some bastardization of Karl Popper's principle of falsification—and philosophers, neither armed nor unarmed with any real scientific knowledge at all, join to do discredit to both their subjects. And neo-Darwinism suffers unjustly' (Ruse, 1981, p. 828).

One of the most persistent and noted critics of the synthetic theory has been Sir Karl Popper. He has claimed that the theory as a whole is untestable, hence is unfalsifiable, and therefore metaphysical rather than scientific. The validity of this criticism depends both on the criterion he uses for the demarcation of science from non-science and on the accuracy of Popper's analysis of the theory.

Not only have his comments been the most far-reaching in their implications, they have also been influential on other evolutionary commentators. Criticism of the synthetic theory has appeared in a number of Popper's writings (1960, 1972, 1976, 1978). Comparison of his earlier and later works reveals a growth in his understanding of evolution, and therefore an increase in the value of his comments. A cogent summary of his arguments appears in the essay 'Darwinism as a metaphysical research programme', published in his autobiography (1976). Some of his comments on the synthetic theory are: Darwinism is an example of 'situational logic' with inherent tautological features; the conjectures of Darwinism are untestable; it does not really predict the evolution of variety and cannot therefore really explain it; its explanation of adaptation 'is feeble'; and, in contrast to the physical sciences, specific predictions cannot be made about evolutionary change.

Although Popper has identified problem areas in the theory, his analysis of the biological aspects tends to be superficial and rather inaccurate. For example, he equates adaptation and fitness (1976, p. 171) although these terms are not synonymous (Mayr, 1982). Popper also made specific criticisms that are incorrect, for example he accuses neo-Darwinians of being 'almost blind' to many difficulties including that:

> '... it may be a little hard to understand why natural selection should have produced anything beyond a general increase in rates of reproduction, and the elimination of all but the most fertile breeds' (Popper, 1972, p. 271).

It is not correct that biologists have been blind to difficulties such as why natural selection does not always maximize the number of offspring. It has been discussed many times including Lack (1966) on clutch size of birds, and Salisbury (1942) on seed size in plants. Indeed, the whole development in ecology of the concept of r and K selection is related to this problem.

The claim that the synthetic theory is tautologous has been retracted by Popper in his most recent paper 'Natural selection and the emergence of mind' (1978). In this paper he has rejected much of his former criticism and is now more interested in whether the theory is really both necessary and sufficient to explain evolutionary phenomena:

> '... I have in the past described the theory as "almost tautological", and I have tried to explain how the theory of natural selection could be untestable (as is a tautology) and yet of great scientific interest. My solution was that the doctrine of natural selection is a most useful metaphysical research programme...
>
> 'I still believe that natural selection works in this way as a research programme. Nevertheless, I have changed my mind about the testability and the logical status of the theory of natural selection; and I am glad to have an opportunity to make a recantation' (Popper, 1978, pp. 344–345).

In this paper he makes the important point that some of the criticisms of evolution are contradictory. In particular it cannot be maintained both that it is a tautology, and that it explains 'too much'—a tautology explains nothing (see later), so cannot simultaneously explain too much. Despite Popper's change of mind, his description of the theory as a metaphysical (rather than a scientific) research programme, has been taken seriously by both scientists and philosophers; for example Campbell (1978), Caplan (1981), Ferguson (1976). Halstead (1980), Lee (1969), Little (1980), Sparkes (1981), Tuomi (1981), and Wasserman (1978). It is also worth discussing Popper's description, as it sheds light on the problem of how to determine the status of theories, and what we should reasonably expect from a scientific theory.

The following section briefly outlines Popper's philosophy of science, as well as some of its difficulties. Our aim is to establish a criterion for distinguishing science from non-science in order to evaluate the criticisms of the synthetic theory. Some of the controversy surrounding Popper's categorization of the synthetic theory as metaphysical is due to a misunderstanding of what Popper means by this term. He does not mean pseudoscientific, or meaningless, or untrue. It is essential to interpret Popper's comment in the light of his philosophy of science where he has made the problem of demarcating scientific theories from non-scientific ones central to understanding the nature of scientific knowledge.

> 'There will be well-testable theories, hardly testable theories, and non-testable theories. Those which are non-testable are of no interest to empirical scientists. They may be described as metaphysical' (Popper, 1965, p. 257).

Popper (1959, 1965), argues that the scientific character of a theory is determined by its potential to be falsified by observation. He rejects the idea that a theory can be judged by the existence of empirical verification alone, since any theory sufficiently vague will always have confirming instances. Popper claims that scientific activity is distinguished by its critical method of *testing* theories rather than trying to prove them.

Accordingly the more specific predictions that a theory gives, the easier it is to test its claims, and hence the more likely it is that contrary evidence will be found, and the theory falsified. Thus the *potential* that a theory has to be falsified can be used to assess its scientific character. In the Popperian world, a metaphysical theory is one which cannot be tested. When Popper said that Darwinism is metaphysical, he was not making a comment about the truth of the theory. Indeed he stated that he saw no scientific alternative and suspected that much of the theory is true. Rather he was claiming that evolution by natural selection cannot be tested and so was not falsifiable.

Popper's philosophy of science was a response to the view which represented the scientific method as an inductive process. In such a model, scientific theories are supposed to be proven, or verified, by their consistency with repeated observations. Hume, and later philosophers including Popper, pointed out major flaws in the model of induction. No matter how many observations support a theory, they will not prove the theory correct. A well-known example of this point is the claim 'all swans are white' which was consistent with innumerable observations . . . until the first black swans were seen in Australia. Hume's moral is that no collection of particular observations can ever justify a general statement. It is impossible to prove that a theory is true, or to justify its scientific status, simply because it is consistent with repeated observations. Popper made an important contribution when he pointed out that theories can be *disproved* by observation (as happened with 'all swans are white'), even though theories can never be proved by observations.

Recognizing the impossibility of ever proving a theory correct through verification, and the importance of theory in guiding observation, Popper has made a very different analysis of the process of scientificdiscovery. He argues that for any given problem, a number of different conjectures or hypotheses can be made to explain it. These initial conjectures have a component of intuitive insight and inspiration in their origin, which makes their inception unpredictable. What makes a theory scientific is that it must be sufficiently specific so as to prohibit some possible observations. A theory that 'swans are either white or they are not white' is clearly true, but of no interest in science because there are no possible observations that would falsify it. Only by fulfilling this condition of *potential falsifiability* is a theory testable and hence, scientific. The more a theory prohibits, the more able it is to be tested.

There are several ways in which a theory may fail to be falsifiable. Suppose we predict that a conjunction of the planets in 1984 will lead to worldwide unrest. Because this prediction is vague, it is very likely that some event will be able to be interpreted as confirmation of the prediction, and hence, also of astrological theory. The hypothesis does not specify the exact time of duration of the 'unrest', nor does it indicate whether this 'unrest' is political, emotional, seismological, climatic, a worldwide period of insomnia, a mass migration, or something else.

Another feature of some theories that makes falsification impossible and thus disqualifies them from Popper's 'scientific' category, is the presence of inbuilt, criticism-deflecting arguments. This addition of *ad hoc* conditions to a hypothesis also decreases its status as a scientific theory (because they prohibit fewer possible observations). If, against our vague prediction, 1984 is a peaceful year in all respects, we might refuse to accept this as a refutation. Instead, we might say that unbeknown to us when we made the prediction, sunspot activity would arise and counter the effects of the planetary conjunction, this being the reason for the very calm terrestrial period observed. Such an amendment protects the hypothesis from the possibility of being refuted, and so reduces the likelihood that the theory, of which it is part, can be falsified.

Popper is careful, however, to say that his criterion of falsifiability is not one that separates theories according to their truthfulness. In fact he believes that most scientific theories have their origins in untestable metaphysical theories and myths. Popper originally put the modern theory of evolution in such a pre-scientific category, although he stressed the value of the theory by calling it a 'research programme' despite its scientific inadequacy. It is a research programme in that it provides a general framework for generating and testing specific proposals such as, for example what may happen to bacteria in a 'penicillin infested' environment (Popper, 1976, p. 172).

The major criticism of Popper's work is that Popper fails to describe how scientists actually work. Kuhn (1970), Lakatos (1970, 1974, 1976), and Feyerabend (1970a, 1970b, 1975) are some of the authors who have produced analyses of scientific methodology that differ from Popper's on the question of the acceptance or rejection of scientific theories. We will consider the schemes of Kuhn and Lakatos. Both believe that the central theory in a research programme is protected, by various mechanisms, from being tested critically.

Kuhn has analysed the history of a number of scientific discoveries to demonstrate how closely the course of experimentation is tied to the corresponding theory. Kuhn claims that normal scientific research proceeds within the bounds of a common conceptual framework (a paradigm), which is shared by the workers involved in a particular field. The specific paradigm held will dictate to the scientists which questions are worth asking and what problems need to be solved in order to explore the full implications of the theory, and to extend its scope and accuracy. Kuhn claims that a paradigm is not displaced in the course of normal science because it is falsified, but only when mounting anomalies within the theory, combined with the appearance of novel discoveries, lead to the production of rival paradigms. Within normal scientific activity, the criterion of falsifiability of individual predictions is considered appropriate, and the outcome of such tests directly determines the fate of the hypothesis concerned. The larger paradigm which guides the course of normal scientific research, it is claimed, cannot be tested in this way. Indeed according to Kuhn a scientist's choice between rival paradigms is initially made as an act of faith.

Lakatos has started from a position similar to that of Popper but like Kuhn has again developed the idea that scientific theories are protected from conflicting empirical data. Accordingly he claims Popper's criterion for demarcation must be altered. Lakatos describes how scientific theories are arranged in hierarchies in which core theories, presumably equivalent to Kuhn's paradigms, are protected from refutation by the presence of auxiliary hypotheses.

> 'The basic unit of appraisal must not be an isolated theory or conjunction of theories but rather a "*research programme*" with a conventionally accepted (and thus by provisional decision "irrefutable") "*hard core*" and with a "*positive heuristic*" which defines problems, outlines the construction of a belt of auxiliary hypotheses, foresees anomalies and turns them victoriously into examples, all according to a preconceived plan' (Lakatos, 1976, p. 9).

So long as a research programme is progressing it is the 'positive heuristic' that guides research rather than the anomalies. In this view, the presence of protective auxiliary theories, or a 'negative heuristic', would not necessarily be interpreted as bad science.

Both Lakatos and Kuhn have shown that at least some scientists do not test a core theory by trying to falsify it. Popper would agree that this is so, but would not condone it (his comments on Kuhn's and Lakatos's analyses are in Popper, 1974, pp. 999 and 1144). As the history of science so amply bears out, the complexity of the structure of a hard-core theory may in the short term make simple and decisive tests difficult, or even impossible. Kuhn's analysis leads to the uncomfortable conclusion that ultimately scientists follow no rational procedure in choosing between rival theories. In contrast, Lakatos believes that a rational reconstruction can be given for the growth of scientific knowledge, despite his belief that core theories are protected from direct testing, and hence from possible falsification.

Popper's description of scientific methodology seems to be in direct contradiction to Kuhn's but it is worth noting that the two authors have different aims. Popper is particularly interested in the theory of knowledge whereas Kuhn's approach is more historical and therefore basically descriptive of science. Kuhn is describing how many scientists have worked (the sociology or psychology of scientists), whereas Popper is interested in the 'demarcation' of science from non-science and is describing how he thinks scientists should work. Popper would claim that the best scientists have worked in the way he advocates and this is illustrated in the following comment:

> 'It is the great scientists which I have in mind as my paradigm for science' (Popper, 1974, p. 977).

Popper's philosophy is, of his own admission, an over-simplification or idealization of scientific methodology. Frequently scientists do not try, or are not able, to design an experiment that could falsify a paradigm or core theory. Though this does not mean that they should not aim to; it is normally difficult to design an experiment that could conclusively disprove a theory.

There are examples in evolution that are typical of the difficulties that scientists experience in choosing between theories, but which ultimately do not prevent them from achieving some sort of rational consensus. Over the past decade a major controversy in population genetics has concerned the origin of genetic polymorphism (diversity) in populations. The *pan-selectionist* position has been that all polymorphisms are maintained in the population by selection. The opposing *neutralist* school of thought is that *most*, but not all, of the variation at a gene locus is the product of neutral mutations, which may then become fixed in the population (or lost) by the process of random drift. They are not selected for or against since their effect on the genotype's chances of survival are neither beneficial nor detrimental. (There is also a possible pan-neutralist position that all variants are neutral. This is often used as a basis for calculation but not as a serious explanation of macroevolution.)

The neutralist's argument is supported by the success of the predictions that can be made from it. These include: that the rate of change in nucleotide sequence is not limited by genetic load (Haldane's dilemma); a protein with a stable function will accumulate nucleotide changes at a steady rate over long periods of time (the molecular clock) and with a variability that can be estimated; it can predict cases when the rates of nucleotide substitution will alter; it explains the high level of polymorphism in populations, relates it to the population size and estimates the variability; it predicts that evolution will be faster both in proteins with a higher heterozygosity and in the parts of the molecule with little functional importance; it leads to the recognition of factors (molecule size, interactions with other molecules, etc.) that will affect rates of evolution; it predicts the loss of function of genes that are not expressed; it predicts that the third site of a triplet should change faster and this faster rate will also occur for many non-expressed sequences; it predicts that the ratio of nucleotide changes in the third site of a triplet to changes in the first two sites should be higher in slower evolving proteins; and finally it predicts that over-all, the average proportion of amino acids in proteins will reflect the number of coding triplets for each amino acid. (Some references to this work will be found in Kimura and Ohta (1974) and Terzaghi *et al.* (1984).)

The pan-selectionists however, do not believe that their theory has been falsified. For example, they interpret the evidence for a molecular clock as meaning that selection rates have been roughly constant over extremely long periods of time and devise reasons why this should have been so (see Van Valen (1974) and Zuckerkandl (1976)). Though the differences between the neutralist and selectionist position have not been resolved directly, the neutralist theory has been increasingly accepted because of its success in making predictions which have been supported experimentally. In contrast the pan-selectionists have found it difficult to make predictions and have more often had to resort to *ad hoc* explanations after new data have been obtained (and it has not been shown that their different *ad hoc* assumptions are compatible in one over-all model). Neither argument can be falsified outright to the satisfaction of both parties. Nevertheless most scientists have been able to decide between the two on the basis that the neutralist theory has made predictions that have been tested and support the neutralist position that a large proportion of nucleotide changes are neutral.

1.3.1. The criterion of falsifiable predictions

The outcome of the neutralist selectionist controversy indicates the type of criterion that scientists use in preferring one theory over another. Popper must be recognized as having made a major contribution to our understanding of the process of scientific understanding by pointing out the fallacy in the method of induction and the asymmetry between verification and falsification. Even more important is the emphasis he has placed on the role of tests, on the value of precise predictions, and on science as a critical method. However, his criterion of falsifiability of *theories* is not corroborated by Lakatos's analysis. Lakatos admired Popper's approach and believed that a rational reconstruction could be given for the growth of scientific knowledge, despite his conclusion that core theories are protected from direct testing and so from direct falsification. Lakatos has proposed a modified version of Popper's criterion of falsifiability, suggesting that the success of a theory can be measured by its ability to produce deductive predictions.

> 'A research programme is said to be *progressing* as long as its theoretical growth anticipates its empirical growth, that is, as long as it keeps predicting novel facts with some success...' (Lakatos, 1976, p. 11).

Many writers have pointed out the importance of predictions in science. The value of predictions is that they can be tested against experimental results and can thus provide a means of evaluating and improving one's explanations. Although a prediction is derived from an hypothesis, *testing a prediction is not the same as testing the hypothesis*.

Imagine what would happen if a prediction, made from the sub-hypotheses in Figure 2, turned out to be wrong. Brady (1982) gives three reasons that could account for an incorrect prediction. We will divide them into four.

1. Experimental error. There is some error in the measurements, or some experimental variable not accounted for. That is, the clause 'all other things are equal' (*ceteris paribus*), is false. 'The refuting instances... are empirical statements and therefore open to all kinds of empirical mistakes' (Popper, 1974, p. 1035).
2. Inadequate analysis. Inappropriate mathematical methods were used to make the prediction (several statistical methods were developed to allow reliable predictions from the components of Figure 2; recently game theory (Maynard Smith, 1982a) has allowed more accurate predictions in some areas; for phylogenetic trees there is still no agreement as to whether statistical or graph theory methods are the most reliable).
3. Theory incomplete. Figure 2 is incomplete and one more sub-hypothesis is necessary before accurate predictions are possible (we would be surprised if Figure 2 was complete, particularly in the components of the sub-hypotheses).
4. Incorrect hypothesis. At least one hypothesis in Figure 2 is wrong.

If careful checking does not change the experimental result, then most researchers will start to consider other possibilities which may have been suggested in the preceding series of experiments. We would not go as far as Lakatos in saying that the 'core' theories are protected from questioning. Rather scientists select an area they think is most likely to be in error, and almost certainly the areas selected for questioning will not be the same for all workers. The history of continental drift is interesting in this context (see Stevens, 1980). There is no mechanism to get all scientists to agree as to what is the 'core' theory that is 'not being questioned'. Scientists, like individuals in a population, vary.

Although tests of predictions do not directly prove or disprove an hypothesis, they do provide directions for further testing or analysis. Testable predictions are an important way to evaluate an hypothesis and its ability to make accurate explanations. We do not accept the criterion that a theory must be directly falsifiable. In its place we propose to use the ability of a theory to lead to testable predictions as the criterion of a theory's scientific adequacy. Sparkes (1981) has given a useful diagrammatic model of the way in which prediction relates to explanation, and hypothesis to observation, which is presented in Figure 4.

We have not yet answered the question of the scientific status of the synthetic theory of evolution. The following sections deal with the specific criticisms that

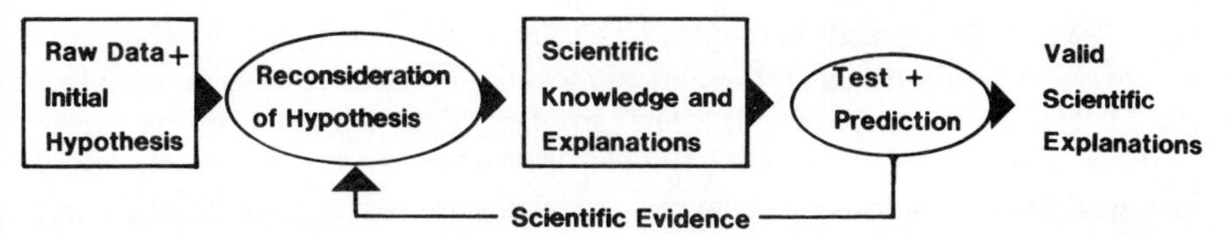

Figure 4. One view of the interrelationship of hypotheses and experiment (after Sparkes, 1981. This first appeared in *New Scientist*, London, the weekly review of science and technology)

were listed earlier. In Lakatosian terms, the synthetic theory acts as a major research programme in biology. The three hypotheses of common ancestry, microevolution, and Darwinism are 'hard-core' theories and as such are locked in an array of auxiliary hypotheses as illustrated for microevolution in Figure 2. Throughout, we use the criterion that a scientific theory, supported by sub-hypotheses, should yield testable predictions. In this way we can evaluate the criticisms in order to decide the question of the scientific status of evolutionary theory.

1.4. CRITICISM 2. THE THEORY OF DESCENT (Hypothesis I)

This is the cornerstone of all evolutionary theory. If evolution had not occurred, then there is obviously no need for a mechanism to explain it. There have been many general criticisms by biologists about the apparent lack of objectivity of particular phylogenetic schemes, (some of these are quoted in Ragan and Chapman, 1978) but these criticisms have not usually gone as far as suggesting that the approach is non-scientific. Popper (1976 and earlier) initially doubted that the theory of descent would lead to falsifiable predictions, but later (as we have already seen and without given reasons) he reversed his opinion (Popper, 1978). Recently a group of taxonomists calling themselves 'transformed' or 'pattern' cladists (see Platnick, 1980; Patterson, 1980) have argued that it is better not to assume that evolution has occurred, and to concentrate on techniques of finding 'natural order' among organisms (see quotation in section 1.1). We will give two responses to these criticisms, one on the role of mathematics in science, and the other on the question of falsifiability of evolutionary trees. The position taken by pattern cladists is similar to that known as 'instrumentalism' (see Popper, 1965, pp. 99ff). This position is that theories are 'instruments' for calculation, but they deny that theories give any knowledge of the real world.

There is an aspect in which the pattern cladists are correct, and this is that tree reconstruction methods are mathematical procedures that work independently of any biological model. The pattern cladists' claim is weakened because in forming a rooted tree (Figure 3A), they define an 'ancestral' and 'derived' state for each character and to define an 'ancestral' state implies that evolution has occurred. Nevertheless it is possible to avoid using 'ancestral' states by a more general tree building method that leads to unrooted trees (Figure 3B). (Note that we are using 'tree' with its mathematical meaning in which cladograms are trees (M. D. Hendy, in preparation).) The problem is then known as the Steiner problem in graphs (Graham and Foulds, 1982). However, even these methods assume that the relationships between organisms are best represented by a tree, rather than any of the alternatives of Figure 3.

Mathematical operations are general methods in that they can be used in different applications within science, but powerful though they may be, they are no substitute for scientific models. Here an analogy will be helpful. Last century, Mendeleev discovered the periodic table showing 'natural order' in the chemical properties of the known elements. If chemists had not tried to explain this 'natural order', they would have hindered the development of most of modern chemistry. Thus we would claim that the pattern cladists by not trying to explain 'natural order' among organisms, are rejecting the normal scientific approach, rather than finding something different about evolutionary theory. They (the pattern cladists) could argue that we were relying on an inductive argument and although scientific models had been important in other areas of science, this was not proof that they would be useful in this case. Our response would be that the theory of descent has indeed been useful in stimulating research, even in quite different areas such as in measuring the variation in populations.

The other aspect of the criticism of the theory of descent is the supposed lack of objectivity and falsifiability. Biologists could claim with considerable justification that the theory of descent has led to many qualitative predictions that have been confirmed. An example would be that as new features of organisms were measured (perhaps protein sequences), most trees derived from these new features agreed reasonably well with trees derived from comparative anatomy and morphology. Another example would be the support for the age of the earth that comes from radioactivity dating, fission track analysis, and sedimentation studies. The confirmation of qualitative predictions is normally considered a valid, though weaker, form of testing a theory. Such qualitative predictions do not seem to have satisfied the entrenched critics of evolutionary theory. The situation is changing rapidly as new mathematical techniques are being developed for the quantitative comparison of evolutionary trees. It has long been recognized (see Sneath and Sokal, 1973) that comparing trees derived from different sets of data (but from the same taxa) offers, in principle, a method for testing the reliability of the techniques used. Trees are 'congruent' if the same trees are found with different sets of data (Mickevich, 1978). To this extent the existence of an evolutionary history is falsifiable, but this is another area where it has been difficult in practice to make quantitative analyses of predictions. Several methods have been used for comparing rooted (or directed) trees (Rohlf, 1982) but these have not allowed the assignment of the probabilities of two trees being similar just by chance.

Our own interest (Penny *et al.*, 1982) has been in using protein and nucleotide

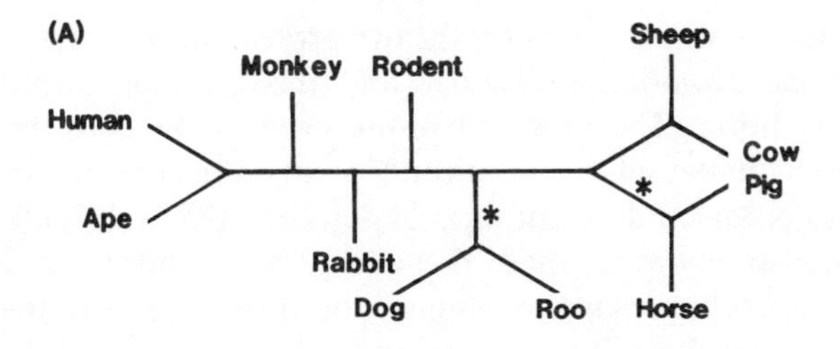

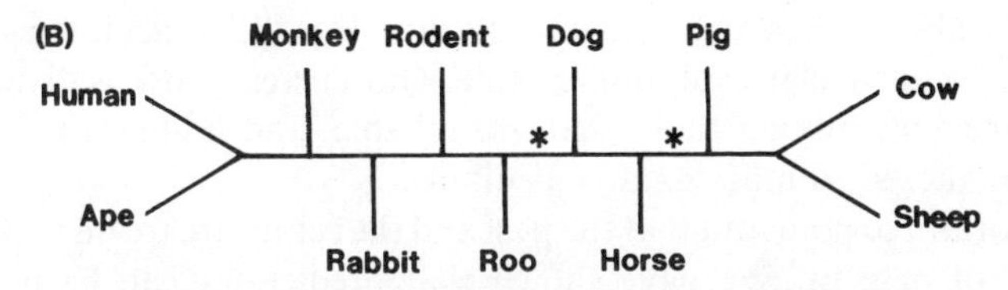

Figure 5. An unrooted minimal tree from cytochrome c and one from hemoglobin β (Penny *et al.*, 1982). (The root can be placed between kangaroo and the rest of the tree.) There are four differences between the two trees, that is, there are two links on each tree for which there is no equivalent link on the other tree, these links are marked. For eleven species, there is only about one chance in 200,000 of randomly selecting two binary trees that are this similar (Penny *et al.*, 1982; Hendy *et al.*, 1984)

sequence data to test the scientific status of the theory of descent. There had been suggestions early in the history of molecular biology that DNA sequences contained information about the evolutionary history of organisms (Crick, 1958; Zuckerandl and Pauling, 1962), but the methods to test these predictions were slow to develop. The three requirements are: (1) being able to obtain the 'best' tree for each set of protein sequences (Hendy and Penny, 1982); (2) being able quantitatively to compare these trees; and (3) being able to assign probabilities to the results of the comparisons. The method used for comparing trees is to remove an internal link of a tree (this divides the species into two sub-sets), then test if there is an equivalent link on the second tree that divides the species into the same two subsets. This is repeated for each internal link.

Figure 5 shows an example of the method which indicated that two trees are very similar, even though they are derived from different proteins. The result strongly supports both the theory of descent and the suggestion (Crick, 1958; Zuckerkandl and Pauling, 1962) that the genetic information of an organism contains a record of its evolutionary history. We would claim that predictions can be made from the theory of descent and that these can now be tested quantitatively. The results invalidate alternatives such as those of Figure 3C–F. Consequently the theory of descent meets our criterion of a science, though it scarcely needs to be mentioned that it is desirable to apply the tests to more data.

There is the question of why the trees for the two proteins are not identical if they have had the same evolutionary history and this question introduces the mechanism of evolution. The observed results seem to be consistent with a mechanism whereby mutations occur at random, and thus it is possible for the same mutation to occur on different lines of descent—this is difficult to detect during tree reconstruction. It had previously been shown in computer simulations (D. Peacock, personal communication) that the 'real' tree may be slightly longer than the shortest tree found from the simulated data. However, more quantitative work is needed before firm conclusions from tree comparisons can be made about the mechanism of evolution. It is still difficult to estimate the reliability of a particular evolutionary tree. Our current work is dividing a tree into sub-trees, identifying those that are reliable, and finding others that are likely to be altered as more data is available.

It is important to point out that the past and the future are treated differently in the theory of descent. We have shown that predictions can be made about historical events, and that these predictions can be tested as more information becomes available. For example, it is possible to predict which evolutionary trees are likely (or unlikely) to be found, if say, collagen sequences were determined for a group of mammals. Predictions can be about unique events in the past, and can be tested as new data are collected.

> '...prediction is not an inference from present to future, but *from known to unknown.* Structure hypotheses...enable us to forecast the *existence* (or the absence) of certain things and properties, whether in the future, the present, or the past' (Bunge, 1959, quoted in, Ferguson, 1976, p. 1102).

The theory of descent is currently the only scientific theory for the origin of species that is supported by the evidence, but as with other areas of science, there are non-scientific alternatives. An example from physical science is: 'atoms and molecules do not really exist, the creator just makes the results of experiments look as if atoms and molecules exist'. A biological example is 'the world was created an hour ago, but was created in such a way that its time of creation was not identifiable; people were created knowing families and friends, and where they lived; organisms were created with gene sequences that looked as if they were derived by an evolutionary process so that we were not able to identify the time of creation'. This assumes a 'great deceiver' as creator. In a Popperian sense it is not possible to claim that the scientific theory is automatically the correct one and that the non-scientific theories are necessarily wrong. Nor can science prove that such theories are false. Popper (1972, p. 21) advocates that a rational person should prefer the best tested theory as a basis for making decisions.

The theory of descent does lead to falsifiable predictions, and thus conforms to the criterion for a scientific hypothesis. We do have sympathy for earlier criticisms about the apparent lack of objectivity of particular evolutionary schemes. However, as Popper has pointed out, there are untestable theories, difficult to test theories, and well-tested theories. The theory of descent is moving from the difficult to test stage (that allowed only qualitative predictions), to being a well-tested theory that gives quantitative predictions. It should be stressed that the scientific status of a theory is not a fixed entity. Another example of a theory whose status has changed is continental drift. This was difficult or impossible to test for nearly 50 years. Now it is one of the most thoroughly tested and supported geological theories.

1.5. CRITICISM 3. IS NATURAL SELECTION A TAUTOLOGY? (Hypothesis II)

The most frequent criticism of the synthetic theory is that the principle of natural selection (see Figure 2) is a tautology. A tautologous statement is necessarily true because it expresses a repetition of the same idea such as, bachelors are unmarried men, or triangles are three-sided figures. Stated either as 'the survival of the fittest' or as 'the differential survival and reproduction of a genotype', the critics feel that the principle is, like a tautology, a repetition conveying no information about the empirical world. Writers who have described the principle in this way include: Brady, 1979; Barker, 1969; Bethell, 1976; Bohm, 1968; Cannon, 1958; Dobzhansky, 1968; Grene, 1974; Haldane, 1935; MacBeth, 1971; Manser, 1965; Peters, 1976; Platnick, 1977; Popper, 1972, 1976; Stebbins, 1974; Waddington, 1960.

It should be pointed out that while Dobzhansky, Haldane, Stebbins, and Waddington are included among those who describe the principle as a tautology, they do not feel that this makes it a meaningless statement.

> 'Genotypes whose carriers differ in Darwinian fitness in a given environment are transmitted from generation to generation at different rates. This last statement is frankly tautological, and yet it is illuminating' (Dobzhansky, 1968, p. 115)

> 'Natural selection, which was at first considered as though it were a hypothesis that was in need of experimental or observational confirmation, turns out on closer inspection to be a tautology, a statement of inevitable although previously unrecognized relation. It states that the fittest individuals in a population (defined as those which leave the most offspring) will leave the most offspring. Once the statement is made, its truth is apparent... only after it was clearly formulated, could biologists realize the enormous power of the principle as a weapon of explanation' (Waddington, 1960, p. 385).

However, the majority of writers view the tautologous character of natural selection less favourably. Most argue that the principle of natural selection is tautological because there is no independent criterion for measuring fitness. The criticism is that the only way to measure a genotype's fitness is in terms of its success in survival and reproduction, that is, in terms of its fitness.

> '... the trouble about evolutionary *theory* is its tautological or almost tautological character: the difficulty is that Darwinism and natural selection, though extremely important, explain evolution by "survival of the fittest" ... Yet there does not seem to be much difference, if any, between the assertion "those that survive are the fittest" and the tautology "those that survive are those that survive". For we have, I am afraid, no other criterion of fitness than actual survival, so that we can conclude from the fact that some organisms have survived that they were the fittest, or those best adapted ...' (Popper, 1972, pp. 241–242).

To these critics of evolutionary theory, the comments of Dobzhansky and others (to this list may be added Fisher (see Popper, 1978) and Simpson, 1949) are astonishing. These latter writers seem to be saying that the principle of natural selection is an empty repetition with no empirical meaning (the definition of a tautology) and yet, are still contending that it has scientific worth for explaining evolution. To resolve this puzzle, a number of underlying problems need to be resolved. Are tautologous statements ever acceptable in science? If they are, what is their nature? Is the principle of natural selection of this type? On the other hand, if there is no room for tautologies in science, what do Dobzhansky and others mean when they say that the principle of natural selection is tautological but still useful?

Tautologies are acceptable as definitions, the second part of the statement may repeat the meaning of the first part, as in the examples about bachelors and triangles. Their tautological nature does not prevent them serving a useful purpose. Thus, tautological statements are permissible when used as definitions.

The principle of natural selection is, however, more than a definition. It is a proposition offering an empirical observation as to why some genotypes are preserved over others. Brady (1979, 1982) points out that in scientific propositions distinct elements are brought together in a causal relationship, and, since cause and effect cannot be the same, the two sides of a scientific proposition cannot be identical. Therefore since the principle of natural selection is stating a cause and effect relationship; it cannot legitimately be called a tautology.

If the two sides of a proposition are in a causal relationship, then one should be able to make predictions of the sort: if this, then that. In the case of the principle of natural selection, it should be possible to say: if genotype n is fit, then genotype n will survive. However, if the only way to measure the fitness of n is to see whether or not it survives, then it will be impossible to test predictions derived from this proposition. It is claimed by many critics that the reason the principle of natural selection is untestable is because of its tautological nature (see the quotation in Section 1.1 from Peters, 1976). This analysis is incorrect for the principle is not intrinsically untestable whereas a tautology is. Rather, the difficulty in testing is due to practical problems in calculating and measuring the variables that determine fitness. These problems have to do with the application of the principle, rather than the principle *per se* and they will be explored more fully in the following section.

At this point it is necessary to return to Figure 2, and to examine its sub-hypotheses in more detail. We would claim that all of the parts of Figure 2 have been independently tested, and all are strongly supported on the available evidence. Populations do have the potential to increase their numbers, natural resources *are* limited, there is genetic variation in populations, and so forth. There are many known mechanisms for genetic variants affecting the probability of survival of an organism. There are countless examples from biochemical genetics, and even in non-living systems (Orgel, 1979). It could be objected that, with most of these biochemical examples, there was no *prediction* that a particular mutant would or would not survive, because the discovery of an altered macromolecule was made after the mutant had been selected. However, this is a trivial objection because it is always possible, in principle, though in practice tedious, to establish a clone, test whether a particular group of enzymes is present, make a prediction about the survival of the clone in a particular environment, and then to carry out the test.

There is no difficulty in principle in deriving testable predictions from the hypothesis that inherited factors affect the expected number of offspring. In practice there are at least two problems. The first concerns measuring the effect of traits that *contribute to fitness*, but do not determine it. It is easy to predict the expected survival of a genotype that either is, or is not, lethal. Outside the laboratory there may be only small differences in the expected survival of two genotypes. Then it is difficult or usually impossible, to make exact quantitative calculations of the expected survival. Qualitative predictions are usually possible such as, plants with fewer stomata will lose less water and so tend to survive better in dry climates. Such qualitative predictions can still be tested by statistical methods.

The difficulty in making quantitative predictions does not change the scientific status of the assertion that differences in genotype can affect probability of survival of an organism. In the same way, quantum theory is not metaphysics simply because we are able only to calculate properties of the simplest molecules. Improved methods of calculation will allow more precise predictions and consequently more thorough testing. There is a major difficulty in relating genetic variation to phenotypic effects (Lewontin, 1980) but this difficulty does not stop qualitative predictions (Maynard Smith, 1978) from being made. A mutation may double the size of leaves in a particular environment and there is no problem in calculating the effects on water loss, light interception and leaf temperature—even though the biochemical mechanism for the change of leaf size is not known.

The second practical difficulty arises again in cases when there are only small differences in expected survival of genotypes. The difficulty is the stochastic (probabilistic) nature of survival and reproduction of individuals. In any particular case, there is no logical necessity that those individuals leaving the most offspring are the most fit. They may or may not be. Both selection *and*

random factors determine the survival of any particular individual's genotype. For instance, two thistle plants growing on a river bank may leave widely different numbers of offspring if one is undercut by the river and swept away, but clearly a biologist would be unwise to make claims about the relative fitness of the two genotypes. Genetic drift, mutation and migration will also lead to changes in genotype frequencies. The term 'differential reproduction' refers not to the survival of any particular individual's offspring but to the expected survival over many generations.

If the principle of natural selection is not tautological, why do so many say that it is? In some cases, the confusion arises from using the term 'fittest' with two quite distinct meanings: (1) those that leave the most offspring (the modern meaning); and (2) those most suited to their environment (Spencer's use of the term). In other cases, like Waddington and Dobshansky, they are misled by the fact that the principle is difficult to test, the unhappy result of our current inability to quantify 'fitness'. The 'tautology argument' is unfortunate for it distracts our attention from the real problem—why fitness is so hard to determine, and what can be done about the difficulty.

The problem is made more intractable by writers who feel that evolutionists should be able to make predictions about the survival of individual organisms. Hull admirably summarizes this viewpoint:

> 'If one only knew enough about the genetic make-up, the embryological development, and the physiology of the organisms concerned, as well as the vagaries of the environment, one could assign a certain degree of fitness to each of these organisms and hence be able to make reasonable predictions about their chances of survival. With this information, one could in turn predict subsequent changes in the population' (Hull, 1974a, p. 68).

Obviously, as Hull points out, we do not have all the required information and the temptation has been to slip into the tautological fallacy in the application of the principle, and to let survival become the criterion for fitness.

> 'One is tempted to rush by all such pragmatic considerations [of the type mentioned in the preceding quotation]. Perhaps biologists do not know all the relevant variables and could not combine them meaningfully if they did, but surely nature does the summing for us' (Hull, 1974a, p. 68).

This temptation to 'let nature do the summing for us' is of course the solution that leads to charges of a tautological criterion for fitness. The conclusion of this section is that, properly phrased, natural selection is not a tautology. We think that distinguishing between Hypotheses II and III does help eliminate this problem.

1.6. CRITICISM 4. EXPLANATORY AND PREDICTIVE POWER (Hypothesis III)

A number of writers have complained that the synthetic theory can explain too much. They argue that it is difficult to imagine any biological phenomenon that could not be explained by natural selection. Such writers include Brady, 1979; Cannon, 1958; Hairston, 1979; Lewontin, 1972, Little, 1980; Løvtrup, 1976; 1974; Medawar, 1974; Platnick, 1978; Polyani, 1974; Popper, 1976; Rosen, 1978; and Wasserman, 1978. Lewontin speaks for many critics when he says:

> 'Can one really imagine observations about nature that would disprove natural selection as a cause of the difference in bill size? The theory of natural selection is then revealed as metaphysical rather than scientific. Natural selection explains nothing because it explains everything' (Lewontin, 1972, p. 181).

The criticism has a nice ring to it, and intuitively one feels that it has some truth to it. However, most of the critics listed simply repeat it, parrot-fashion, without evidence or explanation. If they want to give an example they repeat *ad nauseam* the above hypothetical bird bill example from Lewontin. The critics do not divulge exactly what they mean when they say that natural selection 'does not explain anything or it explains everything...' (Waddington, 1968).

Maynard Smith (1972a) has already answered this extreme form of the criticism by doing what the critics say is impossible. He has given an example of a biological phenomenon that could not be explained by the synthetic theory:

> '... if someone discovers a deep-sea fish with varying numbers of luminous dots on its tail, the number at any one time having the property of always being a prime number, I should regard this as rather strong evidence against neo-Darwinism. And if the dots took up in turn the exact configuration of the various heavenly constellations, I should regard it as adequate disproof. The apparent absurdity of these examples only shows that what we know about existing organisms is consistent with neo-Darwinism' (Maynard Smith, 1972a, p. 87).

There are two complementary ideas here. The first is that we do *not expect* to find organisms that would cause us to reject natural selection, but secondly there are innumerable possible organisms that could not be explained by natural selection. The reply to Lewontin's bird-bill example must be: 'Yes, we can imagine innumerable such organisms that could not be explained by natural selection.' Natural selection is falsifiable in principle in that it is inconsistent with many imaginable organisms (see later).

However, there is a less extreme, but more serious, criticism that is related to the idea that the synthetic theory explains too much. It is that biologists invoke the principle of natural selection uncritically and without recourse to experimental evidence, to explain any and every biological phenomenon in terms of adaptation. Some critics go on to claim that explanations of this nature are not based on any general rule, will not yield testable predictions, and so cannot be regarded as scientific. The less extreme criticism will be considered after a general discussion of Hypothesis III.

The mechanisms of microevolution (natural selection and chance) are proposed, through changes in gene frequencies, to lead to the emergence of new taxa that are better adapted to their environment in comparison to their immediate ancestors. The theories of common ancestry and microevolution are synthesized to form a third 'hard core' or central theory, that of Darwinism. This states that the processes of microevolution are sufficient to account for the phenomenon of macroevolution. This hypothesis, like the other two, is based on a number of sub-hypotheses.

1.6.1. Variability in natural populations

The prediction from macroevolution is the amount and types of genetic diversity found in natural populations must be sufficient for the changes that have occurred during evolution. Many geneticists in the early part of this century would have argued that there was little diversity in natural populations. Mayr and Provine (1980) discuss the importance of early population studies which did find considerable genetic diversity. These studies were important to the development of the synthetic theory. More variation was found when proteins from different individuals were examined by electrophoresis (Lewontin, 1974). Molecular biology has now discovered new types of genetic diversity. In addition to single nucleotide substitutions there are small insertions and deletions, partial or complete duplication of genes, a complex gene structure in eukaryotes, and transposable genes (Terzaghi *et al.*, 1984). This is strong qualitative support for Darwinism but more specific studies are required. Most of the evidence comes from studies on the variability of structural genes and their regulation, simply because their biochemistry is known. During the next decade we can look forward to quantitative tests of predictions on the amount of variation in structural genes, and to estimates of the types and amount of variability of genes regulating development. If microevolution is sufficient to account for macroevolution, then it is necessary that there be extensive variability in genes that regulate development.

1.6.2. Function: A product of selection

Complex features of organisms must give an advantage to those organisms. By 1838 Darwin realized that natural selection explained the fit between the structure of an organism and its environment. However, he also realized that many imaginable organisms could not be explained by natural selection, even though they were consistent with creationism. An organism that would be consistent with creationism, but not with natural selection, is as follows. It could be an elephant with a complex structure on its back that acted only as a nest for a fish-eating bird, or any other bird that gave no benefit to the elephant. There are countless examples of this type that would be 'good design' but are prohibited by natural selection. There is no *a priori* reason why such structures would not be found when over a million species have been described. This does not mean that none will ever be found—only that Hypothesis III does make powerful but qualitative predictions (see also Popper's response to Lakatos, 1974, p. 1004).

1.6.3. Gradualism

Any morphological or biochemical change must be able to be generated by a series of intermediates that could be found in the normal variation of a population. There is no prediction of constant rates (see below) and it is clear that Darwin considered gradualism on an ecological or biological time scale (between generations), not on a geological time scale (see Peckham, 1959, pp. 211, 501, 517; Penny, 1983; Rhodes, 1983). During the past century many biological phenomena have been shown to be consistent with gradualism. Much of Darwin's later work is best understood as demonstrating how particular aspects of biology are consistent with gradualism, and explained by selection. For example it could be assumed that the beautiful and elaborate flowers of orchids were especially created for the delight, edification, and awe of humankind. In contrast, Darwin showed that the orchid flower was consistent with gradualism and selection. The flowers fulfilled a very important function in pollination, and they were formed by modifications to existing flower parts. His work on plant movements and on 'emotions' in humans and other animals is in the same category (Ghiselin, 1969). Goldschmidt (1940) however, gives a list of 17 features for which no satisfactory explanation existed at that time. The example of the origin of feathers is now less of a problem (see later). Other of these features have been partly solved as protein sequences have become available, for example snake venoms. Molecular evolution is strengthening gradualism both from the data it supplies and the mechanisms it uncovers.

The first step is to show that gradualism *could* lead to explanations for the evolution of a particular feature. Such explanations should produce new predictions that can be tested. An example is as follows. Many evolutionary tree building methods reconstruct hypothetical ancestral sequences for proteins or nucleic acids. This is consistent with gradualism in that small changes lead to the differences observed in living organisms. But in addition it is necessary to show that such ancestral sequences would have the appropriate biological activity. This could be done by synthesizing the molecule and testing for say, haemoglobin activity. However, it is likely that the first tests will be computer ones that calculate a three dimensional structure from ancestral sequences and compare them with present day proteins.

1.6.4. Adaptation

Some phenotypes are better able to use environmental resources than are others. For any particular feature that is claimed to be adaptive, it needs to be shown that the change can be generated through a series of intermediate forms (gradualism), and that each intermediate form is of advantage to members of the population or at least is not disadvantageous, (natural selection). Adaptation will be discussed in more detail later.

1.6.5. Complexity

There must be a mechanism that leads to an increase in functional information coded in the genome. It has been one of the successes of molecular biology that it has demonstrated a mechanism of gene duplication, followed by divergence, that explains the origin of many new biochemical features (Terzaghi *et al.*, 1984).

1.6.6. Variable rates

Darwin was careful to stress that the mechanism he proposed would *not* lead to a constant rate of morphological evolution (Maynard Smith, 1982b, p. 126; Penny, 1983; Rhodes, 1983), partly because of the irregular nature of genetic variability, mutation substitution and environmental changes. There has been considerable confusion between Darwinism and orthogenesis. Orthogenesis does predict slow, continuous changes over geological time periods, and then the fossil record could appear as phyletic gradualism.

1.6.7. Absence of planning—lack of perfection

Early biologists assumed that a creator made organisms that were perfectly adapted to their environment. Darwin realized that under his theory there was no reason for organisms to be perfectly adapted. One of his favourite examples was the observation that some plants and animals flourished when transported to a new country, and could even replace the native species. This would not happen if the native organisms were perfectly adapted to their environment. Evolution is opportunistic in that it can only act on the genetic variation that is present in a population. There are 'genetic' and 'functional' constraints on what is possible. Many potentially functional genotypes are difficult to derive from existing genotypes by a series of functional intermediates (see Gould, 1980). Consequently we find that different solutions occur to the same problem such as adaptations of plants to water stress.

All of these hypotheses are controversial. Gradualism and variable rates are both contested by the 'punctuated equilibriists' (Gould and Eldredge, 1977; Stanley, 1979). However, widest dissatisfaction concerns the explanations derived from the hypothesis of adaptation. In both Darwin's original theory and the synthetic theory, one of the main concerns is to explain the complexity and diversity of life. In both theories the concept of change through adaptation is the principle means of doing this. When organisms are compared in the light of the theory of descent from a common ancestor, the idea of adaptation assumes great significance. This explanation of biological complexity and diversity, or macroevolution, has replaced the earlier explanation proposing a divine designer. Critics believe 'adaptive' explanations are scientifically inadequate for the following reasons.

1. The terms adaptation and adaptedness are inadequately defined. It is claimed that biologists explain any and every morphological, physiological, and behavioural trait as an adaptation, without ever making clear how one determines what is an adaptation and what is not (Gould and Lewontin, 1979; Lewontin, 1979). The term 'adaptation' is used with several different meanings. Mayr (1982) points out that it is used to refer to the process of becoming adapted; to the end stage of having achieved adaptedness; to non-genetic physiological changes in individuals (for example humans adapting to high altitude); to changes in gene frequency from continued selection pressure; to single components of the phenotype; or to an over-all fit between an individual and its environment (more or less synonymous with one of the meanings of fitness).

   There is a similar confusion in the definition of 'adaptedness'. The environment is constantly changing (Van Valen, 1973) and an organism's adaptedness cannot be measured against a scale of absolute adaptedness. Adaptedness is also a comparative measure, either between a specific feature of different individuals, or between whole organisms or populations. However, unless we can specify which features of the environment determine adaptedness, it is very difficult to make an accurate comparison.

   Gould and Vrba (1982) have gone some way to help this problem by suggesting that different terms be used to distinguish between features built by natural selection for their current function, and those features now enhancing fitness no matter how they arose. Of course the problem remains to tell which is which! The difficulties in terminology cause confusion and make it difficult to determine whether the hypothesis of adaptation meets our criterion of leading to falsifiable predictions.
2. The second criticism is that adaptive explanations are *ad hoc* and appeal to no general rule. Explanations of the type, 'mammals survived the period in which dinosaurs became extinct because mammals were better adapted to the increasingly cooler temperatures', and that 'giraffes have long necks because they were selectively advantageous over short necks in the giraffes' environment' appear to claim no generality beyond the particular set of facts which they seek to explain.

This criticism can be met in two ways. Firstly, it is possible to make and to test predictions about unique events. Secondly, there is often a pattern of response in different organisms that can be used to form a general rule. These will be discussed in turn.

The history of any taxon is unique and some aspects of its evolution may lack generality. This does not mean that such explanations are unfalsifiable in principle.

> '... the fact that all laws of nature are hypotheses must not distract our attention from the fact that not all hypotheses are laws, and that more especially historical hypotheses are, as a rule, not universal but singular statements about one individual event, or a number of events' (Popper, 1960, p. 107).

The following is an example of how such an explanation can lead to falsifiable predictions. It concerns the familiar conjecture that bird feathers were an adaptation for flight. (This hypothesis raised difficulties for gradualism because it was difficult to imagine what advantage there would be to individuals in the very early stages in the evolution of feathers.) A prediction from this hypothesis would be that none of the non-flying ancestors of birds had feathers. The discovery of feathered, but non-flying, animals (related to dinosaurs and presumably ancestral to birds) is considered to have disproved the hypothesis. In this case the prediction was not only falsifiable, but falsified, and it is now suggested that the earliest function of feathers was insulation. It is therefore possible to make predictions from hypotheses about specific adaptations. As such, they are in the realm of science, as judged by our criterion. More precise versions of such hypotheses should be sought in order to be able to make quantitative predictions.

Our second comment is that in many cases there is an attempt to extend the adaptive interpretation to a general case. Examples will be found in Clutton-Brock and Harvey (1979) who compare many primate species and seek generalizations such as, 'leaf-eating primates have smaller ranges than fruit-eating primates of the same size'. Many other examples are possible, such as that 'plants with C4 photosynthesis survive better in warm dry habitats than C3 plants'. Many of these predictions are qualitative and/or statistical but if the predictions are falsifiable, then it meets our criterion of being a scientific hypothesis. This leads into the third criticism of many adaptive explanations.

3. It is impossible to calculate the effects of all relevant variables in order to determine the effect of some phenotypic feature; this means that 'nature is left to do the summing'.

Biologists use Darwinism to give explanations as to why some species are preserved while others become extinct; why individuals of the same species have different forms of a certain character; and how complex morphological, physiological, and behavioural characters have evolved from simpler forms.

All of these types of explanations require a comparison of relative adaptedness and its contribution to fitness. Such explanations will only be scientific if there is, at the very least, an accurate way of judging adaptedness. The temptation has often been to say, that because it survived, it must have been adapted to its environment. Again, evolutionists have been tempted to let nature do the summing up.

> 'If one simply cannot measure the state variables or the parameters with which the theory is constructed, or if their measurement is so laden with error that no discrimination between alternative hypotheses is possible, the theory becomes a vacuous exercise in formal logic that has no points of contact with the contingent world' (Lewontin, 1974, p. 11).

Brandon (1978a,b) recognizes that an adequate definition of relative adaptedness is important to the synthetic theory. He points out that there is a fundamental law (D) which works in evolutionary explanations such that:

> 'If a is better adapted than b in environment E, then (probably) a will leave more (sufficiently similar) offspring than b in E' (Brandon, 1978b, p. 187).

Unless relative adaptedness *and* its contribution to fitness can be measured, it will remain impossible to make quantitative predictions about the probability of survival of genotypes. The difficulty is that it is not a simple matter to decide how to analyse an organism into parts, and then to determine whether a particular feature is a product of selection, or functional constraints, or pleiotropic action, or allometric effects, or any combination of these factors.

The problem of defining relative adaptedness prevents quantitative predictions that could then be thoroughly tested. However, as we have shown above, it is still usually possible to make qualitative predictions, or predictions of a statistical nature. These are not so easy to test as thoroughly as quantitative tests, but as long as they can be disproved, they belong to the realm of science rather than metaphysics. Hypotheses about adaptation are in principle scientific because such hypotheses can, as the feather example shows, be falsified. We would agree that there are many cases of poor science masquerading under the cover of adaptive explanations and also of scientists protecting hypotheses, rather than trying to test them critically. We would also agree that there are many cases where natural selection has been invoked, when there is no evidence to suppose that anything but chance factors are involved. Better science should eliminate these cases, the problem is with the scientists rather than with the theory being non-scientific.

It has been mentioned previously that neutralists have argued that much of the inherited variability in a population has no selective advantage or disadvantage, and that many sequence changes are neutral. There have been suggestions that such neutral changes are 'non-Darwinian' (for example, Gould, 1982). This is a misunderstanding. The problem, as recognized since the eighteenth century (Burkhardt, 1977), is how to explain those features that cannot be due to chance alone. For while Darwin demonstrated that, in principle, such features could be explained by natural selection, he was careful to point out that not all features of an organism need serve a useful function, either now or in the past (Huxley, 1982; Terzaghi *et al.*, 1984). Darwin never excluded chance factors. He talked about *some* features being useful or injurious, a *tendency* for an inherited variant to increase (see entries in Barrett *et al.*, 1981). Indeed one objection to his theory was that it relied too much on 'blind chance', such as variation being unrelated to the needs of the organism. Our conclusion is that neutral or random events are a part of microevolution and that *the real question is whether microevolution is sufficient to account for macroevolution*. We advocate that from an operational viewpoint it is better to assume that a particular feature has no advantage unless it can be shown that the feature cannot be explained by chance.

The question in hand is the scientific status of Darwinism, whether microevolution is sufficient to account for macroevolution. Qualitative predictions are possible in several areas including: the amount of inherited variability in populations, the existence of a mechanism for the origin of new proteins (which serves as a model for gradualism at other levels of development), and in explanations that a particular feature is an adaptation that increases fitness. It is also apparent that Darwin saw these sub-hypotheses as being falsifiable (Williams, 1973). Thus our third hypothesis is an example of a theory that can only be tested qualitatively, and until the predictions can be made quantitative, it is not possible to decide whether our current understanding of microevolution is sufficient to explain all, or only part, of macroevolution.

1.7. CRITICISM 5. LACK OF EVOLUTIONARY LAWS

The nature of evolutionary laws has been questioned by a number of critics of the synthetic theory, including Goudge (1961), Scriven (1958), and Smart (1963), all of them claiming that there can be no evolutionary laws. They have, for the most part, felt that this is so because the theory describes a unique, historical process:

> 'The fact that the evolutionary hypothesis is not a universal law of nature, but a particular (or more precisely, singular) historical statement about the ancestry of a number of terrestrial plants and animals is somewhat obscured by the fact that the term "hypothesis" is so often used to characterize the status of universal laws of nature...' (Popper, 1960, p. 107).

Since much of evolutionary theory is more concerned with describing a generalized mechanism of evolutionary change through natural selection than with outlining unique phylogenies, Popper's statement is as unjustified as claiming that there can be no laws of astronomy because the history of the sun is unique. To be fair to Popper, however, he is aware of this distinction. It is the possibility of formulating generalized rules of evolutionary change, which are comparable to those found in the physical sciences, that concerns him.

While laws of molecular biology, such as those regarding inheritance, do function in the synthetic theory, it is true that there are no deterministic evolutionary laws of selection, speciation, or extinction, that compare with the deterministic macroscopic laws of chemistry and physics. However, since there are fundamental features of biological entities which distinguish them from strictly physical phenomena, it is hardly surprising that there are differences in the structure of the respective theories of evolution and Newtonian mechanics. As Lewontin (1974) points out:

> 'The metaphysical introduction of ideal bodies moving in ideal paths, so essential to the proper development of physics and so consonant with the habits of thought of the seventeenth century, was precisely what had to be destroyed in the creation of evolutionary biology. Darwin rejected the metaphysical object and replaced it with the material one. He called attention to the *actual* variation among *actual* organisms as the most essential and illuminating fact of nature. Rather than regarding the variation among members of the same species as an annoying distraction, as a shimmering of the air that distorts our view of the essential object, he made that variation the cornerstone of his theory' (Lewontin, 1974, p. 6).

Thus the probabilistic component is important in biological theories. Biologists seek generalizations about the types of responses that populations have been able to make in their environment (see Clutton-Brock and Harvey, 1979). Frequently there are many alternatives that may solve a problem (such as for plants growing in dry areas), and the particular solution will usually depend on the features of the initial population. We see the problem as making it difficult to obtain quantitative predictions, rather than being a fundamental objection.

1.8. CRITICISM 6. SOCIAL AND POLITICAL FACTORS

The last type of criticism is that the success of Darwinism can be explained as merely the application of prevailing social and political ideas to biology (see the extract in Section 1.1 from Radl, 1930). It must first be pointed out that it is almost universally accepted that science is not independent of the knowledge of a society at a particular stage of its development. As an example, it was not possible to establish a model for the motion of the planets without the observations that led to the recognition that a class of objects (planets) had a quite different behaviour from the stars. Similarly the problem of the origin of species was not acute as long as biologists assumed that most species arose by spontaneous generation.

Despite this limitation, we do not find the explanation of Radl at all convincing for the success of evolutionary theory. This is because the initial success of evolution in biology (as opposed to sociology) was in convincing people of the 'theory of descent', rather than in convincing them that natural selection was the main mechanism. It has been estimated that at the end of the nineteenth century nearly all biologists accepted the theory of descent, but that no more than a third accepted natural selection as a sufficient mechanism (Provine, 1982). A major factor in this non-acceptance of natural selection is that such a 'blind chance' mechanism was not convincing to people expecting a highly ordered and deterministic Newtonian universe. Within biology there was no immediate success of what we call Hypothesis III, and the reason was that the mechanism was a radical departure from expected deterministic ideas (see also Mayr, 1982).

1.9. DISCUSSION AND CONCLUSIONS

The question of the scientific status of evolutionary theory has attracted the interest of many biologists and philosophers. In order to examine the problems, we have found it necessary first to consider modern treatments of scientific methodology. There is a diversity of views on scientific methods and we have chosen to consider the question whether it is possible for scientists to choose rationally between major competing theories. While recognizing that cultural factors are always influencing both the problems that are considered soluble and the types of solutions that seem feasible, we believe that there is an increasing acceptance by scientists that they must take a critical attitude to their theories.

Our analysis leads us to conclude that theories often cannot be tested directly. It is predictions that can be tested rather than the theories themselves. This inability to test theories directly is due in part to the fact that several hypotheses are often used in making a prediction, and in part because it is often unclear whether the problem is with the experiment or with the theories.

In this chapter we have used the criterion of falsifiable predictions to consider the scientific status of modern evolutionary theory. It is our experience the analysis is simpler if the problem is broken into three: evolutionary history, microevolutionary studies, and the question of whether microevolution is sufficient to explain the observed features of all organisms (Darwinism).

For a long time the theory of descent has led to qualitative predictions, and these have been supported, for example from a general agreement between classical phylogenies with those derived from protein and nucleic acid sequences. Such qualitative agreement is encouraging but has not convinced all workers. It is now possible to make quantitative tests of predictions from the theory of descent. This has been done with sequences of macromolecules but there is no reason why the methods should not be applied to other biological data. Our conclusion is that the theory of descent does meet the criteria of falsifiable predictions.

We have defined microevolution to include an ecological aspect and a genetic aspect. The most frequent criticism is that the principle of natural selection is a tautology. This would imply that natural selection was merely the empty repetition of a definition, and consequently had no empirical use. Our conclusion is that this criticism is trivial because it is due to semantic confusion. Rewording the principle as: 'some genetic variants in a population alter the probability of survival of descendants' eliminates the problem because such an assertion can be falsified. The tautology criticism tended to draw attention away from a real problem which is how to make quantitative predictions about expected survival, rather than just qualitative predictions.

The third aspect of our sub-division is 'Darwinism'. This is the hypothesis that the mechanisms discovered in microevolution are sufficient to explain the observed features of organisms. It is this aspect of evolution that leads to the most difficulties. Some difficulties are minor, such as not making the distinction between what is possible, and what is expected. The claim that there is no conceivable observation that could disprove natural selection is quite simply wrong. There are literally millions of logically possible observations that would be inconsistent with evolution by natural selection.

We expect that molecular studies will prove particularly valuable in allowing more quantitative tests of Hypothesis III, and of evolutionary theory in general. We have already referred to the use of protein and nucleic acid sequences in testing the theory of descent. This will be a help in testing models in other areas because as we have already commented, it is difficult to make precise tests about the possible selective advantage of a feature, if the phylogeny of the organism is uncertain.

Another area where molecular studies are becoming important is in measuring the amount of genetic variation in populations. These studies have shown that natural populations do have much higher levels of genetic variability than many authors had thought possible (see Lewontin, 1974). Most of the quantitative work has been on protein variation, there is a need for quantitative studies of DNA sequences, both in the amount and types of variation present (see Terzaghi *et al.*, 1984). Studies of this kind would allow a more precise testing of whether the observed molecular information is needed before this model can be tested for its usefulness in explaining the development of new organs.

Evolutionary theory will continue to develop and be subjected to more exacting tests of its predictions. The problems are scientific, rather than with any inherent philosophical problems with evolutionary theory. The greatest difficulty lies in testing Hypothesis III, the idea that microevolution is sufficient to account for all evolutionary phenomena. Many of these tests will come from molecular biological studies, there is a need for imaginative, but critical, scientists to use molecular information for evolutionary studies. The later chapters in this book outline many interesting problems that await quantitative analysis.

ACKNOWLEDGEMENTS

We would particularly like to thank D. G. Graham for many critical discussions, and also Drs A. S. Wilkins, P. Penny, B. and E. Terzaghi, and D. G. Lloyd for their helpful comments on the manuscript.

1.10 REFERENCES

Barker, A. D. (1969) An approach to the theory of natural selection. *Philosophy*, **44**, 271–290.

Barrett, P. H., Weinshank, D. J., and Gottleber, T. T. (1981) *A Concordance to Darwin's Origin of Species*, (*1st Edn*), Cornell University Press, Ithaca.

Bethell, T. (1976) Darwin's Mistake. *Harpers*, **252**, 70–75.

Bohm, D. (1968) Addendum on order and neo-Darwinism,' in Waddington, C. H. (ed.) *Towards a Theoretical Biology*, Edinburgh University Press, Edinburgh, Vol. 2.

Brady, R. H. (1979) Natural selection and the criteria by which a theory is judged. *Syst. Zool.*, **28**, 600–621.

Brady, R. H. (1982) Dogma and doubt. *Biol. J. Linn. Soc.*, **17**, 79–96.

Brandon, R. N. (1978a) Evolution. *Phil. Sci.*, **45**, 96–109.

Brandon, R. N. (1978b) Adaptation and evolutionary theory. *Stud. Hist. Phil. Sci.*, **9**, 181–206.

Burkhardt, R. W. Jr (1977) *The Spirit of System: Lamarck and Evolutionary Biology*. Harvard University Press, Cambridge, Mass.

Campbell, M. (1978) The theory of natural selection—its status and adequacy. *Method. Sci.*, **11**, 129–145.

Cannon, H. G. (1958) *The Evolution of Living Things*. Manchester University Press, Manchester.

Caplan, A. L. (1979) Darwinism and deductivist models of theory structure. *Stud. Hist. Phil. Sci.*, **10**, 341–353.

Caplan, A. L. (1981) Popper's philosophy. *Nature*, **290**, 623–624.

Clutton-Brock, T. H., and Harvey, P. H. (1979) Comparison and adaptation. *Proc. R. Soc. Lond.*, **205B**, 547–565.

Crick, F. H. C. (1958) On Protein Synthesis. In, *The Biological Replication of Macromolecules*, 12th Symp. Soc. Exp. Biology. C.U.P. Cambridge, pp. 138–163.

Dobzhansky, Th. (1968) Adaptedness and fitness, in Lewontin, R. C. (ed.) *Population Biology and Evolution*, Syracuse University Press, Syracuse, N.Y.

Ferguson, A. (1976) Can evolutionary theory predict? *Amer. Natur.*, **110**, 1101–1104.

Feyerabend, P. K. (1970a) Consolations for the specialist, in Lakatos, I., and Musgrave, A. (eds.) *Criticism and the Growth of Knowledge*, Cambridge University Press, Cambridge, pp. 197–230.

Feyerabend, P. K. (1970b) Against Method, outline of an anarchistic theory of knowledge, in Radner, M., and Winokur, S. (eds.) *Minnesota Studies in Phil. Sci.*, University of Minnesota Press, Minnesota, Vol. IV, pp. 17–130.

Feyerabend, P. K. (1975) How to defend society against science. *Radical Philosophy*, **No 11, Summer**, pp. 55–65.

Ghiselin, M. T. (1969) *The Triumph of the Darwinian Method*. U.C. Press, Berkeley.

Goldschmidt, R. B. (1940) *The material basis of evolution*. Yale University Press, New Haven, Conn.

Goudge, T. A. (1961) *The Ascent of Life: A Philosophical Study of the Theory of Evolution*. University Toronto Press, Toronto.

Gould, S. J. (1980) *The Panda's Thumb: More Reflections on Natural History*, W. W. Norton, N.Y.
Gould, S. J. (1982) Darwinism and the expansion of evolutionary theory, *Science*, **216**, 380–387.
Gould, S. J., and Eldredge, N. (1977) Punctuated equilibria, the tempo and mode of evolution reconsidered, *Paleobiology*, **3**, 115–151.
Gould, S. J., and Lewontin, R. C. (1979) The spandrels of San Marco and the Panglossian paradigm: a critique of the adaptationist programme, *Proc. R. Soc. Lond.*, **205B**, 581–598.
Gould, S. J., and Vrba, E. S. (1982) Exaptation—a missing term in the science of form, *Paleobiology*, **8**, 4–15.
Graham, R. L., and Foulds, L. R. (1982) Unlikelihood that minimal phylogenies for a realistic biological study can be constructed in reasonable computing time, *Math. Biosc.*, **60**, 133–142.
Grene, M. (1974) *The Understanding of Nature*, Boston Studies in the Phil. Sci., Reidel, Dordrecht, Vol. 66, D.
Hairston, N. G., Jr (1979) Fitness, survival and reproduction, *Syst. Zool.*, **28**, 392–395.
Haldane, J. B. S. (1935) Darwinism under revision, *Rationalist Ann.*, pp. 19–29.
Halstead, B. (1980) Popper: good philosophy, bad science? *New Sci.*, **87**, 215.
Hendy, M. D., and Penny, D. (1982) Branch and bound algorithms to determine minimal evolutionary trees, *Math Biosc.*, **59**, 277–290.
Hendy, M. D., Little, C. H. C., and Penny, D. (1984) Comparing trees with pendant vertices labelled, *SIAM J. Appl. Math.*, **44**, (in press).
Hull, D. L. (1974a) *Philosophy of Biological Science*, Prentice-Hall, Englewood Cliffs, N.J.
Hull, D. L. (1974b) Review of Ruse, The Philosophy of Biology. *Studies in Hist. and Phil. Sci.*, p. 73.
Hull, D. L., Tessner, P. D., and Diamond, A. M. (1978) Planck's principle, *Science*, **202**, 717–723.
Huxley, A. F. (1982) Address of the president, *Proc. Roy. Soc. Lond.*, **214B**, 137–152.
Huxley, J. (1974) *Evolution, The Modern Synthesis*, (3rd Edn.) George Allen and Unwin, London.
Kimura, M., and Ohta, T. (1974) On some principles governing molecular evolution, *Proc. Natl. Acad. Sci. USA*, **71**, 2848–2852.
Kreitman, M. (1983) Nucleotide polymorphism at the alcohol dehydrogenase locus of *Drosophila melanogaster*, *Nature*, **304**, 412–417.
Kuhn, T. S. (1970) *The Structure of Scientific Revolutions*, (2nd Edn.), University of Chicago Press, Chicago.
Lack, D. (1966) *Population Studies of Birds*, Clarendon Press, Oxford.
Lakatos, I. (1970) Falsification and the methodology of scientific research programmes, in Lakatos, I., and Musgrave, A. (eds.) *Criticism and the Growth of Knowledge*, Cambridge University Press, Cambridge, pp. 91–195.
Lakatos, I. (1974) Popper on demarcation and induction, in Schilpp, P. A. (ed.) *The Philosophy of Karl Popper*, Open Court, La Salle, Illinois, Book 1.
Lakatos, I. (1976) History of science and its rational reconstructions, in Howson, C. (ed.) *Method and Appraisal in the Physical Sciences*, Cambridge University Press, Cambridge.
Lee, K. K. (1969) Popper's falsifiability and Darwin's natural selection, *Philosophy*, **44**, 291–302.
Lewontin, R. C. (1972) Testing the theory of natural selection, *Nature*, **236**, 181–182.
Lewontin, R. C. (1974) *The Genetic Basis of Evolutionary Change*, Columbia University Press, N.Y., pp. 156–168.

Lewontin, R. C. (1979) Sociobiology as an adaptionist programme, *Behavioural Science*, **24**, 5–14.
Lewontin, R. C. (1980) Models in natural selection, in Barigozzi, C. (ed.) *Vito Volterra Symposium on Mathematical Models in Biology*, Springer-Verlag, Berlin.
Little, J. (1980) Evolution: myth, metaphysics, or science, *New Scientist*, **87**, 708–709.
Lovejoy, A. O. (1936) *The Great Chain of Being*, Harvard University Press, Cambridge, Mass.
Løvtrup, S. (1976) On the falsifiability of neo-Darwinism, *Evol. Theory*, **1**, 267–283.
Løvtrup, S. (1981) Macroevolution and punctuated equilibria, *Syst. Zool.*, **30**, 498–500.
Løvtrup, S. (1982) The four theories of evolution, *Rivista di Biologia*, **75**, 53–66; 231–272: 385–409.
MacBeth, N. (1971) *Darwin Retried*, Gambit, Boston.
Manier, E. (1980) Darwin's language and logic, *Stud. Hist. Phil. Sci.*, **11**, 305–323.
Manser, A. R. (1965) The concept of evolution, *Philosophy*, **40**, 18–34.
Maynard Smith, J. (1972) *On Evolution*, Edinburgh University Press, U.K.
Maynard Smith, J. (1978) Optimization theory in evolution, *Ann. Rev. Ecol. Syst.*, **9**, 31–56.
Maynard Smith, J. (1980) *Science, ideology and myth*, J. D. Bernal Memorial Lecture, Royal Society, (unpublished).
Maynard Smith, J. (1982a) *Evolution and the Theory of Games*, Cambridge University Press, Cambridge.
Maynard Smith, J. (ed.) (1982b) *Evolution Now*, MacMillan, London.
Mayr, E. (1978) Evolution, *Sci. Amer.*, **239(3)**, 38–47.
Mayr, E. (1982) Adaptation and selection, *Biol. Zbl.*, **101**, 161–174.
Mayr, E., and Provine, W. B. (eds.), (1980) *The Evolutionary Synthesis: Perspectives on the Unification of Biology*, Harvard University Press, Cambridge, Mass.
Medawar, P. (1974) Comment on 'On chance and necessity', in Ayala, F. J., and Dobzhansky, T. (eds.) *Studies in the Philosophy of Biology*, MacMillan, London, p. 363.
Mickevich, M. F. (1978) Taxonomic congruence, *Syst. Zool.*, **27**, 143–158.
Olding, A. (1978) A defence of evolutionary laws, *Brit. J. Phil. Sci.*, **29**, 131–143.
Orgel, L. E. (1979) Selection *in vitro*, *Proc. Roy. Soc. London*, **205B**, 435–442.
Ospovat, D. (1981) *The development of Darwin's theory: Natural history, natural theology and natural selection, 1838–1859*, Cambridge University Press, Cambridge.
Patterson, C. (1980) Cladistics, *Biologist*, **27**, 234–240. (Reprinted in Maynard Smith, 1982b.)
Peckham, M. (ed.) (1959) *The Origin of Species by Charles Darwin, A Variorum Text*, University of Pennsylvania Press, Philadelphia.
Penny, D. (1983) Charles Darwin, gradualism and punctuated equilibria, *Syst. Zool.*, **32(1)**, 72–74.
Penny, D., Foulds, L. R., and Hendy, M. D. (1982) Testing the theory of evolution by comparing phylogenetic trees constructed from five different protein sequences, *Nature*, **297**, 197–200.
Peters, R. H. (1976) Tautology in evolution and ecology, *Amer. Nat.*, **110**, 1–12.
Platnick, N. I. (1977) Review of Mayr, 1976, *Syst. Zool.*, **26**, 224–228.
Platnick, N. I. (1978) Adaptation, selection and falsifiability, *Syst. Zool.*, **27**, 347–348.
Platnick, N. I. (1980) Philosophy and the transformation of cladistics, *Syst. Zool.*, **28**, 537–546.
Polanyi, M. (1974) Genius in science, in Cohen, R. S., and Wartofsky, M. W. (eds.) *Methodological and Historical Essays in the Natural and Social Sciences*, Reidl, Dordrecht, pp. 57–61.

Popper, K. R. (1959) *The Logic of Scientific Discovery*, Hutchinson, London.
Popper, K. R. (1960) *The Poverty of Historicism*. (2nd end. Routledge and Kegan Paul, London.
Popper, K. R. (1965) *Conjectures and Refutations; The Growth of Scientific Knowledge*, Routledge and Kegan Paul, London.
Popper, K. R. (1972) Evolution and the tree of knowledge, and Of clouds and clocks, in *Objective Knowledge: an evolutionary approach*, Oxford University Press, Oxford.
Popper, K. R. (1974) Replies to my critics, in Schilpp, P. A. (ed.) *The Philosophy of Karl Popper*, Open Court, La Salle, Ill. Book II.
Popper, K. R. (1976) Darwinism as a Metaphysical Research Programme, in *Unended Quest: An Intellectual Autobiography*, Fontana, London, pp. 167–180.
Popper, K. R. (1978) Natural selection and the emergence of mind, *Dialectica*, **32**, 339–355.
Provine, W. B. (1982) Influence of Darwin's ideas on the study of evolution, *BioScience*, **32**, 501–506.
Radl, E. (1930) *History of Biological Theories*, Oxford University Press, Oxford.
Ragan, M. A., and Chapman, D. J. (1978) *A Biochemical Phylogeny of the Protists*, Academic Press, New York.
Rhodes, F. H. T. (1983) Gradualism, punctuated equilibrium, and the *Origin of Species. Nature*, **305**, 269–272.
Rohlf, F. J. (1982) Consensus indices for comparing classifications, *Math. Biosc.*, **59**, 131–144.
Root-Bernstein, R. (1981) Views on evolution, theory, and science, *Science*, **209**, 289.
Rosen, D. E. (1978) Darwin's demon, a review of *Introduction to Natural Selection* by C. Johnson, *Syst. Zool.*, **27**, 370–373.
Ruse, M. (1977b) Karl Popper's philosophy of biology, *Phil. Sci.*, **44**, 638–661.
Ruse, M. (1981) Darwin's Theory; an exercise in Science, *New Scientist*, **25 June**, pp. 828–830.
Salisbury, E. J. (1942) *The Reproductive Capacity of Plants: Studies in Quantitative Biology*, G. Bell and Sons, London.
Scriven, M. (1959) Explanation and prediction in evolutionary theory, *Science*, **130**, 477–482.
Simpson, G. G. (1949) *The Meaning of Evolution*, Yale University Press, New Haven, Conn.
Sneath, P. H. A., and Sokal, R. R. (1973) *Numerical Taxonomy, The Principles and Practice of Numerical Classification*. Freeman, San Francisco.
Smart, J. J. C. (1963) *Philosophy and Scientific Realism*, Humanities Press, New York.
Sparkes, J. (1981) What is this thing called science?, *New Scientist*, **89**, 156–158.
Stanley, S. M. (1979) *Macroevolution: Pattern and Process*, Freeman, San Francisco.
Stebbins, G. L. (1974) Adaptive shifts and evolutionary novelty: a compositionalist approach, in Ayala, F. J., and Dobzhansky, T. (eds.) *Studies in the Philosophy of Biology, Reduction and related problems*, MacMillan, London, pp. 285–306.
Stevens, G. (1980) *New Zealand Adrift*, A. H. and A. W. Reed, Wellington.
Terzaghi, E., Wilkins, A. S., and Penny, D. (1984) *Molecular Evolution: An Annotated Reader*, Jones and Bartlett, San Francisco.
Tuomi, J. (1981) Structure and dynamics of Darwinian evolutionary theory, *Syst. Zool.*, **30**, 22–31.
Van Valen, L. (1973) A new evolutionary law, *Evol. Theory*, **1**, 1–80.
Van Valen, L. (1974) Molecular evolution as predicted by natural selection, *J. Mol. Evol.*, **3**, 89–101.
Waddington, C. H. (1960) Evolutionary adaptations, in *The Evolution of Life*, Yax, S. (ed.) Univ. Chicago Press, Chicago.

Waddington, C. H. (1968) *Towards a Theoretical Biology, NBS Symp. 2*, University of Edinburgh Press, Edinburgh.

Wasserman, G. D. (1978) Testability of the role of natural selection within theories of population genetics and evolution, *Brit. J. Phil. Sci.*, **29**, 223–242.

Williams, M. B. (1970) Deducing the Consequences of Evolution, A Mathematical Model, *J. Theor. Biol.*, **29**, 343–385.

Williams, M. B. (1973) Falsifiable Predictions of Evolutionary Theory, *Philos. Sci.*, **40**, 518–537.

Zuckerkandl, E. (1976) Evolutionary processes and evolutionary noise at the molecular level II. A selectionist model for random fixations in proteins, *J. Mol. Evol.*, **7**, 269–311.

Zuckerkandl, E., and Pauling, L. (1962) Molecular disease, evolution, and genic diversity, in Kasha, M., and Pullman, B. (eds.) *Horizons of Biochemistry*, Academic Press, New York, pp. 189–225.

Evolutionary Theory: Paths into the Future
Edited by J. W. Pollard

Chapter 2

Cladistics: Theory, purpose, and evolutionary implications

PHILIPPE JANVIER

Centre National de la Recherche Scientifique, E.R.A. 963
Laboratoire de Paléontologie des Vertébrés,
Université Paris VI
4 place Jussieu
75230 Paris Cedex 05
France

2.1. INTRODUCTION

Phylogenetic systematics, now often referred to as 'cladistics', 'cladism', 'Hennigism', or 'Hennigian method', was introduced among biologists by the German entomologist Willi Hennig (1950), with a book which is a milestone in the history of comparative biology and evolutionary science. Hennig's goal was an objective method for reconstructing phylogenies, that is the genealogies of organisms, and for establishing classifications which reflect those phylogenies. His method was a response to the widespread habit, among biologists and palaeontologists, of classifying organisms on the basis of heterogeneous criteria (morphological, ethological) and according to their over-all resemblance. Since the eighteenth century, most classifications reflected the personal opinion of 'specialists', more than the real history of the groups, and they depended largely on the importance given to one particular character by the specialist in question, hence the often ludicrous debates about classifications which contributed to the popular image of the 'naturalist' as out-dated biologist. By tying classifications to unique, non-arbitrary events, i.e. the history of life itself, Hennig overthrew the weight of established authority in this field of life science and he also achieved, to some extent, Darwin's prediction that classifications would become 'genealogies' (Tassy, 1983).

Although the interest for Hennig's ideas was recognized as early as the 1950s by Kiriakoff (1955), Günther (1956), and Brundin (1956), discussions and polemics about phylogenetic systematics did not become general until two decades later. This delay was probably due to language barriers, and was overcome by the translation of a much revised version of Hennig's book into English by Davis and Zangerl (1966), as well as Brundin's (1966, 1968) popularization of Hennig's ideas. These publications played a major role in revolutionizing methods in phylogenetics and systematics.

Besides his *magnum opus*, Hennig wrote a number of earlier notes (Hennig, 1936, 1948, 1949) in which his ideas are clearly foreshadowed, and several later notes in defence of his method (Hennig, 1953, 1974, 1975). He died in 1976, just as a worldwide debate on cladistics began. This debate can be followed through the last decade's issues of the journal *Systematic Zoology* and, in a more inglorious form, in recent volumes of *Nature*. Serving as an open battlefield, *Systematic Zoology* will certainly stand out in the history of science as one of the major catalysts for this renewal of biological thought and I encourage the interested student to read the series of papers in the journal which chronicles the evolution of cladistics from classical Hennigian phylogenetic systematics to the so-called 'transformed cladistics' (or 'modern cladistics', 'pattern cladistics'). According to Brundin (1972), Hennig's contribution will be regarded as 'one of the major methodological advances on causal biology after Darwin'. The history of phylogenetic systematics can also be found summarized remarkably thoroughly by Dupuis (1978), and the reader is referred to that paper for extensive data on the early debates about Hennig's work. On a much wider canvas, original views on the place of cladistics in the history of biology are expressed in Nelson and Platnick (1981).

The goal of this paper is to present the general principles of Hennig's phylogenetic systematics, as well as their subsequent transformation by some later theoreticians of cladistics. I shall also try to forecast possible future developments of cladistics by considering the material and human factors which condition natural science in western countries.

The fundamentals of cladistics have already been explained many times, either in semi-popular form (Bonde, 1975; Patterson, 1978, 1981a, 1982a; Janvier *et al.*, 1980; Goujet, 1981), or in excellent textbooks (Cracraft, 1979; Eldredge and Cracraft, 1980; Nelson and Platnick, 1981; Wiley, 1981), and I shall concentrate on some controversial aspects of cladistics, which may be sources of discussion in the 1980s. These are the use of parsimony in phylogeny reconstructions, and the question of the evolutionary implications of cladistics. Some possible scenarios are also hypothesized for the future of cladistics, considering its bearings on various fields such as biogeography, geology, physiology, or biochemistry. Such crystal-gazing may be risky, because the fruitfulness of cladistics in the next 20 years will also depend on socioeconomical factors which, unfortunately, are out of place here, but I shall take this risk, in the hope that these forecasts may influence those who govern educational systems.

2.2. COMPARATIVE BIOLOGY AND CLADISTICS

Phylogenetic systematics, or cladistics, belongs to one of the two main fields of biology, termed *comparative biology* by Nelson (1970). Comparative biology is concerned with the diversity of organisms, the distribution of characters in them, and the possible meaning of this pattern (Wiley, 1981). In contrast, *general biology* deals with the processes generating diversity, and one of the major failures of modern biology is that it often neglects the comparative aspect of problems. In other words, too much attention is paid to the question 'why?' and to little to the question 'what?'. The goal of comparative biology is to study what things really are, or how they are related to each other. Subsequent speculation about why it is so belongs to general biology. Nelson and Platnick (1981) and other cladists consider the current emphasis on process is inherited from Wallace (1855) and Darwin (1859) who, by proposing a model of the evolutionary process (natural selection), initiated the decline of pattern analysis. There may be an earlier, pre-Darwinian, background to this loss of interest in comparative biology: the failure of great comparative anatomists such as Cuvier to deduce genealogies from structural analysis.

2.2.1. Some basic terms of comparative biology

The term *phylogeny* was coined by Haeckel (1866) and meant, in its original sense, the history of the palaeontological development of species. Today, phylogeny expresses the interrelationships of the members of a group, as a result of evolution by descent and transformation of heritable characters. Reconstruction of phylogeny is based on the search for characters shared by two or more organisms. Since different characters must have appeared at different times in the history of a group of organisms, they show a hierarchy in their distribution, or a generality which is correlated with age. Vertebrae, for instance, are inferred to have appeared earlier than the five-fingered tetrapod limb in the history of the chordates, because vertebrae have greater generality, or a wider distribution. Thus tetrapods are all vertebrates, but some vertebrates (fishes) are not tetrapods. The chronology of the occurrence of characters in a group leads to the construction of a *phylogenetic tree* in which ancestral species are inferred at branching points.

Systematics is the study of the diversity of organisms and attempts to rationalize it on the basis of any criterion of relationship. Various 'systems' have been proposed by naturalists since Aristotle, and the one constructed by Linnaeus's (1758) is still largely the basis of present day systematics. In non-Hennigian systematics, three major criteria are used for grouping organisms and ranking these groups: over-all similarity, ecological similarity, and, in some cases, phylogenetic relationships when they are expressed by conspicuous characters (hairs of mammals, feathers of birds, etc.). For this reason, systematists who use all these criteria call themselves '*eclectic*' *systematists* (Simpson, 1975). However, most eclectic systematists, such as Simpson, Mayr and their followers, give much weight to the evolutionary process and try to produce classifications which express descent. Thus, they name ancestral groups and descendent groups, and the rank of the latter often expresses their degree of divergence, as estimated by criteria which include the systematist's personal opinion or 'ingenuity' (Simpson, 1961). Since such classifications are supposed to express the evolution of groups, eclectic systematists are also called *evolutionary systematists*. Systematists who only use the criterion of over-all similarity, whatever methods they may use to manipulate the data, are called *phenetic systematists*. Finally, systematists who consider that classifications must reflect only phylogenetic relationships, and who disregard morphological divergence, are called *phylogenetic systematists*. Since phylogenetic relationships are best expressed by a branching diagram called a *cladogram* (*klados* = branch in Greek) and since phylogenetic systematists only take natural groups (= *clades*) into consideration, Mayr (1969) coined the term '*cladists*' to designate these systematists.

Taxonomy is the technical aspect of systematics. Simpson (1944, 1961) defines it as the theoretical basis of the principles, rules and laws of classification. The term taxonomy, coined by de Candolle (1813), designates all activities concerned with the use of the name attached to a group of organisms, or *taxon* (plural *taxa* or *taxons*), as well as the description of these groups and the history of their discovery. The species is generally considered as the only non-arbitrary category, yet it has many different definitions which reflect controversies about its limits. Species are relatively well limited in nature, e.g. by sterility with other species, but the defining criteria of species are still hotly discussed by biologists. Infraspecific categories (subspecies, variety, etc.) and particularly supraspecific categories (genera, families, orders, classes, etc.) do not have the same reality as the species. They are man-made groupings based on criteria chosen by the systematist. Attempts at finding criteria to define supraspecific categories have failed. Dubois (1982), for instance, recently proposed defining the genus on the basis of the ability of congeneric species to produce viable hybrids, but I doubt that this definition will be acceptable to all biologists.

2.2.2. The principles of phylogenetic systematics

In its original sense, as Hennig defined it, phylogenetic systematics rests on the postulate that descent with modification, or evolution in the modern sense, has taken place. Therefore, the characters observed in any organism may show either a primitive or an advanced (derived) state. As we shall see, some cladists now reject this evolutionary postulate, considering it unnecessary. This new form of cladistics, often referred to as '*transformed cladistics*' (or '*modern cladistics*', or '*pattern cladistics*'), is difficult to understand for biologists brought up in the

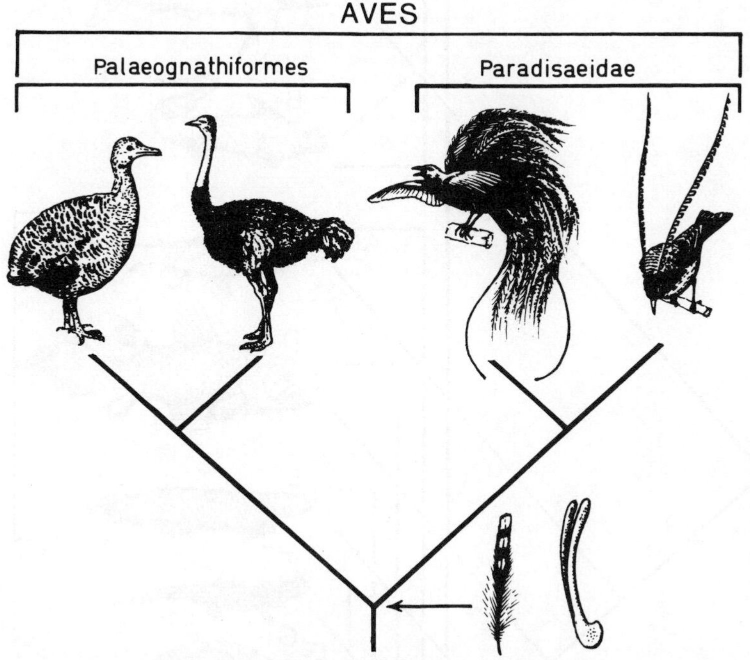

Figure 1. An example of a monophyletic group: birds (Aves). All extant birds share several synapomorphies (unique shared homologies) such as the feathers and the wishbone, which are met with in no other extant vertebrates. These characters show that all birds share the same common ancestral species, although that species is unknown as a fossil. The class Aves contains, in its turn, a number of smaller monophyletic groups. Here, four species derived in various ways (A, *Eudronia elegans*; B, *Struthio camelus*; C, *Astrapia mayeri*; D, *Pterydophora alberti*) illustrate two of the many monophyletic subgroups; the Palaeognathiformes and the Paradisaeidae

evolutionary tradition, and is sometimes erroneously regarded as a form of scientific creationism.

The cornerstone of Hennig's phylogenetic systematics is that resemblances between organisms do not always have the same bearing on relationships. Organisms, or groups of organisms, can resemble each other because they share either advanced or primitive characters. Hennig called the advanced, or derived, characters *apomorphies*, and the primitive characters *plesiomorphies*. When such characters are shared by several organisms, they are *synapomorphies* or *symplesiomorphies*. Hennig considered that synapomorphies alone indicate relationship between organisms, and that all the organisms sharing a synapomorphy form a *monophyletic* group, which has its own history. The feathers, wings and wishbone of birds, for instance, are synapomorphies making the group Aves (birds) monophyletic (Figure 1). It is the same with the hairs of

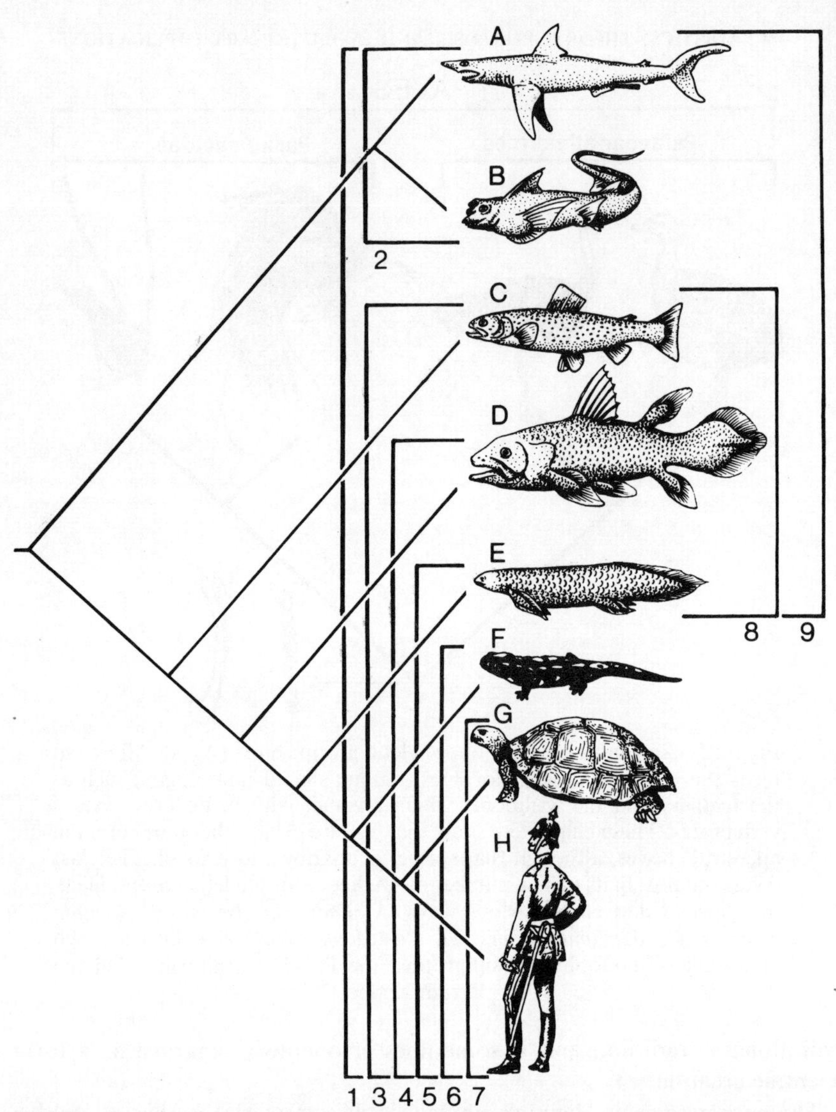

Figure 2. Monophyletic and paraphyletic groups in the classification of the jawed vertebrates (Gnathostomata), here represented by eight monophyletic taxa: A, Elasmobranchii (sharks); B, Holocephali (chimaeras); C, Actinopyterygii (ray-finned fishes); D, Actinistia (coelacanths); E, Dipnoi (lungfishes); F, Urodela (salamanders); G, Chelonia (turtles) and H, Mammalia (mammals). The interrelationships of the taxa, as shown by synapomorphy distribution, are illustrated by the cladogram to the left. On the left side of the animals are indicated the nested monophyletic groups that can be found in this cladogram represented by brackets: (1, Gnathostomata; 2, Chondrichthyes; 3, Osteichthyes (in a cladistic sense, that is including the tetrapods); 4, Sarcopterygii; 5, Choanata; 6, Tetrapoda; 7, Amniota). On the right side are indicated some classical paraphyletic groups (8, Osteichthyes (in a classical sense, that is restricted to bony fishes); 9, Pisces ('fishes')). The phylogenetic classification based exclusively on monophyletic groups mirrors the interrelationships expressed by the cladogram

mammals, the prismatic calcified cartilage of elasmobranchs (sharks and chimaeras), or the shell of turtles. A synapomorphy may secondarily disappear or become modified within a monophyletic group it defines; ontogeny or congruence with other synapomorphies often reveal these cases of *reversion* (whales, for instance, have no hair, but they share many synapomorphies with other mammals having hair and, thus, must have lost this character). A monophyletic group or taxon includes a common ancestral species and *all* the species descending from it, and is recognized by at least one apomorphy which appeared in the common ancestor. Synapomorphies must be homologous characters, that is characters inherited from a common ancestor. We shall see later that the criteria for determining *homologies* are much discussed and that 'transformed cladists' now propose, with good reason, to synonymize synapomorphy and homology.

Symplesiomorphies, or shared primitive characters, indicate common ancestry, but not immediate and exclusive common ancestry. Groups or taxa defined by symplesiomorphies were called *paraphyletic* by Hennig. For instance, fishes (Pisces), as a group defined by the presence of paired fins (Figure 2), are a paraphyletic group because the paired limbs of tetrapods (four-legged vertebrates) are modified paired fins, and because some of the fishes share with tetrapods synapomorphies that are not present in other fishes (jaw bones of bony fishes, choanae of lungfishes). Thus, the paired fins of fishes are a symplesiomorphy for fishes, but a synapomorphy for the gnathostomes (jawed vertebrates = fishes + tetrapods; Figure 2). In contrast, a group including tetrapods and those fishes sharing certain synapomorphies with them, such as jaw bones or choanae, would be monophyletic (Osteichthyes in cladist sense, Choanata; Figure 2). Although these examples may seem trivial, paraphyletic taxa are still widely used not only in common speech ('invertebrates', 'fishes', 'reptiles') but also in scientific papers ('Condylarthra', 'Prosimii', 'Agnatha', 'Mesosuchia', 'Fissipedia').

Finally, a third kind of grouping may be made, on the basis of non-homologous characters, called convergent characters or *convergences*. Such groups or taxa are *polyphyletic*. Polyphyletic taxa are now relatively rare, even in evolutionary classifications, but some of them are tenacious. One example is the Pinnipedia (seals, walruses and sealions), shown to be diphyletic long ago (Mivart, 1882; Tedford, 1976), but still cited as a taxon in the literature (e.g. Thenius, 1980). The seals share with otters synapomorphies in the mastoid and occipital regions of the skull, and with *Potamotherium* (a fossil otter-like mammal) additional synapomorphies in muscular insertion areas on the scapula. Sealions and walruses share synapomorphies with bears, in particular in the auditory bulla, and they swim with their front legs only. Consequently, seals, *Potamotherium* and otters form a monophyletic group, Musteloidea, and sealions, walruses, and bears form another monophyletic group, Arctoidea (Figure 3). Another fossil form, *Enalioarctos*, may be inserted in the Arctoidea,

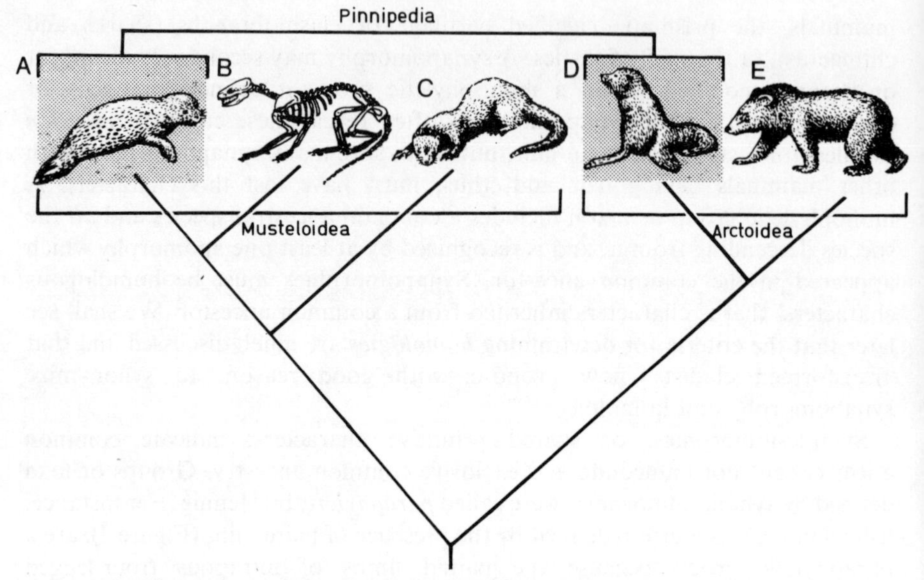

Figure 3. An example of a polyphyletic group: the Pinnipedia.
Seals (A) and sealions (D) have long been placed in the order Pinnipedia on the basis of resemblances in limb structure. Detailed study of their anatomy reveals that seals share synapomophies with otters (C) and sealions share other synapomorphies with bears (E). Thus, the group appears to be polyphyletic (in this case, diphyletic) and seals have to be classified with otters in the Musteloidea, whereas sealions are grouped with bears in the Arctoidea. Some fossil taxa seem to support this view and can be inserted in this cladogram: *Potamotherium* (B), an otter-like Oligocene musteloid, which shares with seals a larger number of synapomorphies than with otters. This polyphyly of the Pinnipedia is, however, not congruent with myoglobin sequences

as an 'intermediate' between sealions and bears. Thus, the similarities between seals and sealions are convergent, or due to similar mode of life (swimming and piscivorous diet). (It is noteworthy, however, that protein sequences are inconsistent with the polyphyly of the Pinnipedia (Patterson, 1981b) and thus throw doubt on the validity of this example!)

Among these three types of group (monophyletic, paraphyletic and polyphyletic), Hennig considered that *monophyletic groups alone have phylogenetic relevance*, because they have their own history. In contrast, paraphyletic groups do not have a proper history or, rather, their own history is also part of that of the monophyletic groups which are rooted into them. For instance, the history of the 'Reptilia' (reptiles) includes the early history of mammals, because there are 'mammal-like reptiles' (therapsids), and also that of birds, because there are 'reptile-like birds' (*Archaeopteryx*). The same is true for polyphyletic groups. The histories of the seals and sealions, for instance, are not linked in any way, except that both are parts of the history of the carnivorous mammals (Carnivora), a monophyletic group.

Here we meet another fundamental point in Hennig's principles: *translation of the phylogeny into a classification.* The history of a monophyletic group is a series of unique events and cannot be modified by subsequent events. What can be modified is the reconstruction of this history by the phylogenetist. There are various ways of reconstructing phylogeny. Eclectic, or evolutionary systematists use an *inductive* method and consider that the accumulation of fossils in stratigraphical order permits one to read the history of a group through the strata. Phylogenetic systematists use a *hypothetico-deductive* method based on comparisons of character distribution patterns among living and, subsequently, fossil taxa. They search for congruent patterns and propose a hypothesis of phylogenetic relationships based on the *congruent distribution of the largest number of synapomorphies* (Figure 4). Such a hypothesis is regarded as the most parsimonious, until it is refuted by a still more parsimonious hypothesis (following the discovery of new characters). The true history of a group is unknowable and can only be approached more and more closely by a succession of hypotheses and refutations. The use of *parsimony* as a criterion for preferring one hypothesis of phylogenetic relationship to another makes the chosen hypothesis accessible to *refutation*, or *falsification.* In this respect, Hennigian hypotheses of phylogenetic relationships, expressed by a cladogram or a phylogenetic tree, are consistent with Popper's definition of scientific hypotheses (Popper, 1959, 1963, 1972; Bock, 1974; Platnick and Gaffney, 1977, 1978a,b), and this approach to phylogeny conforms to Bernard's (1865) method in scientific research. There is now a lively debate about the use of the principle of parsimony in phylogeny reconstruction, and this will be dealt with below. In sum, a Hennigian classification can be regarded as a *prediction*, that can be refuted by new character distribution patterns.

The phylogeny of a monophyletic group, as expressed by a hierarchy of derived characters, summarizes the historical characteristics of the group, thus it is the only non-arbitrary basis for its classification. Critics of phylogenetic systematics claim that since a phylogenetic hypothesis is not the 'true' phylogeny of a group, there is no reason to assume that the classification based on it is more objective than any other. But a phylogenetic classification has the advantage of being refutable, as is the phylogeny on which it is based. In contrast, an evolutionary classification is rarely modified by changes in ancestor–descendant relationships between taxa. Despite the considerable changes in our views on the phylogeny of the Amniota (tetrapods with amniotic eggs: a monophyletic group) since the end of the eighteenth century, their evolutionary classification still comprises three major classes: Reptilia, Aves, and Mammalia. In this respect, evolutionary taxonomy is a frozen language, and it tells us nothing except that evolutionary systematists venerate the authority of their predecessors. This separation between inferred history and classification stems from Lamarck (1809) and, as pointed out by Tassy (1981a), has only been revived by Simpson (1944, 1961). In sum, a classification is a language (Brooks, 1982) and, for

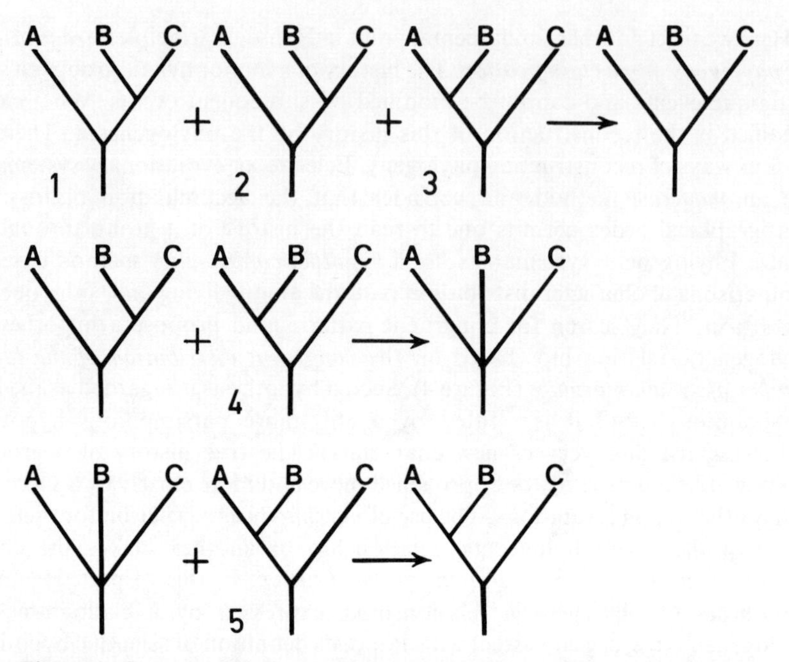

Figure 4. Congruent synapomorphy distribution and the principle of parsimony in phylogeny reconstruction. The analysis of three homologous characters 1, 2, 3, in three taxa A, B, C, may lead to conflicting hypotheses of relationships between these taxa, according to the characters under consideration. Characters 1 and 2 show congruent patterns, with A sister-group of B + C, whereas character 3 leads to a different hypothesis, with C sister-group of A + B. This conflict is solved by choosing the hypothesis which corresponds to the majority of congruent characters (A sister-group of B + C). This hypothesis is the more parsimonious, because it implies only *one* convergence, in the similarity between A and B in character 3. If we chose the hypothesis suggested by character 3, we would have to admit *two* convergences, in the similarities between B and C in characters 1 and 2.

Investigation of another character (4) may reveal the same pattern as character 3. There is then no means to choose (unless one weights characters), and this uncertainty is expressed by an unresolved cladogram showing a trichotomy. Further study of a fifth character (5) may provide one more pattern congruent with 3 and 4. If so, the first hypothesis of relationship between A, B and C is overthrown and parsimony implies the cladogram in which C is the sister-group of A + B

Hennig, this language must express precisely the most objective characteristic of a group: its history. For evolutionary taxonomists, classification may express both the descent and divergence between taxa. For this reason, some systematists (Bock, 1974) consider that evolutionary classications convey more information than phylogenetic classifications. However, degree of divergence is a subjective, largely personal criterion, which does not tell us anything. In order to express this

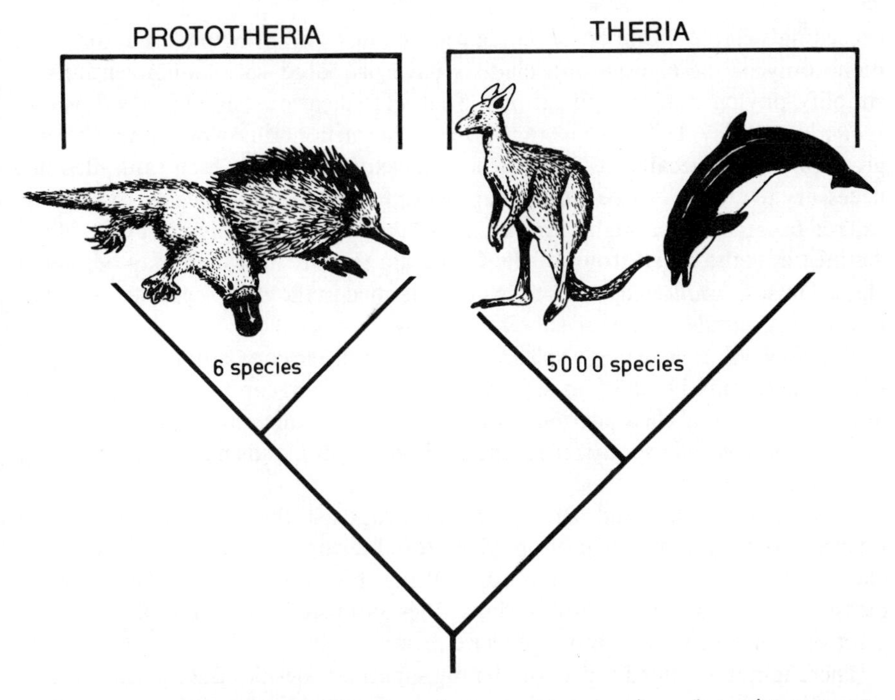

Figure 5. A phylogenetic classification gives equal taxonomic rank to sister-groups. However, they may differ in the number of species they contain. For instance, the class Mammalia (mammals) contains two sub-classes: the Prototheria (living monotremes: platypus and echidna) and its sister-group, the Theria (all marsupials and placentals). Although the Prototheria are represented by six extant species and the Theria by nearly 5,000, these two groups have the same taxonomic rank: they are ranked as sub-classes

divergence, systematists have to give names to primitive (paraphyletic) groups, masking the phylogenetic relationships which are surely the most interesting data.

In a phylogenetic classification, two species or groups which are more closely related to each other than to any other species or group are called *sister species* or *sister-groups*, and are given equal taxonomic rank, however much they differ in divergence or importance in number of species. Thus, when one or a few species are the sister of a very large group, they have to be placed in a supraspecific taxon whose rank is equal to that of the large group. For instance, the Monotremata, or Prototheria (platypus and echidna), represented by only six extant species, are the sister group of the Theria (marsupials and placentals), represented by thousands of species, but in a phylogenetic classification, they have the same rank as the latter (Figure 5).

In response to criticisms from evolutionary taxonomists who considered that phylogenetic classifications would be impossible to use because of the exceedingly large number of categories (ranks: in principle one for each dichotomy of the cladogram), cladists have proposed various conventions to simplify phylogenetic classifications. Such classifications should be *minimally redundant* (Wiley, 1979), for instance when a taxon containing only one species is given high rank because it is the sister of a taxon of equally high rank, it is not necessary to create a succession of supraspecific taxa for this species, the highest ranked taxon being enough. *Latimeria chalumnae*, the single living species of the Actinistia, is the sister group of the Choanata (Figure 2), which may be, say, a class. Thus, *L. chalumnae* can be directly classified in the class Actinistia, without intervening family, order or subclass names.

Phylogenetic sequencing (Nelson, 1974) is another convention according to which the terminal taxa of an asymmetrical cladogram can be given equal rank and be listed, in a classification, in order of their branching sequence. Such a classification would exactly reflect the cladogram yet avoids new names for each dichotomy in it.

Finally, Patterson and Rosen (1977) proposed the term *plesion* as an unspecified rank name for monophyletic fossil groups which are inserted into a classification based on extant groups. With this convention, a phylogenetic classification based on extant groups does not need to be modified by the discovery of a new fossil monophyletic group.

There remains a third aspect of Hennig's method. Besides the exclusive use of synapomorphies in phylogeny reconstruction and the exclusive use of phylogenetic relationships for establishing classifications, Hennig proposed criteria for determining the state (apomorphous or plesiomorphous) of characters. The major criterion is the 'argumentation scheme', now better known as the *out-group rule*: when one wants to analyse characters in the members of a monophyletic group, one has to look for their homologues in its sister-group. If they occur in it, they are plesiomorphous for the monophyletic group under consideration. But if they are absent in the sister-group, and are unique to the monophyletic group under study, they are apomorphous to it. For instance, the five toes of mammals are a plesiomorphous character because they are also present in their traditional sister-group, the Diapsida (lizards, snakes, birds, and crocodiles), although secondary reduction has occurred here and there (birds). This out-group test can be further reinforced by comparisons with turtles (regarded by some phylogeneticists as the sister-group of mammals + diapsids) or lissamphibians (sister-group of all amniotes). Thus, the five toes are not only plesiomorphous for mammals, but also for all amniotes. However, out-group comparison with the sister group of tetrapods, lungfishes, Figure 2 shows that they are apomorphous to the latter (no fish have five toes). The out-group rule has been under heavy fire recently (Nelson, 1978), and there are good reasons to use it with care, but it has also been defended by other cladists (De Jong, 1980; Stevens, 1980; Watrous and Wheeler, 1981; Wiley, 1981). Fortunately, Hennig proposed accessory criteria for character state determination: the geological order of succession of characters, their biogeographical succession, correlation of character transformations, and ontogenetic character sequence. The first accessory criterion, of *palaeontological argument* has been criticized by Nelson (1978) who rejects it as unrefutable and as being a mere 'idol of the academy'. Hennig (1966) himself considered that it should be used only when all the other criteria have failed. Usually, but not always, characters of the geologically oldest representatives of a group are more primitive than their homologues in extant forms. *Archaeopteryx*, the oldest known bird, displays primitive features (teeth, long bony tail, etc.) that are met with in no extant bird.

The secondary accessory criterion, also called the *progression rule*, is linked with biogeography. The more primitive taxa of a group are supposed by Hennig (1966) to be found closer to the centre of origin of the group than the more derived ones. This rule has been discussed by Nelson (1975), Brundin (1981), and Slater (1981) and it is now better understood in the light of vicariance biogeography (see below) than in the dispersalist frame that Hennig inherited from Darwinian biogeography.

The third accessory criterion, or criterion of *transformation series correlation*, consists of comparisons of successive states of different transformation series of characters. For instance, if two characters, or more, show regularly increasing modifications in a series of extant taxa, this can be regarded as an indication of the polarity of other characters in the same taxa. This criterion is merely an extension of the *test of homology congruence* (Patterson, 1982b), advocated by modern cladists—when a majority of homology (=synapomorphy) distribution patterns are congruent, the probability that each pattern provides a good hypothesis of character state polarity is greater. However, to Wiley (1981), transformation series of two characters are more interesting when they conflict, because it indicates that something is wrong with the analysis.

Finally, the *ontogenetic criterion*, that is the order of appearance of characters, or character states, during the ontogeny of an organism, is now regarded by Løvtrup (1974, 1978), Nelson (1978), and Patterson (1982b) as much more than an auxiliary criterion. It is very close to von Baer's law, but not identical to it. According to von Baer's law (1828), the more general characters appear earlier in the development of an individual than the more particular ones. Thus, the most apomorphous characters are those which are the most particular, that is restricted to the smallest number of organisms. The ontogenetic criterion is regarded now as the only reliable one by 'transformed cladists'.

This long introduction to Hennig's method will serve as a basis for subsequent discussion of the heuristic value of cladistics in achieving the goal of comparative biology, and of its future development in scientific research. However, I advise the reader to read Hennig's (1966) book or the 'digest' of his ideas (Hennig, 1965, 1981). Among the books presenting Hennig's phylogenetic systematics, I favour Wiley's (1981), which is remarkably clear and refers to all the literature on the topic. It should become the *vade mecum* of cladists.

2.3. PHYLOGENETIC VERSUS EVOLUTIONARY AND PHENETIC TAXONOMY: THE GOAL AND THE MEANS

How is cladistics better? This question has often been put to me, either by students who had no preconceived opinion, or by professional biologists and palaeontologists who had preconceived opinions! The question can only be answered in the context of the goal of comparative biology, which is to summarize the pattern of the living world in such a way that general biology can rely on it to test hypotheses about processes which have led to this pattern. There are three routes to this goal: cladistics, evolutionary systematics and phenetics. I shall leave aside what Wiley (1981) calls '*traditional taxonomists*': those who consider that any theory of phylogeny and systematics is useless and, consequently, that the goal of science is to provide jobs and careers.

I shall present some points of disagreement between cladistics and the two other methods.

2.3.1. Cladistics and evolutionary systematics

In contrast to phylogenetic systematics, evolutionary systematics has no precise rules, which make it difficult to discuss or criticize. Theoreticians of evolutionary systematics (e.g. Simpson, 1961, 1975; Mayr, 1969, 1974; Bock, 1974; Ashlock, 1980) have presented objections to cladistics but have also pointed out some points of agreement, especially on the use of synapomorphies in phylogenetic reconstruction.

I see five major points of disagreement between cladists and evolutionary systematists: the expression of morphological gaps; of species diversity; and of ecology; the definition of monophyly; and the significance of fossils (as ancestors or indicators of plesiomorphy).

The need to express morphological gaps in classification is a constant concern for evolutionary systematists and probably stems from use of over-all resemblance as a criterion of relationship. Evolutionary systematists consider that there is a larger morphological gap between a therapsid ('mammal-like reptile') and a mammal than between the former and a crocodile. Thus, they classify therapsids in the class Reptilia (a paraphyletic group for cladists; Figure 6) and not in the class Mammalia or as a sister-group of that class. Of course, this classification expresses the gap, but masks the only thing that is real in a ,classification: the relationships due to the history of the group. Consequently, on this point evolutionary systematics cannot reach the goal of comparative biology defined above. Evolutionary systematists will probably protest against this caricature of their goal; they claim that any informed biologist knows the relationships between therapsids and mammals. But what is true for this example is also true for far less self-evident examples, and it would be easy to find groups in which evolutionary classification tells us *nothing* about

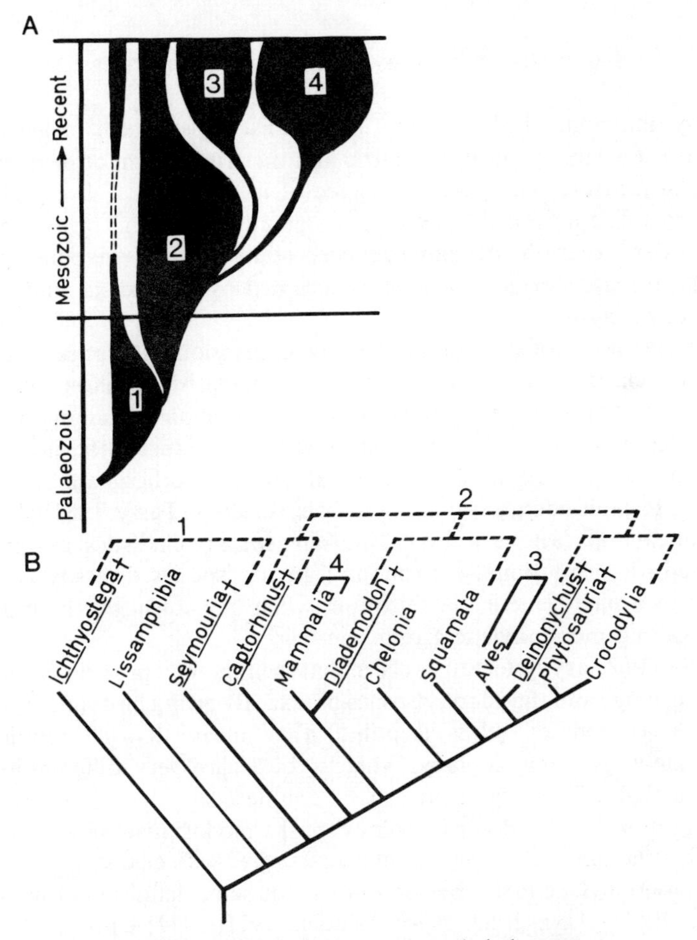

Figure 6. Evolutionary trees and cladograms.

An evolutionary tree, such as the widely accepted tree of the Tetrapoda (A, reproduced from *Vertebrate Palaeontology*, Romer, © 1933, 1945 and 1966 by the University of Chicago, by permission of the University of Chicago Press, with modifications), shows monophyletic groups (3, birds and 4, mammals) and paraphyletic, or 'ancestral', groups (1, amphibians and 2, reptiles). It provides information on the chronological distribution of the groups and on their relative importance in number of species (according to the breadth of the 'branches'). However, it conveys very little information on phylogenetic relationships, except that birds and mammals 'descend' from reptiles which, in their turn, 'descend' from amphibians.

A cladogram of a few selected fossil (†) and extant tetrapods (B), constructed from synapomorphy distribution, provides a maximum of phylogenetic information. Only two of the groups in A can be recognized: the monophyletic groups 3 and 4. Amphibians and reptiles (1 and 2) cannot be characterized by unique synapomorphies and, thus, are paraphyletic. Such a cladogram provides no information on ecology or stratigraphy, but it represents the raw material, the observed reality, from which phylogenetic trees and scenarios can be inferred. This cladogram of the Tetrapoda reflects current views, but it may be refuted in the near future phylogeny (mammals, rodents, etc.). For cladists, assessment of gaps has no interest, for it is largely anthropocentric and the criteria cannot be defined. All attempts to measure such gaps, or degree of morphological divergence, have failed to provide a universal basis; each systematist having his own criterion. In contrast, the criterion of phylogenetic relationship, synapomorphy (or homology, for transformed cladists), is accepted by all cladists, and there are rules for assessing it.

The second point of disagreement is the expression of species diversity or richness in a classification. For evolutionary systematists, a taxon containing a large number of species merits higher rank than a taxon containing only a few species. By contrast, for cladists, phylogenetic relationships are the only guide for ranking taxa. To give the monotremes equal rank to all other mammals (Theria) is shocking to evolutionary systematists. Some cladists (Tassy, 1981b) have seen socio-political implications in this egalitarian aspect of cladistics, as opposed to elitist, gradistic, evolutionary taxonomy! Again, species diversity is as well-expressed in a phylogenetic classification, when one considers the number of nested taxa in a monophyletic group.

Third, evolutionary systematists claim that their method provides information about the ecology and the adaptive zones of taxa. By giving birds (class Aves) the same rank as 'reptiles' (class Reptilia), they mean that all reptiles have approximately the same ecology, whereas birds are very different in being adapted to flight. This argument can be countered in the same way as that concerning morphological gaps: ecology masks phylogenetic relationships.

Fourth, although evolutionary systematists agree with cladists on the need to use only monophyletic taxa, they do not use the same definition of monophyly (Simpson, 1961; Mayr, 1969, 1974; Ashlock, 1971, 1972; Dubois, 1982). For evolutionary systematists, a monophyletic group may be a grade, or a paraphyletic group in Hennig's sense, and they call a monophyletic group in Hennig's sense *holophyletic*. Thus, they only reject polyphyletic groups. Since they do not distinguish the two types of group in classifications, evolutionary systematics cannot express phylogenetic relationship. For instance, in an evolutionary classification of the Amniota, one finds three classes: Reptilia, Mammalia, and Aves. Can you guess which class is monophyletic and what are the relationships between these classes from reading this list of three names?

Finally, evolutionary systematists consider that fossils are the best means of determining evolutionary relationships and of reconstructing the history of a group, whereas cladists consider that fossils can only be a secondary addition to phylogenies based on extant taxa (Figure 6B). Thus fossils cannot contradict a phylogeny based on extant taxa. This question will be discussed in detail in Section 4.2.

Besides these theoretical criticisms, there are many other reasons for evolutionary systematists to attack cladistics. These reasons include simple disturbance in habits; the 'arrogance' of cladists; irritation at cladists' claim to be really scientific; political or scholastic conflicts; use of the principle of parsimony; and the question of character weighting. Many of these criticisms of cladistics can be explained by human factors, but others show a misunderstanding of cladistics or simply a lack of practice in cladistic analysis (Halstead, 1982). Among these criticisms, those against the principle of parsimony and against the absence of character weighting in cladistic analysis are the most interesting, for they seem to summarize the sense of the debate around cladistics in the early eighties.

The *principle of parsimony*, or simplicity, implies that, among several hypotheses of phylogenetic relationship, one chooses that which explains the data (distribution of homologies) in the most simple, or economical manner. For instance, the hypothesis of relationship which implies minimal convergence (characters attaining the same state independently) or reversion (characters apparently returning to a primitive state) will be preferred until it is, in its turn, refuted by a more parsimonious hypothesis (Figure 4). To Popper (1963) the most parsimonious hypothesis is the most easily falsifiable, because it is not burdened with numerous unfalsifiable *ad hoc* hypotheses—secondary statements making the hypothesis fit the data.

The debate about the principle of parsimony and falsifiability has started recently with two papers (Panchen, 1982; Cartmill, 1981) which criticize Engelmann and Wiley's (1977) and Gaffney's (1979) procedures for testing phylogenetic hypotheses. Cartmill's paper, in particular, is well-documented and provides a most interesting base for discussion. It suggests that the author has practised cladistics but is still handicapped by worries about whether nature is simple or complicated. I would like to illustrate the debate about parsimony with the example of the phylogenetic affinities of birds, which have been recently discussed by Gardiner (1982) and Cox (1982). Birds form a monophyletic group (Aves), which is classically considered as the sister-group of crocodiles (Crocodylia). These two extant groups form the Archosauria, a supposedly monophyletic group characterized by e.g. fenestrae in front of the orbits and on the lower jaw. Many fossil groups have also been referred to the Archosauria, including the various 'dinosaurs', the phytosaurs, and the pterosaurs. Striking resemblances, and even plausible synapomorphies, have been pointed out between birds and particular dinosaurs (*Deinonychus*) (Figure 6B) or some primitive archosaurs ('thecodonts'), and the phylogeny of the Archosauria was, until recently, regarded as so reliable that the British Museum (Natural History) presented it as a most attractive exhibit. Gardiner (1982) found a large number of apparent synapomorphies shared by birds and mammals, far outnumbering those shared by birds and crocodiles. But these synapomorphies mainly concern soft anatomy and physiology and, thus, cannot be observed in fossils. Perhaps for this reason, palaeontologists such as Cox (1982) do not like them. The dilemma is that on the one hand, we have apparent mammal–bird synapomorphies which outnumber the bird–crocodile synapomorphies, and on the other hand an evolutionary tree based on fossils showing a gradual passage from primitive archosaurs to birds *via* a series of non-monophyletic groups ('thecodonts', 'dinosaurs'). For Gardiner, this dilemma must be resolved by the principle of parsimony, and his 17 bird–mammal synapomorphies dictate the most economical hypothesis of relationship, that birds are the sister-group of mammals. What his opponents have to do is to falsify this hypothesis by finding at least 18 bird–crocodile synapomorphies, or by showing that the features he lists are not characteristic of birds and mammals. But this example leads us to the question of *character weighting*, which is, in my opinion, the last hope for evolutionary systematists who wish to preserve traditional groups. As Patterson (1982) puts it:

> '...the systematist whose pet group is threatened by the congruence test is unjustified in appealing to weighting of homologies, since weighting by the systematist is an *a posteriori* exercise, designed to prefer one group to another' (Patterson, 1982b, p. 66).

Are some characters more 'important' than others, and how much more 'important'? Cladists are conscious that some characters may appear, after careful study, to be compounded of many unit characters, or that some characters may be linked (controlled by the same gene) and so should be regarded as one single character; but do evolutionary systematists have a method capable of resolving this problem? Who decides the 'quality' of a character? I have experienced this problem many times, and arrived at the conclusion that a character may be considered 'important' by various authors: (1) when it is conspicuous, (2) when the scientists who found it is an authority, (3) when the scientist who found it is a specialist on this character, (4) when it is also to be found in fossils, (5) when it is widespread among taxa or, conversely, when it is shared by only a small number of taxa, and (6) when it marks an 'important' step in adaptation to a particular environment or, conversely, when it is not 'adaptive'. In sum, any judgement on the value of a character can depend on the result one wants to obtain, and the claim to be a cladist does not free one from subjectivity in character weighting. The notion of *adaptive characters* is a rich source of *ad hoc* defences for phylogenetic hypotheses threatened by parsimony. Hecht and Edwards's (1977) method for weighting homologous characters (according to their complexity) bears a heavy burden of preliminary assumptions about how evolution produces characters, and Patterson (1982b) very elegantly showed that characters weight themselves by associations (congruence with other characters) which are beyond the bounds of chance.

The bitterest pill for evolutionary systematists is that cladistics prevents one building up phylogenies according to one's personal feelings. Cladistics provides rules, based on logic, and a comparative biologist who rejects these rules can be compared with a chemist who rejects Mendeleev's periodic table (Bonde, 1977). These rules have not been erected to tyrannize other biologists, but are simply means to reach a goal of comparative biology and, hence, pave the way to general biology. Panchen's (1982) appeal to cladists to reintroduce intuition in classification is a sign of the disarray among evolutionary systematists, some of whom feel that life will not be as easy as before.

2.3.2. Cladistics and phenetic taxonomy

Contrary to phylogenetic systematists, pheneticists do not take into consideration the state (apomorphous or plesiomorphous) of characters, and so

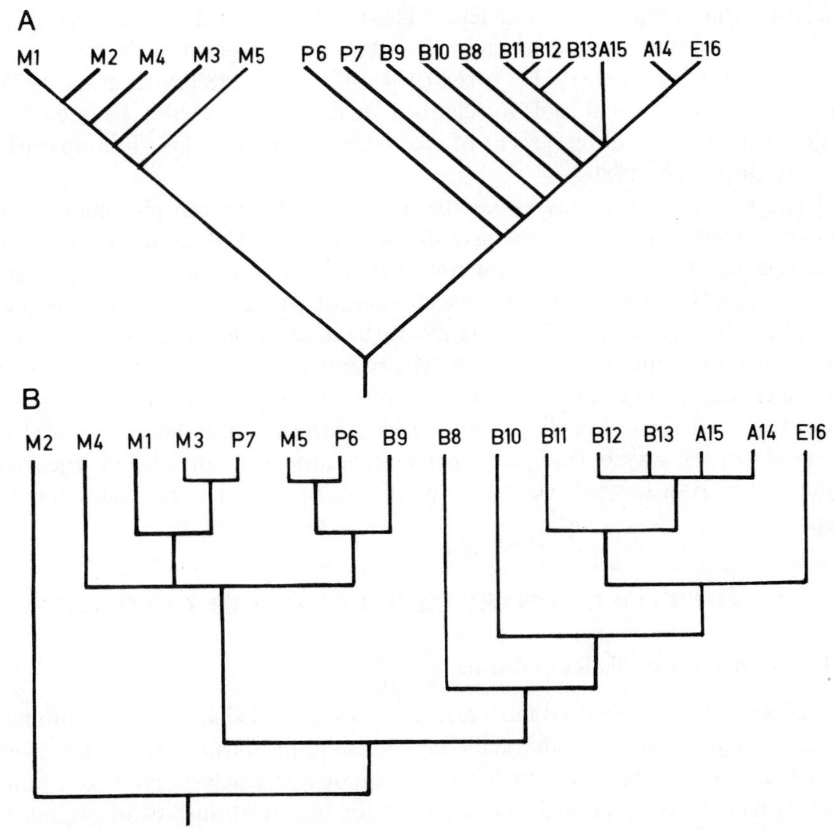

Figure 7. Phenograms and cladograms. This example shows the cladistic (A) and phenetic (B) analysis of morphological characters of 16 populations of five species (M, P, B, A, E) of the fish genus *Menidia* (from Mickevich and Johnson, 1976, redrawn and modified as to the cladogram, by permission of the Society of Systematic Zoology). There are considerable differences between these two results, because the phylogenetic analysis is based exclusively on synapomorphies, whereas the phenetic analysis takes all the characters into consideration, be they apomorphous or plesiomorphous, and counts absence of character as a character

group organisms according to their over-all similarity. The result is that paraphyletic groups whose members share a large number of symplesiomorphies may appear as monophyletic in a *phenogram* (Figure 7). But this does not matter for pheneticists, whose goal is a stable classification with no necessary phylogenetic implication.

The stability of phenetic classification, claimed by Sneath (1961), Sokal and Sneath (1963) or Michener (1963) to be a considerable advantage in comparison with a continually changing phylogenetic classification, is now less self-evident (Farris, 1977; Mickevich, 1978). However, phenetics has been useful in promoting discussion of cladistics among scientists who have mastered the mathematical background of phenetics. This resulted in considerable advances in cladistics, by the establishment of computer programs for phylogeny reconstruction (Wagner, 1961; Farris, 1970, 1973, 1977, 1980; Mickevich, 1978). These new technological tools for cladistic analysis will probably be one of the major factors in the development of cladistics in various fields of comparative biology during the 1980s.

There is not much to say about the disagreement between pheneticists and cladists, except that they do not have the same definition of natural groups. For pheneticists, a natural classification must explain as many characters as possible, but only that (Gilmour, 1961), whereas for cladists, a classification must do that, and also reflect the order of occurrence of the derived characters only. As with evolutionary systematics, there can hardly be any compromise between cladistic and phenetic systematics, since the goal of phenetics is limited to taxonomy. However, 'transformed cladistics' may be regarded as somewhat convergent on phenetics in the way both reject *a priori* evolutionary support, but phenetics differ from transformed cladistics by counting absence of characters as characters.

2.4. THE EVOLUTIONARY IMPLICATIONS OF CLADISTICS

2.4.1. Hennigian cladistics and transformed cladistics

For Hennig (1950, 1966) and most cladists, some knowledge of the evolutionary process is a prerequisite for character analysis, as implied by the use of terms such as 'advanced', 'derived', or 'primitive' to qualify characters. However, some cladists have recently argued that study of the interrelationships of organisms should be freed from the burden of these preliminary assumptions on evolution, and that evolutionary theory should not influence the construction of a cladogram. This modification of Hennig's principles is sometimes called 'transformed cladistics', 'modern cladistics', but, to me, the most appropriate term would be 'pattern cladistics' (Beatty, 1982), as opposed to Hennig's 'phylogenetic cladistics'. Unfortunately, 'pattern cladistics' has given some the impression of a creationist theory, but this misinterpretation, sometimes encouraged by ill-intentioned evolutionary or traditional systematists, stems from ignorance of the purpose of cladistics in general. Clear papers on this subject (Forey, 1982; Patterson, 1982a,b) should help biologists to understand the significance and scope of this transformation.

Transformed, or pattern cladistics may be described as a simplification and a rationalization of cladistics. Thus, characters are no longer called plesiomorphous or apomorphous (terms referring to evolutionary states), but *general* or *particular* (*special*), and synapomorphy is simply synonymized with *homology* (since non-homologous characters are not taken into consideration). The only tests on which transformed cladists rely to determine the hierarchy or the polarity of homologies are congruence and ontogeny ('evolution' in a pre-Darwinian sense). Embryology shows that more general characters appear before particular ones in the development of an organism (von Baer's law). For instance, in the embryonic development of a bird, the notochord appears first (homology of the Chordata), then the jaws (homology of the Gnathostomata), then the fingered limbs (homology of the Tetrapoda), and finally the wishbone, modified forelimbs, and feathers (homologies of the Aves). In contrast, turtles, lizards, and crocodiles never show, during their ontogeny, any uniquely 'reptilian' homology. Thus, the 'class Reptilia' cannot be defined by any unique character. Some claim that reptiles can be defined by the presence of amniote characters and the absence of mammalian and avian characters (Charig, 1982). But pattern cladists insist that groups can be characterized only by the *presence* of unique homologies, and not by their absence. The absence of a character cannot be homologized with anything, unless it is preceded in ontogeny by the presence of a recognizable homology. The congruence test has been illustrated above by the example of the affinities of birds. For a pattern cladist, the congruence of a majority of bird–mammal homologies is enough to propose a hypothesis of relationship, yet evolutionary systematists claim that they can hardly 'imagine' the common ancestor to birds and mammals.

One of the merits of transformed cladistics is to emphasize the difference between a cladogram (pure pattern of homology distribution), a phylogenetic tree (in which ancestors are recognized or hypothetized at branching points) and a scenario (a story based on a tree, including ecological or biogeographical inferences, for example) (Eldredge, 1979; Nelson and Platnick, 1981; Forey, 1982). Thus, several trees are compatible with one cladogram (Figure 8). Transformed cladists do not reject the search for an explanation of the patterns observed in nature, but they refuse to reconstruct this pattern in the light of a theory it has to test. If evolutionary theory can be tested only by the actual pattern of living organisms, this pattern must not rely on evolutionary theory.

Paradoxically, transformed cladistics, by returning to a preevolutionary approach to organismic diversity, will probably be the major factor in future progress in evolutionary science. In 1982, transformed cladistics is restricted to a few scientists in the UK and the USA. Polemics against these views have just

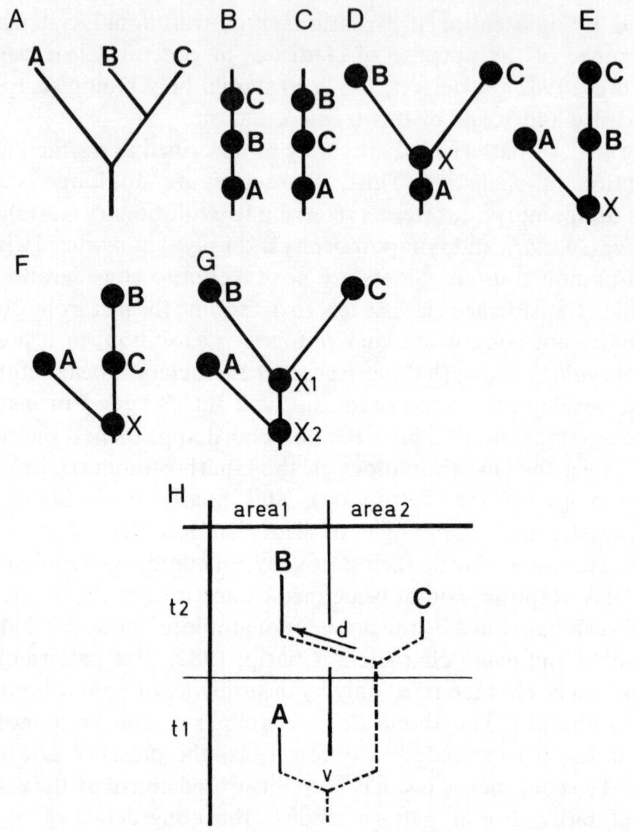

Figure 8. Cladograms, phylogenetic trees and scenarios. A cladogram (A) is a scheme expressing the distribution of synapomorphies (shared unique homologies) of three species: A, B, C. For this cladogram, there are six possible phylogenetic trees (B to G) in which actual or hypothetical ancestors are postulated. Beyond phylogenetic trees, a scenario (H) is a tree (here tree G) placed in a stratigraphical (t1, t2) and biogeographical (areas 1 and 2) framework, with hypotheses about the mode of evolution (natural selection, epigenetics) or biogeography, such as dispersal (d) or vicariance (v) events

started (Beatty, 1982; Charig, 1982; Hill and Crane, 1982) but transformed cladistics is protected by perfect logic and criticisms of transformed cladistics would require thorough knowledge and practice of classical 'phylogenetic' cladistics, which has not yet been acquired by many biologists.

In sum, it appears that cladistics is tending towards the rejection of immediate evolutionary implications. Its relationship to evolutionary science must be to test evolutionary theories, and not to support or illustrate them.

2.4.2. Cladistics and fossils

Palaeontology is classically considered the major science of evolution, and I have met many biologists who claimed that they were ignorant of the phylogeny of the group they studied because it had no fossil representatives! Thanks to the abuse of paraphyletic (ancestral) groups, and the sacrosanct chronological argument, palaeontologists have long given the impression that they could read evolution directly from the rocks. This is now regarded by some cladists as mere superstition (Nelson and Platnick, 1981); meanwhile, some evolutionary systematists show a growing faith in the evolutionary significance of successions of fossils (Gingerich, 1977, 1979; Tintant and Mouterde, 1981). For instance, Chaline and Mein (1979) consider fossil rodents to be the 'Drosophila of palaeobiology'! As a palaeontologist, I would like to defend palaeontology and I think the best way to defend it is to define the limits of its reliability. Ancestor–descendant relationships are, for cladists, beyond those limits. In contrast, there are vast fields of palaeontology, such as analytic comparative palaeoanatomy and analytic palaeobiogeography which are largely under-exploited and should be developed in forthcoming years.

Patterson (1982b) has presented a masterly review of what can be said about fossils from a cladist's point of view. His conclusions are that the unique properties of fossils are: (1) to provide a minimal age for monophyletic groups, (2) to provide new combinations of characters that can reverse decisions on homology and polarity (but rarely or never overthrowing phylogenetic hypotheses based on extant forms), and (3) to amplify biogeographical data by enlarging the distribution of a group in the past. If palaeontologists restricted themselves to these three fields of research in detail, much more useful data would be available to biologists.

We have already discussed (section 2.2) the question of ancestral groups or '*stem-groups*' which are paraphyletic in Hennig's sense, and most evolutionary systematists now recognize that a supraspecific taxon cannot really be ancestral in the same way as a species. But recognition of a fossil species as an ancestor of a fossil or extant monophyletic group still remains the 'hobbyhorse' of most evolutionary palaeontologists. Everybody agrees on the existence of two types of phylogenetic relationship: (1) common ancestry (sister-group relationship) and (2) direct descent (*ancestor–descendant relationship*), but there is only one criterion of relationship: synapomorphies (or homologies). Thus, sister-group relationship cannot be distinguished from ancestor–descendant relationship by the distribution of synapomorphies, since the characteristics of an ancestor are only lack of the apomorphies of the descendant. Consequently, ancestor–descendant relationship cannot be defined by morphology, and evolutionary palaeontologists have to make additional assumptions about stratigraphy, palaeobiogeography, or palaeoecology, which are beyond the bounds of comparative biology. In order to defend the idea that the recognition

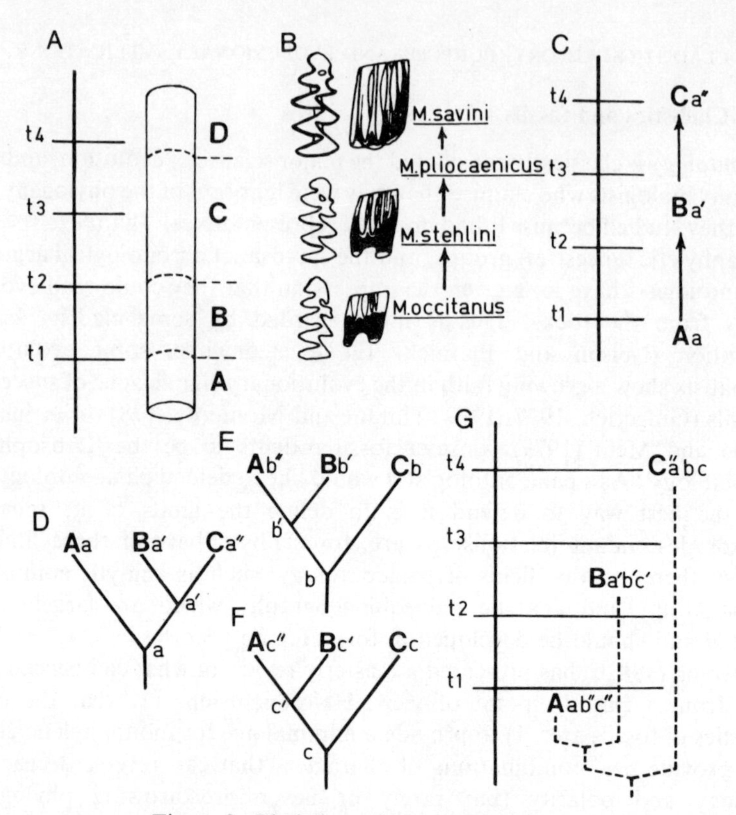

Figure 9. Phyletic evolution and cladistics.

Many palaeontologists consider that stratigraphical succession of fossils provides evidence for gradual transformation of one species into another. This type of speciation is called anagenesis of phyletic speciation. Such a continuum (A) is cut arbitrarily into a number of chronospecies (A, B, C, D) which do not evolve with the same speed. There are numerous examples of models of phyletic speciation, often based on a single character. B shows anagenesis in the rodent genus *Mimomys*, as supported by the structure of the first lower molar tooth (simplified from Chaline and Mein, 1979, by permission of Doin Editeurs, Paris). Other palaeontologists consider that such successions do not generate new species and so call it phyletic evolution or microevolution. From an analysis of three species (A, B, C), based on one character showing more and more derived states (a, a′, a″), as shown in C, one can make only one cladogram (D, see also in Figure 8). But investigation on other characters (b, c) in the same species may show the reverse trend and imply cladograms (E, F) which contradict the first. There may then be a divorce between evolutionary palaeontologists and cladists: the former will give predominant weight to stratigraphy and will minimize the contradicting characters by *ad hoc* hypotheses, whereas the latter will trust the character distribution and modify the relationships of the species, as shown in G. Cladists consider that the conflict between phylogeny and stratigraphic distribution is due to mere incompleteness of the fossil record of ancestors is the 'speciality' of palaeontology, some evolutionary palaeontologists have proposed the *stratophenetic method* (Gingerich, 1979), assuming that the fossil record, when complete, shows ancestor–descendant relationships directly. However, as noted by Patterson (1981b), the only test of stratophenetics would be an impact on the relationships of extant species. This does not seem to have happened so far. For instance, the segments of rodent phylogeny established by palaeontologists are useless as long as no theory of phylogenetic relationships is proposed for the extant taxa of this group.

Cladists, and transformed cladists in particular, restrict the use of fossils to morphological analysis, and they recommend that fossils be inserted into cladograms based initially on extant taxa, and leaving aside the chronological 'message' fossils are supposed to bring (Figure 6B). But some cladists have tried to go beyond the limits of pattern analysis and see how the latter can test hypotheses about *evolutionary processes* inferred from fossils (Eldredge and Gould, 1972; Gould and Eldredge, 1977; Eldredge and Cracraft, 1980; Wiley, 1978, 1979, 1981). There is a tradition, especially among invertebrate palaeontologists, that a good fossil record in a continuous stratigraphical sequence can document the evolution of one entire species into another (*anagenesis*). Such a continuum may have to be arbitrarily cut into a succession of '*palaeospecies*' (Simpson, 1961) or '*chronospecies*' (Mayr, 1942) and is supposed to reflect an evolutionary process, *phyletic speciation* (Figure 9A). Some biologists and palaeontologists (e.g. Eldredge and Gould, 1972) deny that anagenesis produces new species, and therefore call it *phyletic evolution*. Their reasons are that since chronospecies are arbitrarily defined, there is no evidence of speciation. Also, phyletic speciation, as viewed by many palaeontologists, rests on the study of a few (if not one) characters (ornament of ammonites, molars of rodents, etc., Figure 9B) and may be refuted by investigation of other characters (Figure 9C–G). Phyletic evolution may be regarded as the gradual transformation of a character within one species, in response to environmental changes. Such transformations within species are referred to as *microevolution* (Eldredge and Cracraft, 1980), and differ from *macroevolution*, which is speciation proper. All this does not mean that phyletic evolution is uninteresting. It does not tell us anything about speciation itself, but may be a prerequisite to the formation of peripheral isolates.

Those who reject phyletic speciation and consider that *allopatric speciation* (and especially *vicariance*, see below) is the major source of new species are mainly cladists and, for this reason, the debate on speciation processes has sometimes been absorbed into the debate about cladistics. As a matter of fact, one of the three models of allopatric speciation, *punctuated equilibrium* (the others being vicariance and Wiley's (1981) 'model III'), is a result of cladistic analysis of various fossil invertebrate groups (e.g. trilobites) previously interpreted in the light of phyletic speciation. The punctuated equilibrium (Eldredge and Gould, 1972) or peripheral isolate model explains patterns

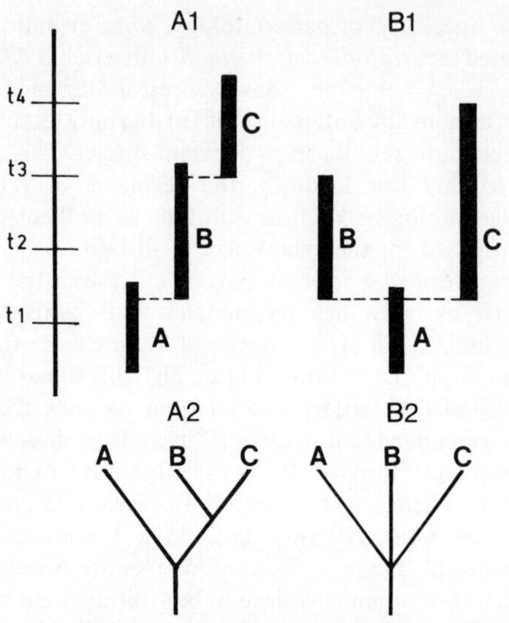

Figure 10. Punctuated equilibria.

The alternative to phyletic speciation is allopatric speciation. One model of allopatric speciation is punctuated equilibrium, or the peripheral isolate model. In palaeontology, successions of species show periods of stasis (black in A1 and B1) separated by gaps (dashed lines in A1 and B1). This succession of equilibria (stasis) and punctuations (speciations) is explained by peripheral isolation of a small population of species A, for instance, which gives rise to species B, and species A then becomes extinct. Such a process is expressed by only one cladogram (A2). When two peripheral isolates succeed at the same time (B1), giving two new species simultaneously, the cladogram shows a trichotomy (B2)

showing a succession of sudden speciation events (*punctuations*) separated by stasis (*equilibrium*) during which phyletic evolution of species may occur. This process of speciation allows two phylogenetic predictions that can be expressed by a cladogram: (1) if a succession of species is due to successive peripheral isolates of the preceding species, a cladogram of this series of species will be dichotomous (Figure 10A), and (2) if several peripheral isolates from a single mother species succeed at the same time, the cladogram of the group will be polytomous (multiple branching, Figure 10B).

The relationship between phylogenetic patterns and evolutionary processes is mostly the preoccupation of palaeontologists, and this may be fruitful in the future, at least in refuting long-lived myths such as phyletic speciation. But the processes inferred from fossils, whatever they may be, must not be taken for more than what they are, that is models applied to occasional cases where the fossil record seems to be more complete than usual. I fear that such hypotheses about process, although more testable than previous ones, still remain beyond the limits of reliability in palaeontology, and the prospect in experimental biology (e.g. epigenetics), as recently outlined by Rosen and Buth (1980) and Løvtrup (1982) look more promising in the search for new evolutionary theories.

In sum, according to modern views, evolution is not implicit in cladistics, but cladistics, as well as other possible tests, should be implicated in evolutionary theories.

2.5. THE FUTURE OF CLADISTICS

The future of cladistics will be conditioned by two sets of factors: sociological and heuristic. Sociological factors have an enormous weight, not in the success or failure of the method, but in controlling its dissemination and development throughout the various branches of comparative biology. Depending on how cladistics will be regarded by those who have the administrative power in natural sciences, it will become generalized in the eighties, or will remain a marginal method for years. The spread of cladistics will certainly occur more rapidly in countries where the power of decision over major reorientations of science is not centralized in the hands of a few. Recently, Boesiger (1980) has pointed out how such centralization delayed the spread of evolutionary thought in France, and this is also true for the spread of mobilistic geology in that country. Consequently, consideration of the future of cladistics must not neglect these non-scientific aspects.

2.5.1. Cladistics, society, and people

In the field of natural science, phylogeny and systematics have no apparent immediate application, thus, their fruitfulness is not sanctioned economically. The use or rejection of cladistics has no bearing on industry—which would not be the case for the rejection of Mendeleev's periodic table!—and researchers in comparative biology are not paid according to the parsimony or internal logic of their theory. In most societies where the educational system includes positions in comparative biology, students and scientists are rarely judged by the parsimony, predictiveness, or refutability of the theories they propose. In most cases, the competence of a student is decided on the basis of his ability to 'verify' previous theories and to produce results which match those of his master. Ability in producing reliable raw data was once regarded as meritorious, but now it seems to be a pity to produce only honest discriptive work. Since the goal of cladistics is refutability, it does not fit a system based on such criteria! Cladists disturb the system by erecting criteria of judgement (parsimony, internal logic, refutability) which are independent of psychological (taste, fashion, fear of being contradicted, etc.) or sociological (authority, coterie, politics) factors. In the name of freedom, traditional, or evolutionary systematists sometimes cite examples of guesses based on intuition, which have not been contradicted but 'supported' by subsequent discoveries. This does not prove anything but that the authors of these guesses are lucky. Lucky guesses are useful in supporting the popular belief that there are geniuses and ordinary people!

There are a few instances where cladists have gained power of decision, and the turmoil it provokes gives an idea of the disarray among biologists and palaeontologists (cf. the debate about the cladistic organization of the exhibits of the British Museum). This may be an interesting field of research for psychologists and sociologists!

When a student or a researcher is judged on the programme of research he proposes, he is often asked what 'scientific problem' he intends to solve. To me, a 'scientific problem' is a set of conflicting *refutable* theories, but, unfortunately, the question whether the proposed programme fits this definition is rarely asked. As far as comparative biology is concerned, there is an urgent need for such conflicting refutable theories, and cladistics is the major source of such theories. The example of the affinities of birds mentioned above would be an ideal subject in this respect. Finally, there is one way for biologists and palaeontologists to bypass the criterion of refutability, by taking refuge in domains where inductive reasoning is largely used, such as palaeoecology, functional anatomy, or narrative palaeobiogeography.

2.5.2. Predictiveness of cladistics

Besides sociological factors, the development of cladistics in the eighties will depend on its ability to predict discoveries in various fields such as molecular biology or geology. *Predictiveness* is not the ability to guess (intuitive belief in a future result) but the ability to propose one or several refutable hypotheses that will *not* be refuted by future discoveries. If this predictiveness appears to be real, and not due to mere chance, cladistics will prove its heuristic superiority. If it fails to predict in spite of good character analysis, then it will have to be replaced by a better method.

The study of *biochemical characters* in a phylogenetic framework will certainly expand in forthcoming years, yet many biologists have been tempted to treat such data by phenetics, probably because the procedures were more easily available, and also because it avoids worries about the polarity of biochemical characters. As pointed out by Wiley (1981), and contrary to current belief, the closer one comes to the genes, the greater is the problem of non-homologies which are structurally identical, and can only be recognized by the test of congruence (congruence of many character distribution patterns). However, successful methods have been used by Mickevich and Johnson (1976) and Fitch (1971) to analyse phylogenetically complex biochemical data (*Wagner tree algorithms, maximum parsimony method*, etc.; see also Felsenstein, 1979, for general discussion of these methods). Cladistic analysis of morphological characters can predict the distribution of some biochemical characters, and this may encourage biochemists to draw conclusions from biochemical characters distributions which are inconsistent with the evolutionary classifications they once learnt in the university! There are numerous examples of biochemical data with a supposedly aberrant distribution, because it is superimposed on classifications or trees containing paraphyletic groups (for instance: substance X is 'present in reptiles, except in turtles', or substance Y is 'present in reptiles, amphibians, and some fishes'). There are also examples of biochemical data which have been found in one species, and then generalized to the whole group to which this species belongs, be it mono- or paraphyletic (for instance: substance X 'is found in mammals (rabbit)' or substance Y 'occurs in reptiles (*Testudo graeca*)'). In contrast, when it is impossible to search for a biochemical datum in all species of a monophyletic group, a relatively well tested cladogram of this group, based on morphological data, will allow a reasoned choice of the species in which this datum can be looked for. The biologist who has practised cladistic analysis can easily understand that the perils of generalizations based on evolutionary classifications may be severe in applied biology!

Biogeography is the field of comparative biology where cladistics can be expected to give the most fruitful results. The prospect of cladistic biogeography has been outlined in a masterly way by Nelson and Platnick (1981), and I recommend that book to students who wish to go into this question. Since Wallace (1876) and Darwin (1859), the distribution of organisms on the earth has generally been interpreted in the light of dispersal from centres of origin, over a fixed continental pattern (Simpson, 1947; Darlington, 1957). When continental drift became well documented geologically, biogeographers have adapted their hypotheses of dispersal to the various past geographies (Darlington, 1970). This biogeography, adopted by evolutionary systematists because it permitted speculations about the distribution of fossil ancestral groups, has been called *narrative historical biogeography*. In contrast, *analytic historical biogeography* is based on the assumption that the phylogeny of a group reflects the geological history of the areas in which its components live. This biogeography includes two modes of analysis: *phylogenetic biogeography*, outlined by Hennig (1966) and Brundin (1966), in which the concepts of centre of origin and disperal routes are retained; and *vicariance biogeography*, of which Croizat (1952, 1958, 1964) is regarded as the precursor, and which is based on replicated patterns of relationships established for a large number of groups living in the same areas. The example in Figure 11 illustrates in a simple way the principle of vicariance biogeography. The pattern of relationship, expressed by a cladogram, for three

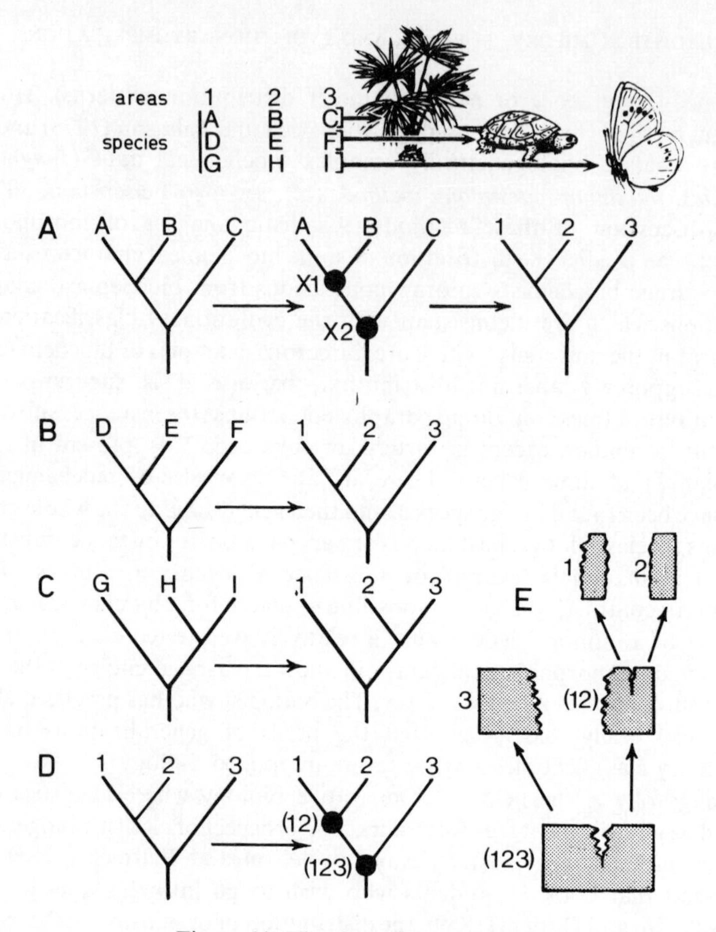

Figure 11. Vicariance biogeography.

Cladistics can provide original hypotheses on the history of the continents by means of vicariance biogeography. This rests on the postulate that allopatric speciation is due to the rise of natural barriers (e.g., break-up of continents), so that the phylogeny of groups is a reflection of the geological history of the areas they occupy. Suppose three separate areas 1, 2, 3 (islands, continents) are inhabited by species belonging to three monophyletic groups (palm trees, turtles, and butterflies, for instance). Each area bears only one species of each group, thus, each species is endemic to an area, with the distribution as shown in the figure. Cladistic analysis of the characters of the group ABC (palm trees, A), will establish a cladogram, or a phylogenetic tree (with hypothetical ancestors X1 and X2) and the corresponding area cladogram in which each species is replaced by the area to which it is endemic. Study of group DEF (turtles, B) may give the same area cladogram. Thus palm trees and turtles show replicated patterns. But study of group GHI (butterflies, C) leads to a different area cladogram. As in Figure 4, parsimony implies that the replicated pattern provides the most reliable area cladogram (D), from which an area tree can be established, with postulated 'ancestral' areas (123) and (12). This area tree is a reflection of a process of break-up of these ancestral areas into the extant areas (E). Area cladograms which are inconsistent with the majority of congruent cladograms (for instance that of group GHI) can be explained by errors in character analysis or by dispersal (butterflies flying accidentally between areas 2 and 3, for instance) species A, B, C (for instance, palm trees) living respectively in areas 1, 2, 3, is supposed to reflect the order of fragmentation of the initial common area (123) on which the ancestral species (X2) lived. The rise of two species (for instance A and B) by fragmentation of the area of distribution of their common ancestor (X1) is called *vicariance*. It is one of the models of allopatric speciation. Given that the species under consideration are restricted (*endemic*) to the respective areas, they can be replaced in the cladogram by the areas 1, 2, 3 (Figure 11A), giving thus an *area cladogram*. The same analysis can be made for other sets of species forming monophyletic groups and living also in the same areas. Species D, E, F (for instance turtles) may lead to the same area cladogram (Figure 11B), thus, there are *replicated patterns*. In contrast, species G, H, I (for instance butterflies) may give a conflicting pattern (Figure 11C), but parsimony implies that the largest number of congruent area cladograms provides the best hypothesis about the history of the areas (Figure 11D,E), until it is refuted by a larger number of contradictory cladograms based on other monophyletic groups. Area cladograms which are inconsistent with the majority can be explained either by mistakes in character analysis, or by dispersal events (that is, distribution of species which does not depend on the history of the areas). In the example above, the resulting area cladogram is that of groups ABC and DEF. The cladogram of GHI would mean, if no mistake has been made in character analysis, that dispersal may have occurred between areas 2 and 3 (butterflies flying from one area to another, for instance).

Vicariance biogeography offers the advantage of hypotheses about the history of areas that can be tested by purely geological data, without being protected by intervening *ad hoc* hypotheses. It can also suggest to geologists inconsistencies between plate tectonic models and biogeography, as, for instance, in the case of the history of the Pacific (Nelson and Platnick, 1981; Nelson, 1981). It is a synthesis of phylogeny (group cladograms), paleontology (extension of areas of endemism in the past), speciation processes (allopatric speciation), and tectonics (hypotheses about area relationships). Its future thus offers a much broader scope than any other derivative of cladistics. Vicariance biogeography is, however, faced with technical problems which make its early development difficult. First, it requires cladistic analysis of many different groups (insects, plants, mammals, etc.) living in the areas under study, and that the various cladograms are equally well tested. This difficulty may soon disappear through meetings of exclusively cladist biologists (e.g. the *Willi Hennig Society*) and through teamwork among scientists working on different groups in the same geographical region. This need for variety is one of the reasons why vicariance biogeography was initiated by a person with encyclopaedic knowledge (Croizat), and then developed in museums (e.g. the American Museum of Natural History) where people working on different groups are gathered in the same institution. Second, it is difficult to compare many cladograms without the help of computer programs, in the search for replicated patterns, but suitable programs will soon be available (Mickevich, 1982) and will certainly accelerate the development of vicariance biogeography in the 1980s. Despite these two difficulties, there are already good examples of vicariance biogeographic analysis, e.g. the now classical study of poeciliid fishes from Central America (Rosen, 1975, 1978, 1979).

2.6. CONCLUSIONS

Cladistics, whether in the form of classical Hennigian phylogenetic systematics or modern 'transformed cladistics', is a method of discovering the pattern of nature and expressing it in a way that makes it refutable. It was originally impregnated with evolutionary tradition and, thus, could hardly be a satisfactory test of evolutionary theories. Its transformation by rejection of any evolutionary presuppositions, mistakingly regarded as a form of neo-creationism, makes it the only reliable test of theories about evolutionary processes. The history of comparative biology can be paralleled to that of astronomy, where progress in understanding movements within the solar system have been due exclusively to successive tests of models by observed patterns. The numerous epicycles of the Ptolemaeic model can be seen as *ad hoc* hypotheses to justify anomalous (unpredicted) movements of the planets. Copernicus's more parsimonious model could predict these anomalous movements without intervening epicycles and has therefore been retained by scientists today. The rediscovery of the heliocentric model is only a question of models (hypotheses) tested by patterns, and speculations on the celestial mechanisms (music of the spheres, angels moving the planets, etc.) played no role in this progress. Even now, knowledge of the structure of the universe makes considerable progress in spite of our knowledge of the process of gravity. If the nature of gravity is found, it will be by deductions from the pattern of the universe. Apparently, evolutionary science has not got that far, yet speculations about processes, such as natural selection, have often proved to be useless in reconstruction of the pattern of evolution. In comparative biology and palaeontology, the future will belong to those who understand the structure of nature, who concentrate their efforts on what is knowable instead of creating new mysteries in the form of untestable hypotheses. In this respect, cladistics can lead to a considerable renewal of comparative biology in the forthcoming years, and will thus have impacts on general biology and even geology.

ACKNOWLEDGEMENTS

For criticism and improvement of the manuscript of this chapter, I am particularly grateful to Colin Patterson (British Museum). I thank also Gareth Nelson and Donn Rosen (American Museum of Natural History) for having opened my eyes on 'transformed cladistics' and vicariance biogeography, and

Daniel Goujet and Pascal Tassy (Museum national d'Histoire naturelle and Université Paris VI) for numerous and fruitful discussions about cladistics.

2.7. REFERENCES

Ashlock, P. H. (1971) Monophyly and associated terms, *Syst. Zool.*, **20**, 63–69.

Ashlock, P. H. (1972) Monophyly again, *Syst. Zool.*, **21**, 430–437.

Ashlock, P. H. (1980) An evolutionary systematicist's view of classification. *Syst. Zool.*, **28**, 441–450.

Baer, K. E., von (1828) *Ueber Entwickelungsgeschichte der Thiere Beobarchtung und Reflektion*, Gebrüder Kornträger, Königsberg.

Beatty, J. (1982) Classes and cladists, *Syst. Zool.*, **31**, 25–34.

Bernard, C. (1865) *Introduction à la médecine expérimentale*, J. B. Baillière et fils, Paris.

Bock, W. J. (1974) Philosophical foundations of classical evolutionary classification, *Syst. Zool.*, **22**, 375–392.

Boesiger, E. (1980) Evolutionary biology in France at the time of the evolutionary synthesis, in Mayr, E., and Provine, W. B. (eds.) *The Evolutionary Synthesis: Perspectives on the Unification of Biology*, Harvard University Press, Cambridge, Massachusetts, pp. 309–320.

Bonde, N. (1975) Origin of 'higher' groups. Viewpoints of phylogenetic systematics, *Coll. Intern. C.N.R.S., Paris*, **218**, 293–324.

Bonde, N. (1977) Cladistic classification as applied to vertebrates, in Hecht, M., Goody, P., and Hecht, B. (eds.) *Major Patterns in Vertebrate Evolution*, Plenum Press, New York, pp. 741–804.

Brooks, D. R. (1982) Classification as languages of empirical comparative biology, in Funk, V. A., and Brooks, D. R. (eds.) *Advances in Cladistics*, The New York Botanical Garden, Bronx, pp. 61–70.

Brundin, L. (1956) Zur Systematik der Orthocladiinae (Dipt. Chironomidae), *Rep. Inst. Freshwat. Res. Drottningholm*, **37**, 1–185.

Brundin, L. (1966) Transantarctic relationships and their significance, as evidenced by chironomid midges, with a monograph of the subfamily Podonominae, Aphroteniinae and the austral Heptagyiae, *K. Svenska VetenskAkad. Handl.*, **11**, 1–472.

Brundin, L. (1968) Applications of phylogenetic principles in systematics and evolutionary theory, in Ørvig, T. (ed.) *Current Problems in Lower Vertebrate Phylogeny*, Almqvist & Wiksell, Stockholm, pp. 473–495.

Brundin, L. (1972) Phylogenetics and biogeography, *Syst. Zool.*, **21**, 69–79.

Brundin, L. (1981) Croizat's panbiogeography versus phylogenetic biogeography, in Nelson, G., and Rosen, D. E. (eds.) *Vicariance Biogeography: a Critique*, Columbia University Press, New York, pp. 94–138.

Candolle, A. P., de (1813) *Théorie élémentaire de botanique*. Detréville, Paris.

Cartmill, M. (1981) Hypothesis testing and phylogenetic reconstruction, *Z. f. zool. Systematik u. Evolutionsforschung*, **19**, 73–96.

Chaline, J., and Mein, P. (1979) *Les Rongeurs et l'Evolution*, Doin, Paris.

Charig, A. (1982) Systematics in biology: a fundamental comparison of some major schools of thought, in Joysey, K. A., and Friday, A. E. (eds.) *Problems of Phylogeny Reconstruction*, Academic Press, London and New York, pp. 367–384.

Cox, B. (1982) New branches for old roots. *Nature*, **298**, 321.

Cracraft, J. (1979) Phylogenetic analysis, evolutionary models and palaeontology, in Cracraft, J., and Eldredge, N. (eds.) *Phylogenetic Analysis and Paleontology*, Columbia University Press, New York, pp. 7–39.

Croizat, L. (1952) *Manual of phytogeography*, Vitgeverij Dr. W. Junk, The Hague.

Croizat, L. (1958) *Panbiogeography*, L. Croizat, Caracas.

Croizat, L. (1964) *Space, Time and Form: the Biological Synthesis*, L. Croizat, Caracas.

Darlington, P. J., Jr. (1957) *Zoogeography: the Geographic Distribution of Animals*. Wiley and Sons, New York.

Darlington, P. J., Jr. (1970) A practical criticism of Hennig–Brundin 'Phylogentic Systematics' and Antarctic biogeography, *Syst. Zool.*, **19**, 1–18.

Darwin, C. (1859) *On the Origin of Species by Means of Natural Selection, or the Preservation of Favoured Races in the Struggle for Life*, John Murray, London.

De Jong, R. (1980) Some tools for evolutionary and phylogenetic studies, *Z. zool. Syst. Evol.-forsch.*, **18**, 1–23.

Dubois, A. (1982) Les notions de genre, sous-genre et groupe d'espèces en zoologie, à la lumière de la systématique évolutive, *Monitore zool. ital.* (*N.S.*), **16**, 9–65.

Dupuis, C. (1978) Permanence et actualité de la systématique: la 'systématique phylogénétique' de W. Hennig (Historique, discussion, choix de références), *Cahiers des Naturalistes*, **34**, 1–69.

Eldredge, N. (1979) Cladism and common sense, in Cracraft, J., and Eldredge, N. (eds.) *Phylogenetic analysis and paleontology*, Columbia University Press, New York, pp. 165–198.

Eldredge, N., and Gould, S. J. (1972) Punctuated equilibria: an alternative to phyletic gradualism, in Schopf, T. J. M. (ed.) *Models in Paleobiology*, Freeman, Cooper & Co, San Francisco, pp. 82–115.

Eldredge, N., and Cracraft, J. (1980) *Phylogenetic Pattern and the Evolutionary Process*, Columbia University Press, New York.

Engelmann, G. F., and Wiley, E. O. (1977) The place of ancestor–descendant relationships in phylogeny reconstruction, *Syst. Zool.*, **26**, 1–11.

Farris, J. S. (1970) Methods for computing Wagner trees, *Syst. Zool.*, **19**, 83–92.

Farris, J. S. (1973) A probability model for inferring evolutionary trees, *Syst. Zool.*, **22**, 250–256.

Farris, J. S. (1977) On the phenetic approach to vertebrate classification, in Hecht, M., Goody, P., and Hecht, B. (eds.) *Major patterns in Vertebrate evolution*, Plenum Press, New York, pp. 823–850.

Farris, J. S. (1980) The information content of the phylogenetic system, *Syst. Zool.*, **28**, 483–519.

Felsenstein, J. (1979) Alternative methods of phylogenetic inference and their interrelationship, *Syst. Zool.*, **28**, 49–62.

Fitch, W. M. (1971) Toward defining the course of evolution: minimum change for a specific tree topology, *Syst. Zool.*, **20**, 406–416.

Forey, P. L. (1982) Neontological analysis versus palaeontological stories, in Joysey, K. A., and Friday, A. E. (eds.) *Problems of phylogenetic reconstruction*, Academic Press, U.K., pp. 119–157.

Gaffney, E. S. (1979) An introduction to the logic of phylogeny reconstruction, in Cracraft, J., and Eldredge, N. (eds.) *Phylogenetic Analysis and Paleontology*, Columbia University Press, New York, pp. 79–111.

Gardiner, B. G. (1982) Tetrapod classification, *Zool. J. Linn. Soc.*, **74**, 207–232.

Gilmour, J. S. L. (1961) Taxonomy, in Macleod, A. M., and Cobley, L. S. (eds.) *Contemporary Botanical Thought*, Oliver and Boyd, Edinburgh, pp. 27–45.

Gingerich, P. D. (1977) Patterns of evolution in the mammalian fossil record, in Hallam, A. (ed.) *Patterns of Evolution, as Illustrated by the Fossil Record*, Elsevier, Amsterdam, pp. 469–500.

Gingerich, P. D. (1979) The stratophenetic approach to phylogeny reconstruction in

vertebrate paleontology, in Cracraft, J., and Eldredge, N. (eds.) *Phylogenetic Analysis and Paleontology*, Columbia University Press, New York, pp. 113–163.
Goujet, D. (1981) Systématique et phylogénie, *Universalia*, **1981**, 356–358.
Gould, S. J., and Eldredge, N. (1977) Punctuated equilibria: the tempo and mode of evolution reconsidered, *Paleobiology*, **3**(2), 115–151.
Günther, K. (1956) Systematik und Stammgeschichte der Tiere 1939–1953, *Fortschritte d. Zool.*, **10**, 33–278.
Haeckel, E. (1866) *Generelle Morphologie der Organismen*, G. Reimer, Berlin.
Halstead, L. B. (1982) Evolutionary trends and the phylogeny of the Agnatha, in Joysey, K. A., and Friday, A. E. (eds.) *Problems of Phylogenetic Reconstruction*, Academic Press, London and New York, pp. 159–196.
Hecht, M. K., and Edwards, J. L. (1977) The methodology of phylogenetic inference above the species level, in Hecht, M., Goody, P., and Hecht, B. (eds.) *Major Patterns in Vertebrate Evolution*, Plenum Press, New York, pp. 3–51.
Hennig, W. (1936) Beziehungen zwischen geographischer Verbreitung und systematischer Gliederung bei einigen Dipterenfamilien: ein Beitrag zum Problem der Gliederung systematischer Kategorien höherer Ordnung, *Zool. Anz.*, **116**, 161–175.
Hennig, W. (1948) Theorie der zoologischen Systematik, in *Die Larvenformen der Dipteren*, **1**, 2–22. Akad Verlag, Berlin.
Hennig, W. (1949) Zur Klärung einiger Begriffe der phylogenetischen Systematik, *Forschungen u. Forschr.*, **25**, 136–138.
Hennig, W. (1950) *Grundzüge einer Theorie der phylogenetischen Systematik*, Deutscher Zentralverlag, Berlin.
Hennig, W. (1953) Kritische Bemerkungen zum phylogenetischen System der Insekten, *Beitr. z. Ent.*, **3**, 1–85.
Hennig, W. (1965) Phylogenetic systematics, *Annual Rev. Ent.*, **10**, 97–116.
Hennig, W. (1966) *Phylogenetic systematics*, University of Illinois Press, Urbana, Chicago, London.
Hennig, W. (1974) Kritische Bemerkungen zur Frage 'cladistic analysis or classification', *Z. zool. Syst. Evol.-forsch.*, **12**, 279–294.
Hennig, W. (1975) Cladistic analysis or cladistic classification?: a reply to Ernst Mayr, *Syst. Zool.*, **24**, 244–256.
Hennig, W. (1981) *Insect phylogeny*, J. Wiley & Sons, Chichester, New York.
Hill, C. R., and Crane, P. R. (1982) Cladistics and the origin of angiosperms, in Joysey, K. A., and Friday, A. E. (eds.) *Problems of Phylogeny Reconstruction*, Academic Press, London and New York, pp. 301–306.
Janvier, P., Tassy, P., and Thomas, H. (1980) Le cladisme, *La Recherche*, **11**, 1396–1406.
Kiriakoff, S. G. (1955) Le système phylogénétique: principes et méthodes, *Bull. Ann. Soc. r. Ent. Belg.*, **91**, 147–158.
Lamarck, J. B. (1809) *Philosophie zoologique*, Dentu, Paris.
Linnaeus, C. (1758) *Caroli Linnaei Systema Naturae, Regnum Animale* 10th ed. British Museum (Natural History), London (reprinted 1939).
Løvtrup, S. (1974) *Epigenetics, a treatise on theoretical biology*, J. Wiley & Sons, New York.
Løvtrup, S. (1978) On von Baerian and Haeckelian recapitulation. *Syst. Zool.*, **27**, 348–352.
Løvtrup, S. (1982) The four theories of evolution. II. The epigenetic theory, *Riv. Biol.*, **75**, 231–272.
Mayr, E. (1942) *Systematics and the Origin of Species*, Columbia University Press, New York.
Mayr, E. (1969) *Principles of Systematic Zoology*, McGraw-Hill, New York.

Mayr, E. (1974) Cladistic analysis or cladistic classification?, *Z. Zool. syst. Evolut.-forsch.*, **12**, 94–128.

Michener, C. D. (1963) Some future developments in taxonomy, *Syst. Zool.*, **12**, 151–172.

Mickevich, M. F. (1978) Taxonomic congruence, *Syst. Zool.*, **27**, 143–158.

Mickevich, M. F. (1982) Quantitative phylogenetic biogeography, in Funk, V. A., and Brooks, D. R. (eds.) *Advances in cladistics*. The New York Botanical Garden, Bronx, pp. 209–222.

Mickevich, M. F., and Johnson, M. S. (1976) Congruence between morphological and allozyme data in evolutionary inference and character evolution, *Syst. Zool.*, **25**, 260–270.

Mivart, S. G. (1882) On the classification and distribution of the Aeluroidea, *Proc. Zool. Soc. London*, **1882**, 135–208.

Nelson, G. J. (1970) Outline of a theory of comparative biology, *Syst. Zool.*, **19**, 373–384.

Nelson, G. J. (1974) Classification as an expression of phylogenetic relationships, *Syst. Zool.*, **22**, 344–359.

Nelson, G. J. (1975) Historical biogeography: an alternative formulation, *Syst. Zool.*, **23**, 555–558.

Nelson, G. J. (1978) Ontogeny, phylogeny, and the biogenetic law, *Syst. Zool.*, **27**, 324–345.

Nelson, G. J. (1981) Summary, in Nelson, G., and Rosen, D. E. (eds.) *Vicariance biogeography, a critique*, Columbia University Press, New York, pp. 524–537.

Nelson, G. J., and Platnick, N. (1981) *Systematics and biogeography: Cladistics and vicariance*, Columbia University Press, New York.

Panchen, A. L. (1982) The use of parsimony in testing phylogenetic hypotheses, *Zool. J. Linn. Soc.*, **74**, 305–328.

Patterson, C. (1978) *Evolution*, British Museum (Natural History), London.

Patterson, C. (1981a) Pattern versus process in nature: a personal view of a method and a controversy, *Biologist*, **27**, 234–240.

Patterson, C. (1981b) Significance of fossils in determining evolutionary relationships, *Ann. Rev. Ecol. Syst.*, **12**, 195–223.

Patterson, C. (1982a) Cladistics and classification, *New Scientist*, **1982**, 303–306.

Patterson, C. (1982b) Morphological characters and homology, in Joysey, K. A., and Friday, A. E. (eds.) *Problems of Phylogenetic Reconstruction*, Academic Press, London and New York, pp. 21–74.

Patterson, C., and Rosen, D. E. (1977) Review of ichthyodectiform and other Mesozoic teleost fishes and the theory and practice of classifying fossils, *Bull. Amer. Mus. nat. Hist.*, **158**, 81–172.

Platnick, N. I., and Gaffney, E. S. (1977) Systematics: a Popperian perspective, *Syst. Zool.*, **26**, 360–365.

Platnick, N. I., and Gaffney, E. S. (1978a) Evolutionary biology: a Popperian perspective, *Syst. Zool.*, **27**, 137–141.

Platnick, N. I., and Gaffney, E. S. (1978b) Systematics and the Popperian paradigm, *Syst. Zool.*, **27**, 381–388.

Popper, K. R. (1959) *The Logic of Scientific Discovery*, Hutchinson, London.

Popper, K. R. (1963) *Conjectures and Refutations—The Growth of Scientific Knowledge*, Routledge and Kegan Paul, London.

Popper, K. R. (1972) *Objective knowledge—an Evolutionary Approach*, University Press, Oxford.

Rosen, D. E. (1975) A vicariance model of Caribbean biogeography, *Syst. Zool.*, **24**, 431–464.

Rosen, D. E. (1978) Vicariant patterns and historical explanation in biogeography, *Syst. Zool.*, **27**, 159–188.

Rosen, D. E. (1979) Fishes from the uplands and intermontane basins of Guatemala: revisionary study and comparative biogeography, *Bull. Amer. Mus. Nat. Hist.*, **162(5)**, 267–376.
Rosen, D. E., and Buth, D. G. (1980) Empirical evolutionary research versus neo-Darwinian speculation, *Syst. Zool.*, **29**(3), 300–308.
Simpson, G. G. (1944) *Tempo and Mode in Evolution*, Columbia University Press, New York.
Simpson, G. G. (1947) Holarctic mammalian faunas and continental relationships during the Cenozoic, *Bull. geol. Soc. Amer.*, **58**, 613–688.
Simpson, G. G. (1961) *Principles of Animal Taxonomy*, Columbia University Press, New York.
Simpson, G. G. (1975) Recent advances in methods of phylogenetic inference, in Lucket, W. P., and Szalay, F. S. (eds.) *Phylogeny of the Primates, a Multi-Disciplinary Approach*, Plenum Press, New York, pp. 3–19.
Slater, J. A. (1981) Discussion of L. Brundin's article, in Nelson, G., and Rosch, D. (eds.) *Vicariance biogeography, a Critique*, Columbia University Press, New York, pp. 139–143.
Sneath, P. H. A. (1961) Recent developments in theoretical and quantitative taxonomy, *Syst. Zool.*, **10**, 118–139.
Sokal, R. R., and Sneath, P. H. A. (1963) *The Principles of Numerical Taxonomy*, Freeman and Co, San Francisco.
Stevens, P. (1980) Evolutionary polarity of character states, *Ann. Rev. Ecol. Syst.*, **11**, 333–358.
Tassy, P. (1981a) Lamarck and systematics, *Syst. Zool.*, **30**, 198–200.
Tassy, P. (1981b) Le crâne de *Moeritherium* (Proboscidea, Mammalia) de l'Eocène de Dor el Talha (Lybie) et le problème de la classification phylogénétique du genre dans les Tethytheria McKenna, 1975, *Bull. Mus. natn. Hist. nat. Paris*, **3**, 87–147.
Tassy, P. (1983) Actualité de la classification zoologique de Darwin, in Conry, Y., and Lecourt, D. (eds.) *De Darwin au darwinisme: science et idéologie*, pp. 261–273.
Tedford, R. H. (1976) Relationship of pinnipeds to other carnivores (Mammalia). *Syst. Zool.*, **25**, 363–374.
Thenius, E. (1980) *Grundzüge der Faunen- und Verbreitungsgeschichte der Säugetiere*. Gustav Fischer Verlag, Stuttgart, New York.
Tintant, H., and Mouterde, R. (1981) Classification et phylogenèse chez les ammonites jurassiques, in Martinell, J. (ed.) *International symposium on 'Concept and Method in Paleontology*', University of Barcelona, Barcelona, pp. 85–101.
Wagner, W. H., Jr. (1961) Problems in the classification of ferns, *Recent Advances in Botany*, 841–844.
Wallace, A. R. (1855) On the law which has regulated the introduction of new species, *Ann. Mag. Natur. Hist.*, **16**, 184–196.
Wallace, A. R. (1876) *The Geographical Distribution of Animals*, MacMillan and Co, London.
Watrous, L. E., and Wheeler, Q. D. (1981) The out-group comparison method of character analysis, *Syst. Zool.*, **30**, 1–11.
Wiley, E. O. (1978) The evolutionary species concept reconsidered, *Syst. Zool.*, **27**, 17–26.
Wiley, E. O. (1979) An annotated Linnaean Hierarchy, with comments on natural taxa and competing systems, *Syst. Zool.*, **28**, 308–337.
Wiley, E. O. (1981) *Phylogenetics: The theory and practice of phylogenetic systematics*, John Wiley & Sons, New York.

Evolutionary Theory: Paths into the Future
Edited by J. W. Pollard

Chapter 3

Hierarchies and history

DONN E. ROSEN
*Department of Ichthyology,
The American Museum of Natural History,
Central Park West at 79th Street,
New York,
New York 10024
USA*

3.1. INTRODUCTION

Ontogeny teaches that each organism is ordered hierarchically during its transformation from zygote to adult. A comparison of different ontogenies teaches that the differences between organisms are part of this hierarchical order. Hierarchical classifications of all organisms can be understood as partial summaries or estimates of ontogenetic histories. Phylogeny is the concept that these hierarchical classifications reflect diversification through time of an underlying, or ancestral, ontogenetic pattern. The intertwined concepts of natural order, hierarchy, and transformation have direct empirical ties to the ontogenetic process.

In fact, all our direct experiences of the world include the notion of change or transformation, the notion of history. The empirical bases for historical inquiry are observations of connectedness and transformation. Offspring are connected to parents; butterflies give rise to more butterflies; young grow into adults. How can butterflies be changed into something else? A proximate cause is a change in ontogeny. A remote cause is a change in the genetic control of ontogeny. How might one study this change that has been called the evolutionary process? The naïve answer is simply to study the details of genetic and ontogenetic mechanisms divorced from any more general problem and to formulate universal statements of which our particular world is merely assumed to be an example. No viable evolutionary theory has emerged, or can emerge, from such study so long as it evades the specific historical question of how butterflies and moths arose as parts of a complex hierarchical ordering of 'horses and tigers and things'. Contemporary evolutionary theories all appear to suffer from what Alfred North

Whitehead called the 'fallacy of misplaced concreteness'—mistaking the abstractions inherently necessary for analytic progress for the empirical reality from which the abstractions are made.

The idea that hierarchy is history is also a transformational interpretation of branching networks of sequential relationships that expand in complexity toward the lower, less inclusive, hierarchical ranks, and decline in complexity toward the higher, more inclusive, ones. The directionality of the branching networks is suggested by ontogeny, which proceeds from the general to the less general.

A corollary to the concept of a uniquely determined hierarchy is that any historical process underlying it should be irreversible. In ontogenetic terms this means that frogs should not transform into tadpoles or arms into limb buds. In phylogenetic terms it means that tetrapods should not give rise to fishes or feathers and hair to scales. Although irreversibility is a property of ideal hierarchical patterns, it is not a logical requirement of ontogenetic change in the real world; instead, irreversibility of ontogenetic transformations is an empirical generalization. Since phylogenetic investigation is really an attempt to interpret historically the hierarchically arranged components of ontogeny, there is no empirical basis for assuming that phylogenetic reversals are possible either. If reversibility were to have occurred among the bits of information that are used to discover hierarchical patterns, the different kinds of information would produce conflicting patterns. Conflicting data do, of course, exist, but more than 200 years of detailed comparative ontogenetic and anatomical investigations have produced increasingly consistent patterns, in part by discovering that at least some of the conflict, the noise in the system, is due to analytical error. In all these investigations ontogeny has had a special status because it can reject hierarchical theories derived from comparative anatomy, but the reverse is not true. Ontogeny can serve as an arbiter of what is noise and what is signal.

The chief preoccupation of comparative biologists, therefore, is to discover congruent hierarchical patterns of relationship among organisms based on data with a favourable signal to noise ratio. These hierarchies are the objects of historical, or evolutionary, interpretation. Hence, inquiry about what evolution has taken place, the phylogeny, requires no special notion of process. Phylogeny is an inference drawn from knowledge of reproduction and development to explain the orderliness of organismic diversity—the diversity of parent–offspring relationships, ontogenetic stages, and kinds of organisms. Modern systematics—the study of congruence among hierarchically ordered natural systems (Nelson and Platnick, 1981)—adds empirical support to the inference that hierarchy is history because of a realized expectation that the ordered pattern of life will, in some degree, mirror the hierarchical pattern of the world in which the organisms arose. Well corroborated hierarchical theories of natural order provide a precise tool for also bringing genetics and development into analytical historical inquiry. These lessons of the last 200 years of biology suggest that congruence theory provides a needed research programme, discussed below, for establishing deterministic relationships among the data of genetics, development, systematics, and earth history.

3.2. DISCOVERING NATURE'S HIERARCHY

Even now little is known about development in most plant and animal species; generally, order in nature is perceived by comparisons of structures from a single life-history stage in these species. Nevertheless, the underlying justification for the comparative method should be the expectation that discovering patterns of hierarchical character distributions among adult organisms will enable us indirectly to recover the information about hierarchical order that is provided directly by ontogeny.

Discovering nature's hierarchical structure begins with an inference about how some feature of an organism might be a transformation of or part of the same transformation as a similar feature in another organism. Such inferences are really theories of homology, as, for example, the relationship between a fish's fin and a tetrapod's limb; but because of the paucity of ontogenetic data theories of homology are generally restricted to an interpretation of features of adult organisms. Nevertheless, ontogenetic transformations can in principle be recovered from the adult stages of several related species where different conditions of some feature, such as keratin are interpreted correctly as parts of the same multistate character assemblage or sequence (e.g., amphibian keratinized skin, reptilian scales, mammal hair, and bird feathers). Whether these conditions truly represent parts of a single kind of ontogeny is checkable by direct observation of the embryology of each species and comparison of these embryologies.

Because it is common in systematic research for ontogenetic data to be lacking, there is often a question if such transformations have been interpreted correctly. To illustrate this an actual example, taken from Berry (1964) and Rosen (1982), employs two sets of ontogenetic data: A—development of upper jaw bones; B—development of upper jaw dentition; in four teleost fish species: T_1—a primitive, elapomorph species; T_2, T_3, T_4—three members of a derived, neoteleost, group: T_2—a primitive neoteleost; T_3—a more advanced neoteleost; and T_4—a representative of the most derived subgroup of neoteleosts. Each of the two anatomical features can be viewed as transforming through three character states, as shown in Figure 1, and represented by the symbols 1^{1-3} and B^{1-3}. These character states correspond with three ontogenetic stages here treated for convenience as prejuvenile, juvenile, and adult. Reference to Figure 1 shows that in the prejuvenile stage of the most primitive of the four species (T_1) the upper jaw bones (A) and dentition (B) are in their untransformed states (A^1 and B^1) and that in the most advanced of the four species (T_4) the upper jaw and dentition are in their completely transformed states (A^3 and B^3) from the prejuvenile stage on.

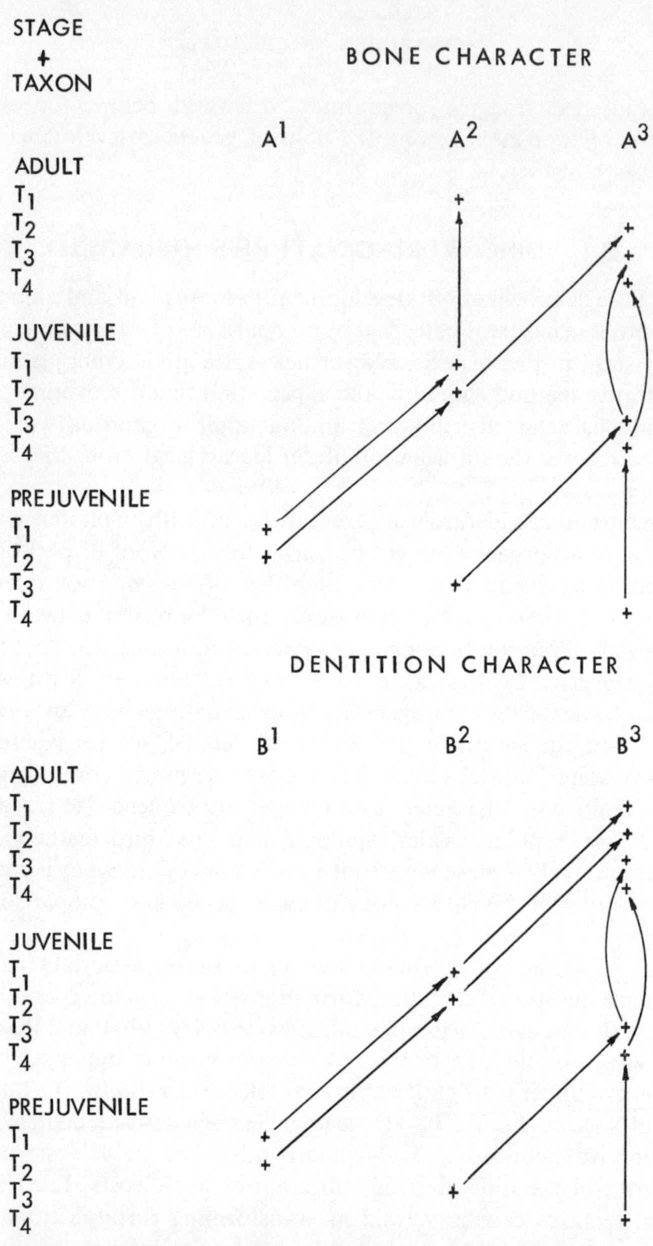

Figure 1. Ontogenetic transformations of the upper jaw bones and dentition in four groups of teleostean fishes: T_1, elopomorph, or primitive teleost; T_2, T_3, T_4, members of more advanced teleost subgroups of the Neoteleostei. Characters: A^1, maxilla not underlain at all by premaxilla; A^2, maxilla partly excluded from gape by premaxilla; A^3, maxilla entirely excluded from gape; B^1, maxilla fully toothed; B^2, maxillary dentition reduced; B^3, maxilla without teeth

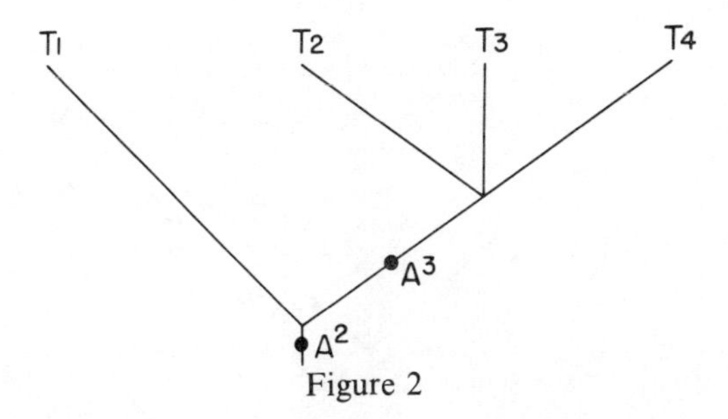

Figure 2

Thus, in T_4, transformation states A^{1-2} and B^{1-2} are absent throughout development whereas transformed character states A^{2-3} and B^{2-3} are attained in the other species T_{1-3} only by passing through less transformed states earlier in ontogeny. These ontogenetic relations can be simplified and represented graphically, as follows:

		Character transformations
Taxa	T_1	$A^1 \rightarrow A^2$
	T_2	$A^1 \rightarrow A^2 \rightarrow A^3$
	T_3	$A^2 \rightarrow A^3$
	T_4	A^3
Taxa	T_1	$B^1 \rightarrow B^2 \rightarrow B^3$
	T_2	$B^1 \rightarrow B^2 \rightarrow B^3$
	T_3	$B^2 \rightarrow B^3$
	T_4	B^3

where A^1 transforms only as far as A^2 in the adult of one species (T_1) and A^3 and B^3 appear as the only character states from early development to adulthood in another species (T_4). Whereas the addition of transformed states (A^3, B^3) to the sequence can be detected in adults, the deletion of states (A^1, A^2, B^1, B^2) cannot. When the adult data are analysed with the knowledge that all other organisms in the group of which T_{1-4} are members possess state A^2, the A data tell us that $T_2T_3T_4$ form a group within which, however, interrelationships are not specifiable (Figure 2).

The B data for adults are completely uninformative (Figure 3).

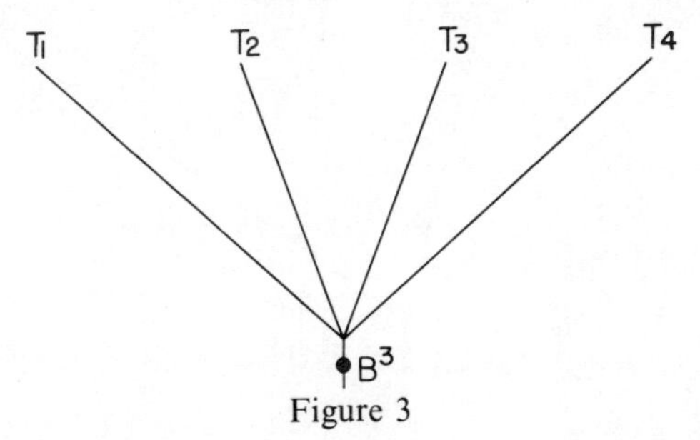

Figure 3

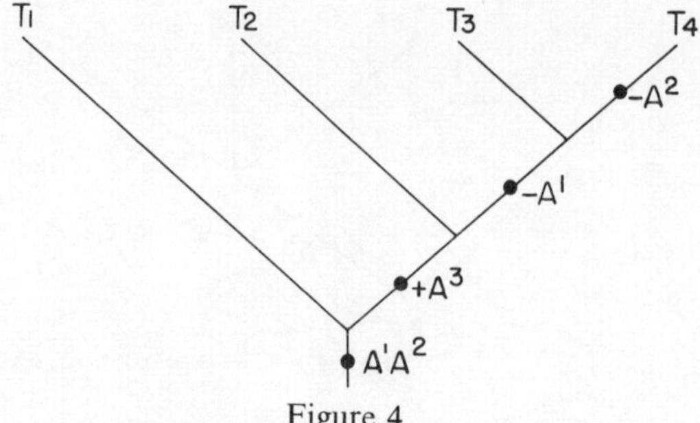

Figure 4

Reference to the ontogenetic data, however, permits a more detailed hierarchical statement by specifying direction of character state transformation and the extent to which some species are represented by all or only parts of the transformation sequence. Ontogenetically the A data not only show the reality of a transformation from A^1 to A^3 but also a decreasing representation of untransformed states and an increasing representation of transformed states, such that T_3 and T_4 form a group, T_2 a more inclusive group with them, and T_1 a still more inclusive group with those three (Figure 4).

The B data, treated similarly, resolve this hierarchical statement (Figure 5).

Notice that though the B data are less informative than the A data, the ontogenetic data are consistently more informative than the comparative data derived only from adults. This is because characters viewed as transformations have observable properties of deletion and extension whereas, viewed in the conventional way as features of adults, characters have only the observable property of data points. Ontogeny, thus, is more informative because it includes more character-state information. This conclusion does not mean that hierarchical relationships cannot be deduced without ontogenetic data. It does mean, however, that in everyday systematic work with adult organisms additional data sources are often required to achieve the level of resolution of ontogenetic data.

Thus, in comparative anatomical argumentation a character-state transformation (i.e., a homology statement) must be postulated, rather than observed,

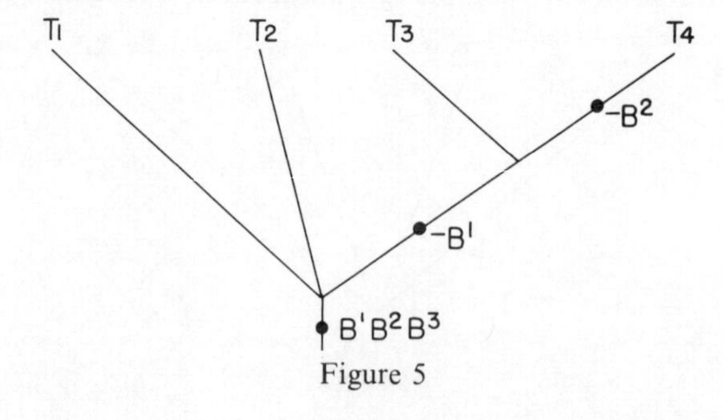

Figure 5

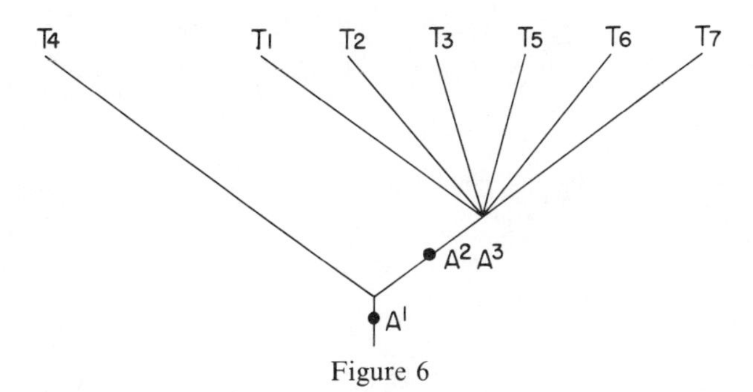

Figure 6

and the polarity of the transformation also must be postulated rather than observed. Character polarity (the direction of transformation) is decided by inferring (rather than observing) that a character lying both inside and outside a group is untransformed and that its alternative (homologous) state occurring only within the group is a derivative condition. The difference between the two approaches to character analysis is therefore that the ontogenetic argument is direct and the comparative anatomy argument indirect with respect to the determination of character polarity.

Although the two approaches differ with regard to hierarchical resolution under conditions when early life history stages have been deleted, they encounter similar problems, at least in theory, when terminal stages are deleted. This problem is illustrated by the transformation $A^1 \rightarrow A^2 \rightarrow A^3 \rightarrow A^4$, where A^4 is at some level of discrimination indistinguishable from A^2. Under those conditions the transformation can be represented by the alternative formulations $A^1 \rightarrow A^2 \rightarrow A^3 \rightarrow -A^3$ or $A^1 \rightarrow A^2 \rightleftharpoons A^3$. For a three-state transformation there are seven ways in which the complete transformation can be represented that allow for the deletion of A^1, A^2, or A^3 from the sequence:

		Character transformations	Discriminated and (real) adult states
	T_1	$A^1 \rightarrow A^2 \rightarrow A^3$	A^3
	T_2	$A^1 \rightarrow A^2 \rightleftharpoons A^3$	$A^2(-A^3)$
	T_3	$A^1 \rightarrow A^3$	A^3
Taxa	T_4	$A^1 \rightleftharpoons A^3$	$A^1(-A^3)$
	T_5	$A^2 \rightarrow A^3$	A^3
	T_6	$A^2 \rightleftharpoons A^3$	$A^2(-A^3)$
	T_7	A^3	A^3

When only the discriminated adult states are used in hierarchical analysis, the result shown in Figure 6 is achieved in which the A^1 state is known to characterize the nearest relatives of group T_{1-7}. When the discriminated adult states are used together with a knowledge of early ontogenetic deletions, the relationships shown in Figure 7 are specified. When complete ontogenetic data is used (a combination

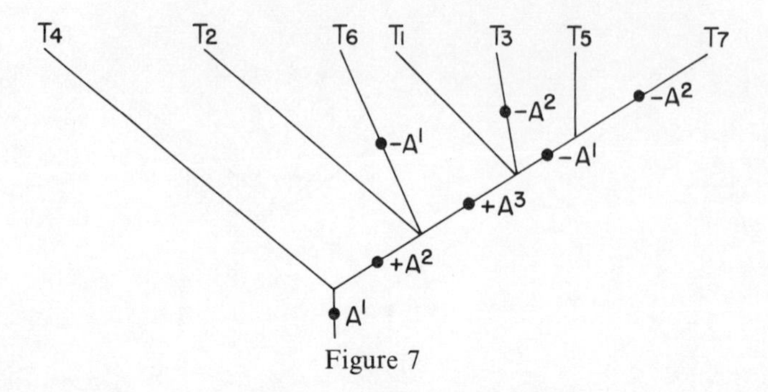

Figure 7

of real adult states and early ontogenetic deletions), a third result is obtained (Figure 8). Notice that in the first two hierarchies, that rely on discriminated adult states, the groupings differ from this solution because the adult states are incorrectly perceived, and that they differ from each other because early ontogenetic information is unavailable to the first solution. In theory, therefore, undetected deletions of final ontogenetic stages can have a crucial effect on how hierarchies are formulated. It only remains to ask: how common might such deletions be? If they are undetectable, the answer is that we can never know. At the present time this theoretically possible source of error has not been investigated except to the extent that we are aware of some particular instances when structural primordia appear in early development and are then later resorbed; in this case the terminal deletion can be detected by studying ontogeny. But we also know that in related species not even the primordia are present. Therefore, the possibility of such error appears to be real and will be exaggerated by the sole use of adult character states.

A final point about hierarchical order concerns the matter mentioned earlier, of character state reversals in evolution or developmental stage reversals in ontogeny. Macbeth (1980) summarized present concepts of reversals in evolution:

> '. . . evolution never goes into reverse in a big way . . . evolution frequently goes into reverse in a small way' (Macbeth, 1980, pp. 402–404).

These two statements can be translated into ontogenetic terms: ontogeny never goes into reverse in a big way by the reversal of early developmental stages; ontogeny frequently appears to go into reverse in a small way by the deletion of terminal developmental stages. Some examples of the latter that are commonly cited are the failure of some sexually mature salamanders to have metamorphosed from the branchiate stage into abranchiate adults or the dedifferentiation of adult tissues into mesenchyme (e.g., during limb regeneration). But reversal in the sense of something deleted, as in the case of failed metamorphosis, is characteristic of all levels of development and it is idiosyncratic to refer to the phenomenon as a deletion in early ontogeny and a

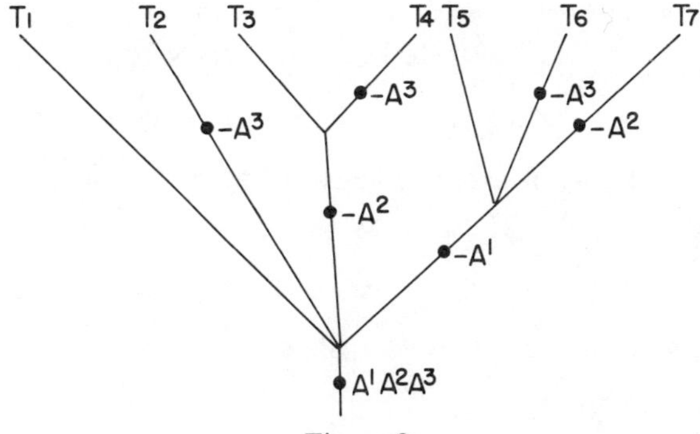

Figure 8

reversal at the final ontogenetic stage. And reversal in the sense of dedifferentiation, especially as it has been applied to vertebrate regeneration, is perhaps an illusion caused by the sudden proliferation of an always present small population of embryonal-like stem cells and the co-ordinated histolysis of damaged tissues. The co-ordination of histolysis and histogenesis is a normal feature of sexual differentiation in at least some teleosts (Rosen and Kallman, 1959). Both terminal ontogenetic deletions and assertions of 'dedifferentiation' are at best ambiguous and at worst inappropriate evidences of ontogenetic reversal. The sense of something reversed is that in a transformation $A^1 \rightarrow A^2 \rightarrow A^3$ there sometimes appear unique relations of the type $A^3 \rightarrow A^2$ or $A^2 \rightarrow A^1$, as in the following five possible transformations:

$$A^1 \rightarrow A^3 \rightarrow A^2$$
$$A^2 \rightarrow A^3 \rightarrow A^1$$
$$A^2 \rightarrow A^1 \rightarrow A^3$$
$$A^3 \rightarrow A^2 \rightarrow A^1$$
$$A^3 \rightarrow A^1 \rightarrow A^2$$

which, in combination with each other or with the normal transformation $A^1 \rightarrow A^2 \rightarrow A^3$, lack hierarchical structure. Terminal ontogenetic deletions, even though undetectable in some instances are, therefore, best treated as deletions, rather than reversals, as given by the notation $A^1 \rightarrow A^2 \rightleftharpoons A^3$.

3.3. CONGRUENCE AND THE MEANING OF EVOLUTION

If the initial objective in studying evolutionary history is to produce a phylogeny (= a hierarchical statement) that correctly reflects the information stored in an organism's complete life history (the holomorph), then the single-character analysis of only the adult stage is unlikely to achieve this goal very effectively in many instances because the analysis of any given trait is subject to observational error. Nevertheless, we know from experience that the method of comparative

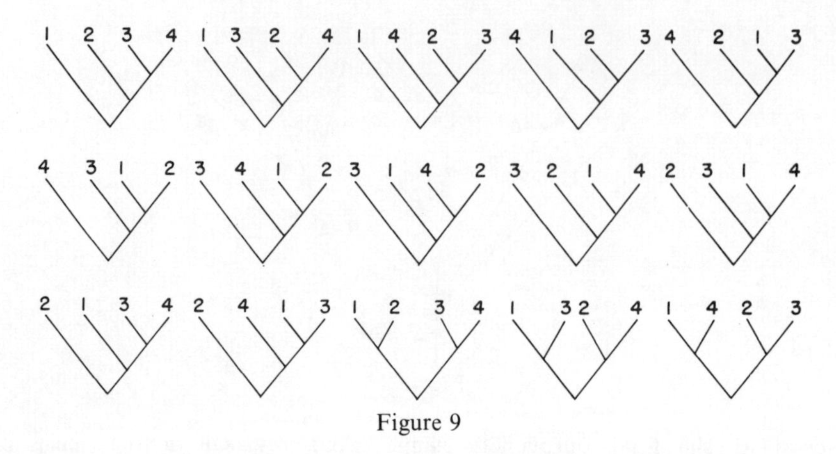

Figure 9

anatomy can work very well. The concepts of Mammalia, Gnathostomata, and Vertebrata have existed for a very long time, have repeatedly withstood attempts at rejection, and were initially formulated without benefit of ontogenetic data. The reason it works as well as it does is that the method involves repeated sampling of the products of many different ontogenetic transformations until a level of agreement among the sampling results is achieved indicating that the agreement is unlikely to have arisen by chance. It is a search for congruent hierarchical solutions, a search that must continue until that acceptable level of corroboration is achieved. Congruence, therefore, is a method of discovering the hierarchy of organisms in the absence of knowledge about the ontogenetic transformations of individual characters that constitute the adult stage.

Accepting that there is an adequate method for constructing hierarchies (see Chapter 2 for a description of cladistic analysis), we can move on to questions of what can be done with hierarchical information. It remains only to reiterate that this method, like all scientific methods is not error-free partly because of different perceptions by scientists and partly because of problems inherent in reading correctly the results of ontogeny reflected in the adult anatomy.

Imagine that several investigators independently attempt to resolve the interrelationships of four species, each using a different set of characters, e.g., morphological, behavioural, biochemical, etc. If nature is orderly and if all character transformations are interpreted correctly, all of the independently derived statements about relationship should agree. There is, after all, just one reality.

What level of agreement or congruence is needed before one can claim to have a corroborated theory of relationship? For four taxa there are 15 possible completely resolved ways in which they might be interrelated (Figure 9). It is clear, therefore, that to find two character transformations that resolve the same relationships among four taxa would be expected to occur once in 15 random searches for such evidence, and for three character transformations, only once in

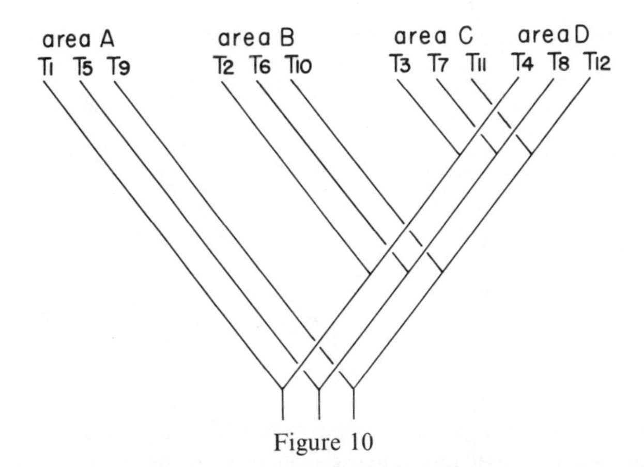

Figure 10

every 225 (15 × 15) searches if such congruence were ruled by chance. The discovery of such congruence minimizes the chance of error inherent in single-character comparative anatomical argumentation, and makes an attested assertion about character transformation. The asserted transformation should agree with ontogeny to the extent that ontogenetic character-state analysis yields the same branching sequence of taxa.

Since all organisms exist in a geographic setting, if each of the four species is endemic to a particular geographic area, another question of congruence might arise: are any or all of four geographic regions (A, B, C, D) areas of endemism for other groups of species and, if so, are these other species related to each other in the same manner with respect to their distribution? Are taxa endemic to areas C and D always each other's closest relatives, is the nearest relative of C–D always in B, and is the nearest relative of B–C–D always in A? (Figure 10). If that were so, one might say that the probability of three groups of four species exhibiting the same set of hierarchical relationships with respect to four areas is 1/255 if ruled by chance. Statistically such occurrences are non-random and illustrate another aspect of order in nature, *viz.*, areas of endemism can be interrelated hierarchically. The geographical distribution of species are the characters in theories of area relationships.

Area relationships can also be deduced from geological data such as stratigraphy, soil chemistry, magnetic anomalies, and surface morphology. Suppose now that geological analysis of the same four areas of biological endemism leads to a corroborated theory of area relationships such that areas C and D are most closely allied, area B is linked most closely to areas C–D, and area A to B–C–D. In an ideal and perfectly orderly world such general congruence might be true without exception. Of course, the world is neither ideal nor perfectly orderly, but to pursue a general theory of natural order I will assume that all the results just described are at least attainable. If attained, the corroboration of nature's hierarchical structure would be compelling. It would

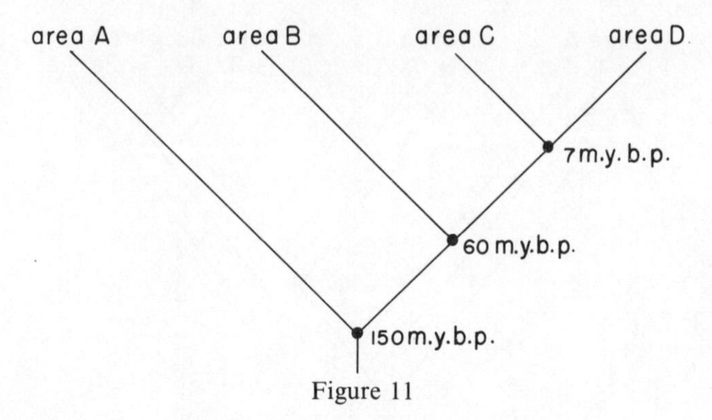

Figure 11

involve congruence among the life-histories and adult character sets of twelve species in three groups, of the interrelationships of these species with respect to four geographic areas, and of the interrelationships of these four areas with respect to a number of different geological data sources. Research in this field of general congruence expressed as agreement among specific hierarchical theories was begun in earnest less than 20 years ago, but significant results have already been achieved (see Wiley, 1981, Chapter 6).

If the hierarchical relationship between the earth and its life is to be viewed in a historical framework, what is needed is a time dimension that is tied explicitly to that pattern of relationship. This is supplied by isotopic decay rates of radioactive elements associated with geographic areas of biological endemism. The importance of these rates in extrapolating biological time lies in the fact that they are independent of biological systems, provide a time scale from the present to more than 700 m.y.b.p., and can be determined with an analytical error as low as 2%. However, radioactive decay rates provide a useful time scale only to the extent that there exists a general theory of congruence to associate them with and that they are themselves consistent with the requirements of relative age imposed by hierarchical structure, as in the example in Figure 11, in which the geological age of isolation of areas A–D corresponds with the branching sequence. In this example ancestral area C–D might have been separated into C and D by penetration of a glacier, area B–C–D might have been divided into B and C–D by mountain building, and area A–B–C–D into areas A and B–C–D by tectonic fragmentation of a landmass.

If we could thereby specify that the organisms in areas C and D were separated and underwent differentiation for a period of no more than 7 million years, as compared with those in area A which existed for as long as 150 million years, then we would appear to have satisfied one of the main methodological requirements of a theory of the evolution of life on earth. At least since the end of the nineteenth century biologists have wanted to know about the sequences of appearance and ages of the parts of a natural hierarchy that includes some 2 million kinds of extant plants and animals.

Comprehensive historical statements of this type, which specify the ways in which the earth and its life have evolved together, involve a minimum of extrapolation and a maximum of empirical content. The success of such theories about what happened during the recoverable history of life on earth depends utterly on our success in embryology, systematics, biogeography, and, in what has been known traditionally as historical geology. This congruence can serve as a basis for further inquiry into the comparative nature of evolutionary changes in morphology, biochemistry, and behaviour and the causal factors responsible for the changes in ontogeny that characterize different taxa. But it hardly needs emphasizing that without a good theory of what happened, without a well-corroborated hierarchical theory, there is nothing called evolution to explain. Without an empirically based theory about the product of evolutionary history, what problem would questions of evolutionary process address?

Dobzhansky (1970) distinguished between two main endeavours of evolutionary biologists when he wrote that:

> '. . . as the study of evolution proceeded, two main approaches were employed. The first concentrated on unravelling actual evolutionary histories, that is, phylogenies of various groups of animals and plants . . . The second approach emphasized studies of the mechanisms that bring evolution about, of causal rather than historical aspects. Genetics, especially population genetics and ecological genetics, has supplied the basic concepts and . . . methods (Dobzhansky, 1970, p. 28).

For a great many biologists today, evolutionary research and population genetics are virtually synonymous, and are included under such umbrella terms as the synthetic theory or neo-Darwinism to indicate that modern evolutionary biology is a melding of Darwin's theory of natural selection with genetics. How important population genetics is to this view is capsuled by the following statements selected by a sympathizer (Michod, 1981):

> 'Evolutionary change is the modification through time of genes and gene frequencies' (Eldredge and Gould, 1977, p. 26).
>
> '. . . biological evolution can be defined . . . as any change from one generation to the next in the proportion of different genes' (Futuyma, 1979, p. 7).
>
> '". . . gene frequency changes *are* evolution" captures in essence the population genetics approach as well as providing the basis of a formal framework within which population genetics as well as other evolutionary biologists work' (Michod, 1981, p. 4).

Which other evolutionary biologists did Michod have in mind? Probably population ecologists, for, with a sense of approval, Michod writes:

> 'There has been a shift in population genetics to incorporate ecological models and principles. This shift began . . . with the synthesis of population genetics, population ecology and evolutionary biology into "population biology"' (Michod, 1981, p. 2).

Sensitive to criticisms that population genetics has made no contribution to understanding causal mechanisms underlying the structure of life's hierarchy or even the origin of its humblest species Michod insists nonetheless that all is well with neo-Darwinism, that the empirically empty concept of fitness (the fit are those that survive; those that survive are fit) can be 'fixed' by adopting an ecological bridge principle (i.e., an ecological extrapolation) about ecological constraints that are independent of survival but for which there are at present no data and no known way of acquiring them (see, e.g., Brady, 1979). Michod (1981, p. 3) seems fairly to represent the fraternity of population biologists, at least from the standpoint of an outsider looking in. Why he is undisturbed by this outcome is evident in how he regards his view of evolution simply as gene frequency changes. Of this view, which he regards as the 'hard core' that gives justification to his models, he writes:

> 'Without this core, the models would have no import in evolutionary biology at all'

But of this core he also writes:

> 'This core is irrefutable by methodological decision . . .Acceptance (of the core) arises more from a process of indoctrination . . . than from any rational decision . . . scientific investigation proceeds by building a *protective belt* around the hard core . . . [this belt consists] of subsidiary hypotheses which buffer the hard core from refutations'.

Indeed a bleak view of science!

What, in my view, has happened to Michod and most other population biologists who regard themselves as evolutionists is that they have either forgotten or have chosen to ignore that it is the hierarchy and its included species that need explaining. They have acted so because they are satisfied with causal theories that are connected to the empirical world only by extrapolation—an extrapolation that seems to them safe enough because their theory is immunized against falsification, on the one hand, and because it is designed so as to explain all things, on the other:

> '. . . there can be no doubt that other genetic changes are equally important, especially in the speciation process. However, an argument could be made that these other genetic changes will, more often than not, be reflected in gene frequency changes' (Michod, 1981, p. 8).

Question: would the theories of population genetics be less relevant in a world designed by creationists, where all of life were created in an instant, fully formed, where no evolution had taken place? Answer: certainly not, since neo-Darwinism does not address the origin of the hierarchy, and the observation that in laboratory and natural populations gene frequencies can fluctuate in no way conflicts with any plan of creation, natural or supernatural. Michod and others simply had not noticed that when population genetics merged with ecology in search of a clearer evolutionary identity, it took another step farther away from, not closer to, its intended goal. It is a step farther away because our understanding of the natural hierarchy of plants and animals is dependent on our ability to comprehend the plan and execution of ontogeny, not a theory of ecology.

Michod derived his notions of a 'hard core' surrounded by a 'protective belt' from Lakatos (1970) whose view of science includes the not inherently unreasonable attitude that under *certain conditions* one should stick to a theory as long as possible:

> 'Without it we could never find out what was is in a theory—we should give the theory up before we had a real opportunity of finding out its strength . . .' (Lakatos, 1970, p. 177).

The *certain conditions* are that enough effort has been expended to test the theory's applicability and usefulness for a specified research objective. In more than a half century, hundreds of investigators in many more hundreds of research papers have tried, and failed, to find a direct link between population genetics and what originally was viewed as the fundamental problem: the origin of species. That population geneticists have since changed their research objective without (to put it politely) any fanfare is reason enough to declare that the neo-Darwinism game is up so far as it concerns the origin of nature's hierarchy or any of its parts. This criticism is not intended to dismiss population genetics as a useful undertaking; its incorporation into population biology undoubtedly adds both depth and precision to ecology.

One might conclude from the history of the neo-Darwinian period that a good reason for research in population genetics is to enlarge our understanding of the genetics of populations, not to discover a mechanism that drives evolution by the gradual accumulation of micro-variations that add up to macro-differences (speciation). Nevertheless, it was the absence of evidence to connect changes in gene frequencies with speciation that seemed to re-enforce the continued use of some version of Darwin's theory. 'Adaptation by natural selection' is the bridge principle used to connect the two. The use of bridge principles is a technique for combining in a coherent theory two or more kinds of data for which an empirical connection has never been demonstrated and, in this case, is undemonstrable because of an immunizing stricture that evolution by this principle is too slow to observe. The embarrassments of this rationalization were eliminated by a stroke of the pen (Michod, 1981, pp. 3, 4 and 20): 'gene frequency changes *are* evolution'! A bridge principle is what has been called a canonical statement (one constructed to agree perfectly with known facts). Since it is *made* to agree with the data, and since any number of other canonical statements are possible, a given bridge principle must be accepted as true axiomatically, i.e., cannot be questioned by those who agree to play the game according to the rules.

Darwinism, neo-Darwinism, punctuated equilibrium (Eldredge and Gould, 1972), epigenetic neo-Darwinism (Løvtrup, 1974), non-equilibrium thermodynamic theories of evolution (Wiley and Brooks, 1982, Chapter 6), and even the various versions of the creationist theory of life all share another shortcoming as explanations of evolution. Each can equally well explain any evolutionary history with a minimum of empirical constraint. None of them uniquely determines one hierarchy. They prohibit no theory of relationship. Creationist theories, of course, make only a small pretence at using data and are therefore beyond further notice, but are mentioned here to make a point with as much force as possible. A theory of process that is not deterministically tied to the pattern it seeks to explain is no explanation of that pattern at all, but only an explanation of patterns of that kind. Whether it is a true explanation of that kind of pattern is not knowable in the context of scientific corroboration because we cannot tell if it succeeds in explaining any one particular pattern of that kind. As noted by Dobzhansky (1970) the term evolution has been used ambiguously. Sometimes it means the process, sometimes the product. The product of evolution is phylogeny. The process that drives it is unknown, but is conjectured by various biologists, according to their training and prejudices, to involve aspects of ecology, genetics, behaviour, and ontogeny. Almost all agree that at least genetics and development are involved in some significant way. But most current versions of evolutionary process are presented in such a vague way and include so many areas of apparent indeterminacy that too many products are possible with the proposed mechanisms. Few, if any, phylogenies are incompatible with the proposed mechanisms; selection theory, for example, will explain an incorrectly devised phylogenetic statement with the same facility that it will a correct one because it can explain all statements of that kind. Hence, current theories of process matter little, if at all, to how we discover nature's hierarchy or recover its singular history. But without this hierarchy there is, in fact, nothing called evolution to explain.

3.4. INFERRING EVOLUTIONARY CHANGE

Once it is accepted that a deterministic theory is what is needed to understand the coherence of complex biological systems, from genetics and development to phylogeny and biogeography, the battle, if not won, is at least aptly joined. What is needed first is a set of constraints or prohibitions so that idle conjecture can no longer serve to guide research programs. For evolutionary theory the underlying constraint must always be a highly corroborated theory of relationships among taxa.

Given such a corroborated theory, a hierarchical pattern exists with which other data might be compared. For purposes of illustration, assume the existence of a hierarchical theory for six taxa. On the depiction of relationship (Figure 12) are plotted the distribution of three phenotypes (P^{1-3}), three ontogenetic stages

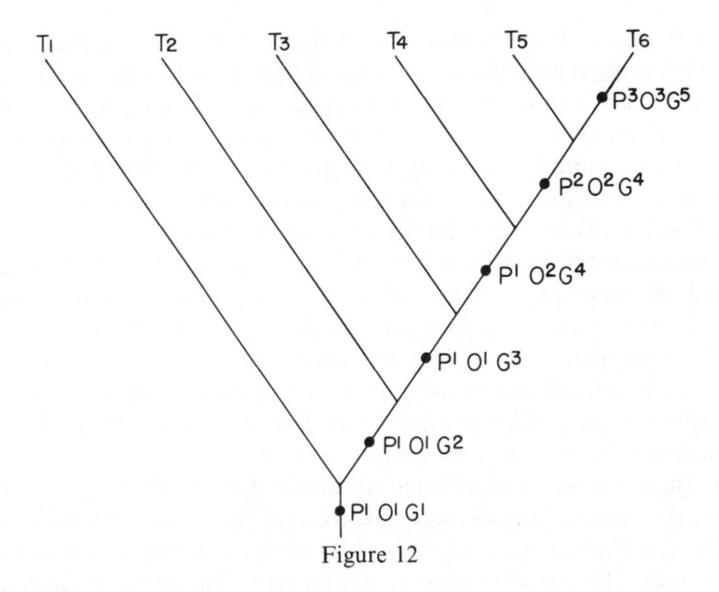

Figure 12

in which these phenotypes first appear during development (O^{1-3}), and five genetic characteristics (G^{1-5}) that are causally responsible for the onset of development of the three phenotypes. This 'character-state tree' can be read as if it were a record of biological history to see what might be learned about the association of phenotypic change with the timing of development and the nature of genetic control.

The first observation that can be made is that each kind of data is perfectly consistent with the given hierarchical order of taxa, i.e., each kind of data, P, O, or G, describes a sequence of relationships of taxa that does not conflict with the original hierarchy, although none of the data are sufficient to specify the entire hierarchy (Figure 13).

The second thing to notice in this theoretical example is that, interpreted historically, particular phenotypes, ontogenetic stages, and the genetic control of development did not evolve synchronously, i.e., G changed twice (G^2, G^3)

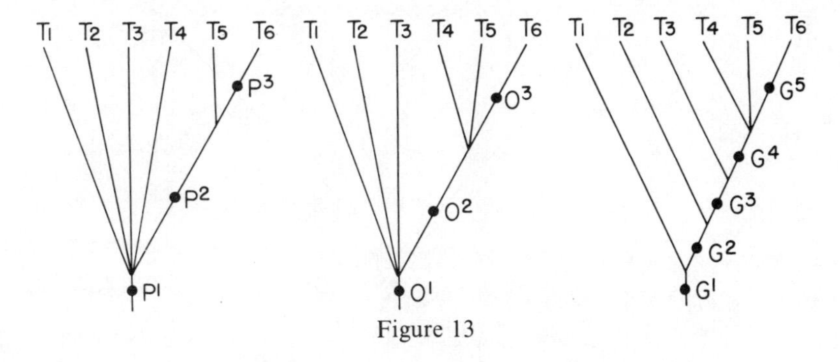

Figure 13

without a change in P or O, O and G changed once without a change in P, P changed (P^2) without a change in O or G, and the three traits changed together in one taxon (T_6: $P^3O^3G^5$). Since the example is theoretical I am, of course, not compelled to explain what such a result might mean. But such a result in the real world might be entirely unexpected. If so, it is clear that a new research programme is indicated to learn: (1) how general such results might be in other groups of organisms; and (2) what these unexpected results might mean in terms of current concepts of genetics and development. Can the timing of ontogenesis get ahead of phenotypic change (P^1O^2)? Can genetic control mechanisms transform faster than the phenotype and the ontogeny ($P^1O^1G^1$, $P^1O^1G^2$, $P^1O^1G^3$)? Systematists have long recognized that plotting character distributions on hierarchical patterns of this sort can provide unexpected insights into the way nature works, and, more important, that there is no limit to the kinds of data that can be examined in this way.

Nor is there reason to expect that all kinds of data will be congruent with hierarchical structure. Consider a recently revived theory of relationships among tetrapods (Gardiner, 1982), simplified here to include only a salamander and three amniotes, the crocodile, mammal, and bird. According to Gardiner, the scheme is well corroborated. Many decisive characters group the salamander with amniotes (i.e., all those features uniquely characterizing tetrapods). Amniotes themselves are uniquely characterized by numerous features of the reproductive and skeletal systems, among others. And birds are uniquely linked with mammals on the basis of features of development, endocrine systems, brain and spinal cord, pancreas, pineal, urine chemistry and kidney, anterior naris, cartilage origin, vertebrae, upper jaw bones, behaviour, gamete development, myoglobin sequence, and haemoglobin. Using Gardiner's scheme as a reliable standard for comparison of other features is not unreasonable since, for four taxa, support of each of the three branches of the hierarchy by as few as five characters at each node is equivalent to a statement of congruence among five independent character transformations (with a probability of 1/50,625 or

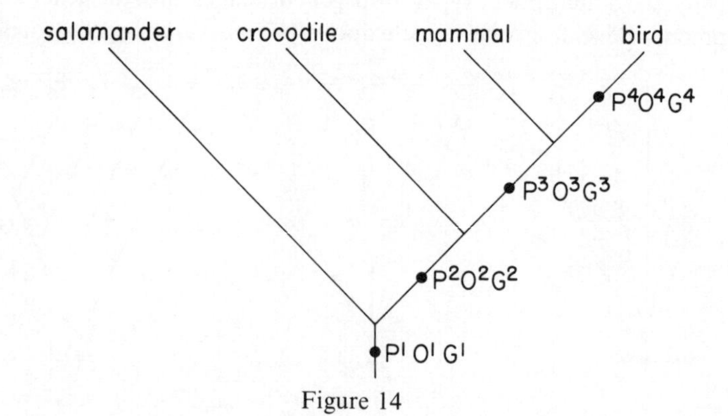

Figure 14

$15 \times 15 \times 15 \times 15$ of arising by chance alone). With this scheme a specific question can be asked about the course of development of various keratinized dermal structures and their genetic control in the four taxa. Four phenotypes are recognizable: uncondensed keratin in the dermis (salamanders); scales (crocodiles); and two kinds of interscale structures, feathers (birds), and hair (mammals). Keratin is sometimes strongly expressed in amphibians, as in the more terrestrial, dry-skinned species such as toads. Scales are also present in some parts of the body of birds (legs and feet) and mammals (tail of rodents), between which the feather or hair papillae arise. The developmental sequences of character transformation in the four taxa might be as shown below:

	Transformation	Adult state
Salamander	P^1	P^1
Crocodile	$P^1 \rightarrow P^2$	P^2
Mammal	$P^1 \rightarrow P^2 \rightarrow P^3$	P^3
Bird	$P^1 \rightarrow P^2 \rightarrow P^3 \rightarrow P^4$	P^4

where P^1 represents the absence of a scale papilla, P^2 the scale papilla, P^3 the interscale hair papilla, and P^4 the interscale feather papilla (Gardiner, 1982, p. 213, discussed the hair–feather transformation sequence). The different end-product phenotypes might therefore be expected to occur with a different ontogenetic timing, P^4 appearing later than P^3. Or, as some recent work suggests, hair and feathers are alternative transformations so that one would not be expected to precede the other ontogenetically; the example will not be affected by this question, however. The object now is to see how a genetical theory of integumentary evolution might be constructed. We might discover, as in the foregoing theoretical example, that the genetic information associated with each phenotype (e.g. simple, additive nuclear DNA substitutions) specifies the same hierarchy as the one we started with and would, therefore, constitute another element of congruence (an additional corroborating instance for each node) (Figure 14).

But suppose, instead, the result shown in Figure 15 was obtained. In this case simple, additive nuclear DNA substitutions would mean nothing in relation to a general theory of genetic control of development because the genetic data, treated as a transformation sequence, yields the theory of relationship shown in Figure 16. Such a discovery does not make the genetic determinations in some sense wrong, it merely makes them irrelevant as components of a general explanation of hierarchical structure, the phylogeny of tetrapods, or the evolution of the genetic control of ontogeny. Incongruence of this sort between genetic data and corroborated hierarchical schemes in the real world would undoubtedly cause geneticists to reconsider the nature of the questions asked about which part or aspect of the genome should be investigated in relation to general theories of genetic control of hierarchical order. But it might also reflect a real problem of our ability to recover that relevant genetic data. Whatever the

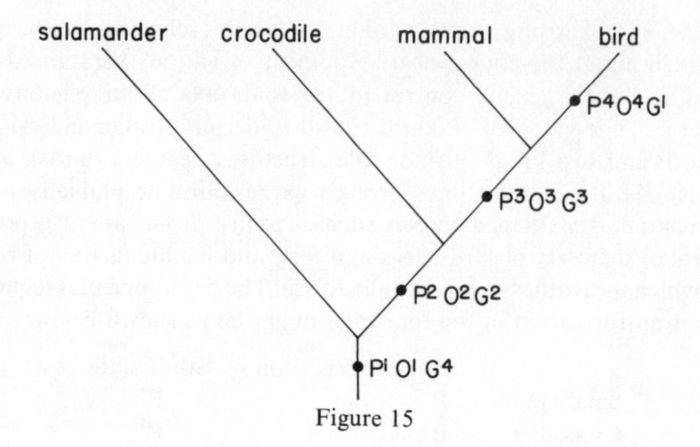

Figure 15

outcome of such study, it is clearly better to know where the problems might lie in 'causal' evolutionary theories than to assume that the bits and pieces of genetic information so far assembled must inevitably explain evolution. There is, after all, no need to placate the ghost of neo-Darwinism; it will not haunt evolutionary theory much longer.

I conclude that something new and constructive can be learned about the evolution of ontogenies and their genetic control when such theories are constrained by the empirical data of natural order. This conclusion acknowledges the importance of progress to date on the nature of genetic systems, ontogeny, and the genetics of populations as general statements. It also recognizes that progress to date to relate this general biology to the evolution of life as we know it is largely wishful thinking couched in the obscure language of bridge principles and thinly veiled extrapolations. Or it evades the question of how they are related by redefining the problem: 'gene frequency changes *are* evolution'.

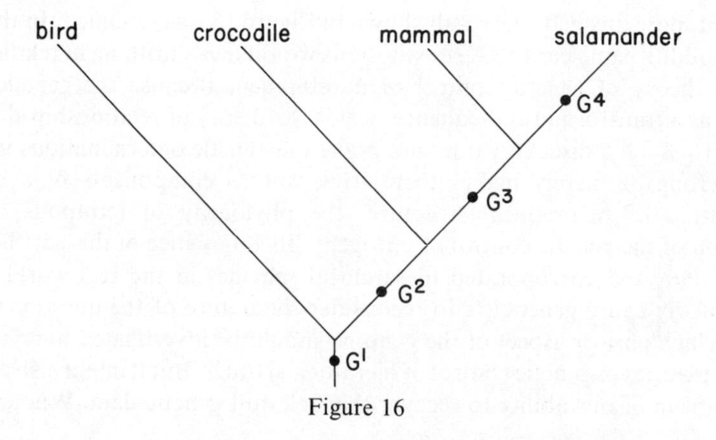

Figure 16

ACKNOWLEDGEMENTS

For helpful discussion and commentary on the typescript, I thank David Hull, Gareth Nelson, Norman Platnick, Darrell Siebert, and G. L. Vankin.

3.5. REFERENCES

Berry, F. H. (1964) Aspects of the development of the upper jaw bones in teleosts, *Copeia*, **1964**, 375–384.

Brady, R. H. (1979) Natural selection and the criteria by which a theory is judged, *Systematic Zoology*, **28(4)**, 600–621.

Dobzhansky, T. (1970) *Genetics of the Evolutionary Process*, Columbia University Press, Columbia.

Eldredge, N., and Gould, S. J. (1972) Punctuated equilibria: an alternative to phyletic gradualism, in Schopf, T. J. M. (ed.) *Models in Paleobiology*, Freeman, Cooper and Co., San Francisco, pp. 82–115.

Eldredge, N., and Gould, S. J. (1977) Evolutionary models and biostratigraphic strategies, in Kauffman, E. G., and Hazel, J. E. (eds.) *Concepts and Methods of Biostratigraphy*, Dowden, Hutchinson and Ross, Stroudsburg, Pennsylvania, pp. 25–40.

Futuyma, D. J. (1979) *Evolutionary Biology*. Sinauer Association, Sunderland, Massachusetts.

Gardiner, B. G. (1982) Tetrapod classification, *Zoological Journal of the Linnaean Society*, **74(3)**, 207–232.

Lakatos, I. (1970) Falsification and the methodology of scientific research programmes, in Lakatos, I., and Musgrave, A. (eds.) *Criticism and the Growth of Knowledge*, Cambridge University Press, Cambridge, pp. 91–195.

Løvtrup, S. (1974) *Epigenetics: a treatise on theoretical biology*. John Wiley & Sons, New York.

Macbeth, N. (1980) Reflections on irreversibility, *Systematic Zoology*, **29(4)**, 402–404.

Michod, R. E. (1981) Positive heuristics in evolutionary biology, *British Journal of Philosophy and Science*, **32**, 1–36.

Nelson, G., and Platnick, N. (1981) *Systematics and biogeography: Cladistics and vicariance*. Columbia University Press, Columbia.

Rosen, D. E. (1982) Do current theories of evolution satisfy the basic requirements of explanation? *Systematic Zoology*, **31(1)**, 76–85.

Rosen, D. E., and Kallman, K. D. (1959) Development and evolution of skeletal deletions in a family of viviparous fishes (Cyprinodontiformes, Poeciliidae), *Quarterly Journal of the Florida Academy of Sciences*, **22(4)**, 169–190.

Wiley, E. O. (1981) *Phylogenetics: the theory and practice of phylogenetic systematics*. John Wiley & Sons.

Wiley, E. R., and Brooks, D. R. (1982) Victims of history—a non-equilibrium approach to evolution, *Systematic Zoology*, **31(1)**, 1–24.

Evolutionary Theory: Paths into the Future
Edited by J. W. Pollard

Chapter 4

Changing from an evolutionary to a generative paradigm in biology

BRIAN C. GOODWIN,
School of Biological Sciences,
University of Sussex,
Falmer, Brighton,
Sussex BN1 9QG

4.1. INTRODUCTION

In this chapter, I shall propose that what is now required in biology is not a new evolutionary theory, but the construction of a new conceptual scheme from which both evolution and development emerge as essential aspects of biological process. The modern evolutionary synthesis which emerged in the middle of this century was a detailed elaboration, with modifications, of the conceptual scheme proposed by August Weismann towards the end of the last century, a scheme which provided Darwinism with a theory of inheritance compatible with the principle of natural selection, and which suggested a specific causal explanation of embryogenesis or development in terms of inherited determinants of morphology (later to become the genes). This conception of embryogenesis as an aspect of inheritance effectively assimilated development into Darwin's vision of biology as an essentially historical subject with evolution as its unifying theme, a view still held by the majority of biologists.

I shall argue that it is this persistent attempt to understand biology in historical terms, and thus to stress the role of particulars, of contingencies, and of genealogies, which is the source of current tension in the subject. In contrast, workers in various disciplines are rediscovering and emphasizing evidence of regularity, of constraint, of order in their empirical material which resists assimilation into an historical interpretation of biological process of the type which characterizes Weismannism and the modern synthesis. Furthermore, theoretical developments, including insights arising from molecular biology, have now reached the point where models of morphogenesis can provide tentative explanations of the origins of constraint and regularity in embryogenesis. Mathematical models embodying principles of spatial organization which satisfy the long-recognized need in developmental biology for high-level relational order are now available in the form of field theories of morphogenesis. Within this perspective, a different way of looking at biological process becomes possible.

In systematics, there are similarly conceptual developments which seek to understand taxonomic relationships not simply as histories or genealogies, but in terms of logical principles of biological order. This theme has been ably and forcefully expounded by such authors as Gould and Eldredge (1977), whose arguments lead to the proposition that species are real entities existing as natural kinds (Eldredge, 1979; Patterson, 1983) rather than arbitrary groupings of characters devised by the taxonomist for purposes of classificatory convenience. These principles can be closely connected with morphogenetic regularities, since the latter can account for the former. From these developments there is now emerging support for a tradition of thought in biology which has always been a part of the subject, but relegated very much to the fringe during the past century of research. This is the view that the central task in biology is the construction of a theory of biological organisation, which could provide a sufficient context for the partial theory of inheritance proposed by Weismann and for the theory of molecular composition of organisms which have both been elaborated in great detail during this century. Such a comprehensive theory could then constitute a logical formulation from which evolution and development emerge as aspects of constrained biological process, or transformation. It is the elaboration of this idea in relation to a specific problem in evolution and development which forms the central content of this essay.

4.2. TYPICAL FORM AND HOMOLOGY AS REVEALED IN TETRAPOD LIMB MORPHOLOGY

The rational morphologists of the late eighteenth and early nineteenth centuries such as Geoffroy St. Hillaire, Cuvier, and Richard Owen shared the belief that the biological realm is one of intelligibility and order, that there are laws of biological form and organization; and they provided much evidence from their study of comparative morphology to support this belief. Among the conceptual tools which they forged from their empirical studies were of the notion of typical form and the concept of homology, with which they conducted their search for evidence of the transformational equivalence of different morphological structures. A classic and well-known example was the demonstration that all tetrapod limbs share certain structural features, so that they may all be understood as transformations of one another. The common morphological characteristics of these limbs are: they start proximally with a single bone, the humerus (forelimb) or the femur (hind limb), then there is a pair of bones (radius

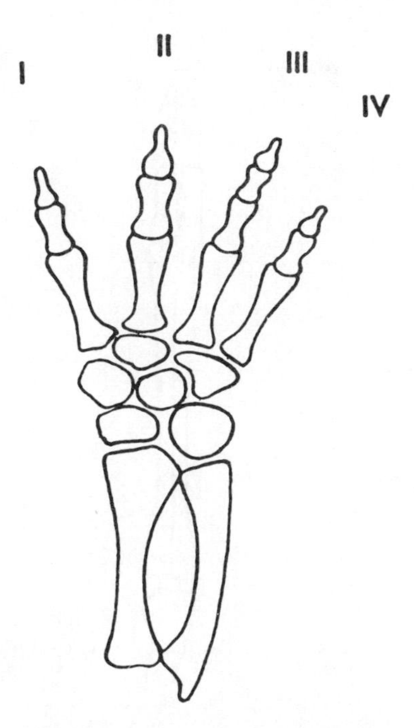

Figure 1. Bone structure of the four-digit limb of the salamander, *Necturus*, distal to the humerus

and ulna, or tibia and fibula), then a group of carpals or tarsals, followed by a set of metacarpals or metatarsals and ending with a series of phalanges which constitute the digits. Figure 1 illustrates this for the case of a four-digit fore limb (*Necturus*; humerus not shown). Limbs of different species show great variation in the size, shape, and number of skeletal elements, most variation occurring in the autopodium (carpals or tarsals plus all distal elements).

The task of the rational morphologists was to discover the order which was hidden in the obvious functional diversity of vertebrate limb morphology. Geoffroy developed a general 'Principle of Connections' which directed this search for morphological order and invariance by the systematic study of the position and relationship of skeletal parts, regarded as structural components independently of any secondary functional diversification. This concentration on relational spatial order at a fairly abstract level of analysis gave rise to Owen's concept of homology. Homology is an equivalence relation on a set of forms which share a common structural plan and are thus transformable one into the other. This is therefore a logical relation, independent of any historical or genealogical relationships which the actual structures may have. Owen showed that vertebrate limbs constitute a homological set of forms by virtue of the

Figure 2. Bone structure of the two-digit limb of the antelope distal to the radius

invariant spatial relationships shown by the limb elements. For example, although the antelope's limb has only two terminal digits it nevertheless shares with all other tetrapod limbs the same basic spatial pattern of a sequence of elements whose numbers differ successively in proximo-distal order by small integers. Thus whereas the four-digit salamander limb has a proximo-distal sequence that proceeds approximately 1, 2, 3, 3, 4, 4, 4, in the antelope (Figure 2) the sequence is 1, 1, 3, 2, 2, 2 (humerus not shown).

Having established this notion of equivalence, one can then ask in what sense the patterns are transformable one into the other, and what is the status of the invariant or unchanging features which make all the forms equivalent to one another. These are difficult questions to answer without a quite highly developed theory of how the limbs are formed in the first place, i.e. without a theory of limb morphogenesis. Although embryology was a rapidly-developing area of biological study during the first half of the nineteenth century, it had certainly not advanced to the point where it could even suggest the mechanisms whereby eggs undergo systematic transformations to adult forms. Furthermore, the rational morphologists tended to believe in divine creation as the ultimate origin of organismic forms, so that whatever principles of order they deduced from the study of morphology revealed the logic of the divine plan. This view was similar to that taken by Newton with respect to the study of natural regularities such as planetary motion, so that there is no intrinsic incompatibility between holding such metaphysical views and developing scientific theories of form and transformation. However, the rational morphologists did not pursue a mathematical description of biological form in terms of transformations and invariants, which are the basic concepts used in the systematic classification of forms in physics and in mathematics itself. And it is not surprising that they failed to do so, since at that time the relevant mathematical concepts were only being developed at the appropriate level of abstraction. Furthermore, as stated earlier, the generative approach to biological form through embryology was insufficiently developed to reveal the morphogenetic principles underlying limb development. This led to a kind of conceptual impasse in the study of biological form, resulting in a division of views which inevitably led to serious conflict. Morphologists such as Richard Owen and Louis Agassiz attempted to explain biological form by adopting a transcendental idealist position which regarded actual forms as variants on perfect or ideal forms. Owen took this to imply that there is one primary ideal limb form, which he called the pentadactyl (five-digit) limb, all other limbs being variations of it. Thus type came to mean ideal type or archetype rather than the abstract concept of typical form used by Cuvier, which meant a class of forms showing a certain constancy of spatial relations in the elements, no member of the class being singled out as special in any way since all are equivalent under transformation. Thus Owen himself idealized the concept of homology and in a sense layed the seeds of decay of the rationalist tradition in morphology.

The great challenge came of course from Darwin, who went in precisely the opposite direction to Owen but actually preserved, within a different context, the latter's concept of a special form, a morphological type which was to be singled out for particular attention. In reaction to Owen's notion of the pentadactyl limb as an ideal form whose variants appear in actual tetrapods, Darwin reified the concept, materialized the form, and gave it a special place in his metaphysical scheme, which elevated historical process to a primary explanatory principle in biology. To this special historical form he gave the name 'common ancestor', which was simply given by historical circumstance and by chance: its origins were not explained, except in relation to antecedent common ancestors, also given, not explained. *The Origin of Species* thus turned out to be a rather misleading title since the arguments presented have virtually nothing to say about how common ancestors originate, but simply how their descendents may become different from one another in a historical process. In place of given ideal forms, we have given historical forms, and we are no closer to a solution of the problem of biological genesis. Homological invariance then took on the meaning of descent from a common ancestor:

> '. . . propinquity of descent, the only known cause of the similarity of organic beings' (Darwin, 1859).

Such a view automatically brought to an end the search for laws of organization in biology as explanatory principles of observed morphological regularity. For Darwin, the

> '...chief part of the organization of every being is simply due to inheritance' (Darwin, 1859).

Thus the tracing of genealogies, of temporal patterns of inheritance, was in itself sufficient as an explanation of biological form, and biology became a historical science. This is so deeply entrenched in modern consciousness that it has become commonsense to 'explain' the presence of six toes on a cat by simply pointing out that the cat's mother has six toes. This is like saying that the reason why the earth now follows an elliptical trajectory around the sun is that last year it followed an elliptical trajectory. Although there is an element of truth in both of these statements, it will be acknowledged that the latter is not very acceptable as a scientific explanation. Why, then, is this type of explanation so widely accepted in biology? There are interesting reasons for this which I shall examine briefly below; but it is clear, I think, that anyone who accepts such explanations as satisfactory has, at the very least, lost sight of more basic questions such as why it is that mutations give rise to only certain categories of morphological disturbance, or that extra toes occur at the ends of limbs and not sticking out of the humerus, say. These are questions relating to organization, to invariance, and to transformation, which cannot be answered in terms of historical processes and inheritance. Let us now see how Darwin's proposals affected the study of biological form as illustrated by studies on vertebrate limb morphology, and what evidence forces us back to a view which once again addresses problems of biological organization rather than those relating simply to inheritance and to adaptation.

4.3. THE PROBLEM OF BIOLOGICAL FORM IN TERMS OF THE EVOLUTIONARY PARADIGM

Darwin's conception that evolutionary change arises from the cumulative effects of selection on small variations in the component parts of common ancestral forms resulted in specific questions which were pursued vigorously by comparative morphologists, whose legacy remains in contemporary textbooks. Since the common ancestral form of tetrapod limbs was accepted to be pentadactyl, one with five digits, it was considered that a four-digit limb such as that of the salamander (Figure 1) must have arisen by a gradual reduction in the size of one of the outside digits, either the anterior or the posterior, with eventual loss resulting in a limb with four digits. A question that arises from this description is then: which digit has actually disappeared? This is illustrated in Figure 3, based on a diagram from a textbook of comparative anatomy by Kent (1965). Are the remaining digits of the salamander limb to be labelled 2, 3, 4, and 5, or I, II, III, and IV? The question can be answered only if each digit has an

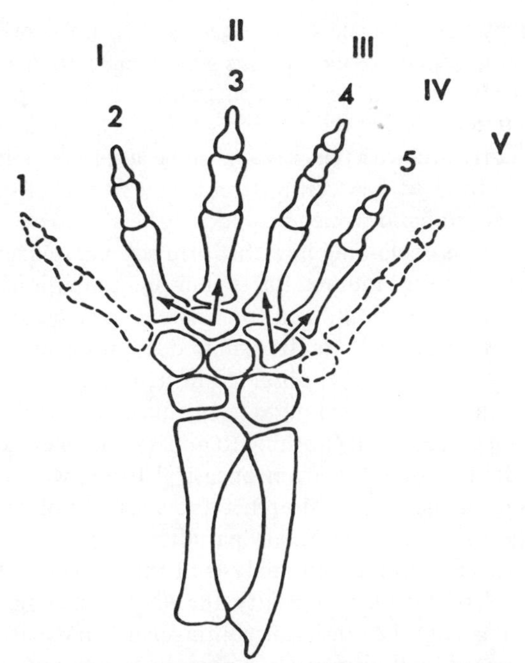

Figure 3. Hypothetical alternatives in the evolution of a four-digit limb from a five-digit 'common ancestor', according to the Darwinian concept of homology and inheritance. The original pentadactyl limb is assumed to have had 1–5 or I–V, either digit 1 or digit V disappearing during evolution to give a four-digit limb, 2–5 or I–IV (after Kent, 1965)

identity and a continuity from generation to generation; i.e. the problem is real only within a certain conception of the inheritance of biological form. This conception postulates the inheritance of individual components of the pattern, and so implies an atomistic theory of inheritance. For Darwin, the continuity was direct: each element of a biological pattern in the adult, such as a skeletal element in a limb, gives rise to the same element in the offspring, with possible variation by some (unspecified) means. This implies Lamarckian inheritance, since it is the parental form which determines the form of the offspring. Weismann (1885) could find no evidence for this type of inheritance in higher organisms and so he altered the model but retained atomistic causality. In his scheme, each element of a pattern is specified by an inherited determinant in the germ plasm, that part of organisms which is continuous from generation to generation and is the vehicle of inheritance. Thus for Weismann, biological inheritance involves the transmission of a collection of material units which are the sufficient causes of biological form. These determinants have become the genes. It is from this type of reasoning that has come the belief that the presence of six toes on the limbs of a cat is explained by the presence of six toes on the limbs of its mother. The 'explanation' is simply that the mother has genes which are a sufficient cause of the appearance of six toes, and these are transmitted to the offspring. The element of truth in this statement is that the *difference* between a normal five-toed cat and one with six toes correlates with the presence in the latter of a polydactylous gene, and this is transmitted with a certain frequency to the offspring. However, identifying an inherited factor which may loosely be said to cause a difference of form is not the same as explaining how the form is generated, either the normal five-toed limb or the six-toed mutant. But if your preoccupation is to explain how evolutionary *change* comes about, then you will tend to be satisfied with accounts of how differences arise and are transmitted to progeny. Furthermore, if whatever similarity of form that is observed in structures is ascribed to descent from a common ancestral form, and this ancestral form itself arises by contingencies, then there is really nothing to understand in evolution apart from the origins of adaptive diversity from ancestral forms which lead back to a primordial organism, the historical archetype. This is biology conceptualized within the framework of an evolutionary paradigm (see Webster and Goodwin, 1982, for a detailed description and analysis of this concept). The subject thus becomes a realm of irreducible complexity, the whole genealogical succession of organisms reflecting only the effects of contingencies and particular historical conditions, transposed by Weismann to a domain of genotypic causes and their phenotypic effects (using the contemporary expressions for germ plasm and somatoplasm). There are then no general principles of biological organization to be grasped because all has been explained away in terms of inheritance and selection.

Belief in the inherited continuity of skeletal elements from a common ancestral pentadactyl limb has resulted in extensive anatomical studies of vertebrate limbs with reduced numbers of digits in an attempt to pursue the type of question illustrated in Figure 3. We find that the two digits of the antelope limb, for example, are labelled III and IV, while the single digit of the modern horse is labelled III. As pointed out above, this reflects a particular view not only of inheritance but also of morphogenesis: heriditary determinants are considered to be sufficient generative causes of the elements of the limb pattern, so that loss or gain of these determinants can result in loss or gain of, say, individual carpals, metacarpals or phalanges. It is necessary now to consider what sort of generative process is likely to be responsible for the appearance of specific limb patterns during embryogenesis, and whether this is consistent with the above assumptions.

4.4. MORPHOGENETIC MECHANISMS OF LIMB FORMATION

Extensive studies have been made on developing limbs in a variety of vertebrate species, most work concentrating on amphibians such as the newt and the axolotl, and on the chick. (For a comprehensive survey of this work, see Hinchliffe and Johnson, 1980.) Early in the study of the amphibian limb it was recognized that the flank tissue of the developing embryo which has the potential of forming a limb has the properties of what is known as a morphogenetic field (Harrison, 1918). A field is a domain of tissue capable of forming a structure, such as a limb, with the capacity of responding as a unitary, self-organizing whole to a variety of disturbances. Among these disturbances are operations involving the removal of tissue, exchange of tissue between parts of the field, addition of extra tissue to it, or transplantation of the tissue to a different site in the body. After any of these perturbations, if carried out within certain time periods during embryonic development, the tissue gives rise to a perfectly normal-looking limb; i.e., it is capable of undergoing internal reorganizations of state such that normal internal relations are re-established and a normal structure is generated.

A number of different models have been proposed to account for such behaviour, but roughly speaking they fall into two rather distinct categories. One stresses the role of detailed genetic information in the specification of the limb pattern, and such models are therefore readily assimilable into the historical perspective of the evolutionary paradigm. An example is the general model proposed by Wolpert (1969, 1971) wherein specific morphological structure arises during development by a process in which the genes within an embryonic cell 'interpret' signals which specify that cell's position within a field. The position-specifying signals ('positional information') act essentially like general co-ordinates of position, carrying no specific information other than that which 'names' a cell spatially relative to other cells in the field. This spatial label, together with another label which defines the cell's developmental memory (i.e., what states it has already passed through as a result of membership in earlier embryonic fields in the development of the individual) allows the cell to respond in some species-specific manner arising from its specific genotype. Different types of limb then arise as a result of different genetic interpretations of the positional signals in the limb field, which signals could be the same in all tetrapod limb fields specificity arising solely from genetically-based interpretations. Such a model clearly satisfies the type of atomistic genetic determinism envisaged by Weismann, and it fits the metaphor of the 'genetic programme' as the directing agency of embryonic development. Each element of structure in the limb is separately encoded in the programme, and can be independently altered by changes in the programme (i.e., in the genes). It is a conceptual scheme which readily accounts for evolutionary change of structure. But because there are no constraints on interpretation, any form is possible and there is no explanation, within such a model, of regularity of form over large taxonomic groups such as one sees in the tetrapod limb. Wolpert (1982) has himself recognized this limitation, and is reconsidering the possibility that the other category of model may play some role in limb morphogenesis.

This alternative class of model makes use of morphogenetic mechanisms in

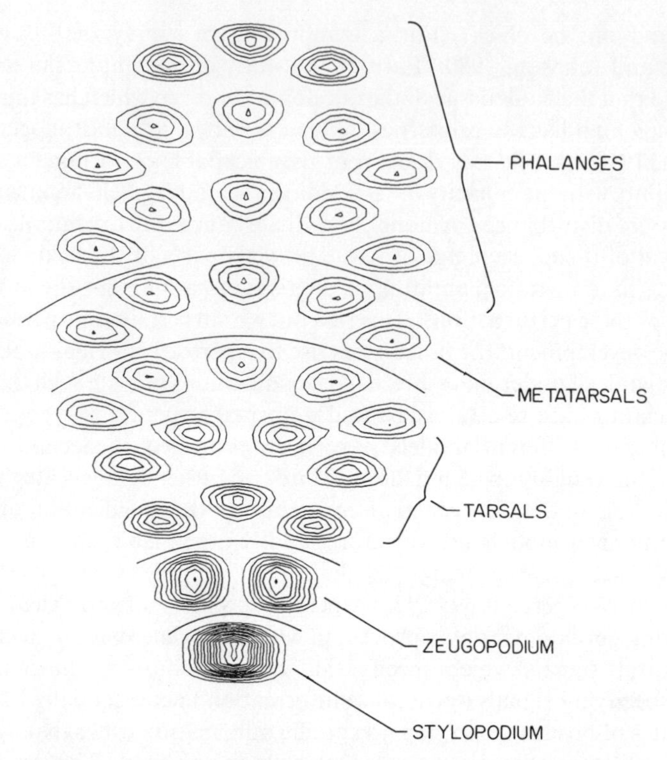

Figure 4. Contour plots of the sequence of field solutions generated by a model of pattern formation for a hind limb with five digits. Each closed curve represents a particular level on a hill which describes a part of the field solution at some proximal–distal position. Thus the five metatarsal elements correspond to a single field solution with five peaks of the function which is selected by a particular 'energy' level at this position in the proximo-distal axis

which there are constraints on possible limb patterns by virtue of limitations on the generative processes involved in the specification of the elements of the pattern. These constraints arise because elements are not generated separately but in groups, and the models differ from one another in the extent to which these groups are themselves constrained relative to each other. The skeletal elements are layed down in a particular order in the developing limb, from proximal to distal, and they start as small centres of condensation of cells which will form cartilage. These chondrogenic sites are of fairly uniform size initially, the long bones of the upper arm forming by growth of the corresponding units during the development of the limb. Focussing on the initial phase of limb formation, the chondrogenic sites are generated in a pattern which has two dimensions of periodicity, proximo-distal and antero-posterior. This is shown schematically in

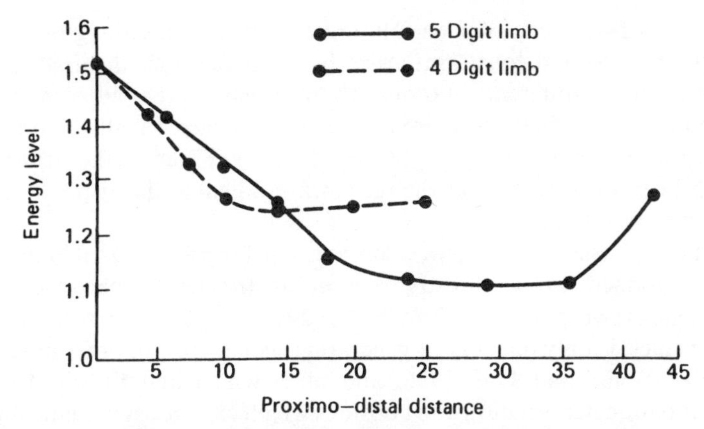

Figure 5. The proximo–distal 'energy' profile which selects or stabilises a particular series of solutions of the field equation in successively more distal levels of the limb bud. The solid curve gives rise to the sequence shown in Figure 4, each solid circle on the curve defining an 'energy' level which corresponds to a particular eigen-solution of the field equation, with a specific number of peaks across the antero–posterior axis. The dashed curve generates the pattern shown in Figure 6, the curve flattening out at an 'energy' level which gives eigensolutions of period 4

Figure 4, which is a computer print-out of a model of early chrondrogenic pattern formation in a generalized and simplified pentadactyl hind-limb. This model is characteristic of the class under discussion in that at each proximo-distal level, the complete set of elements is generated simultaneously as members of a single solution of a field equation. Thus, for example, all five metatarsal sites are layed down together as elements of a single solution, with five-fold periodicity, of a particular equation which describes the properties of the limb field according to a certain set of assumptions described in Goodwin and Trainor (1983). The sequence of solutions, from proximal to distal, differ in their periodicity across the antero-posterior axis from one to five with a distinct order determined by a systematic change in a parameter which selects or stabilizes each of these different solutions in different relative positions in the limb field. Models differ in how this is achieved, but the example shown in Figure 4 makes use of a continuous gradient along the proximo-distal axis, different solutions being stable within particular ranges of the gradient so that there is a transition from one solution to another as the generative process proceeds in proximo-distal direction. The type of gradient involved in this model is shown in Figure 5, where the variable is referred to as an 'energy level' which changes in some systematic manner along the proximo-distal axis.

Not only is there a constraint on the elements in an anterio-posterior array, since they are components of a single field solution; there is also a constraint on the changes of periodicity of each solution as one proceeds along the proximo-distal axis. This second constraint arises from the fact that the gradient which selects or stabilizes different solutions is continuous, so that adjacent solutions cannot differ in periodicity by more than a small integer, usually by one. Hence one gets a sequence such as 1, 2, 3, 4, 5, 5, 5, 5, 3, as in Figure 4, the terminal drop from 5 to 3 being due to a rapid change in the gradient at the distal extremity of the limb bud (see Figure 5).

I should emphasize that the type of proximo-distal gradient used in this model, which is assumed to be already present at the time when the wave of chondrogenesis sweeps over the limb bud and leaves a pattern determined by the gradient values, is only one way of generating such forms. In other models such as those of Wilby and Ede (1976) and of Newman and Frisch (1979), the equivalent parameter is a diffusion constant which changes systematically along the proximo-distal axis and stabilizes different solutions of field equations known as diffusion-reaction equations. These are basically very similar to those used by Goodwin and Trainor (1983) although in the former the variable is a chemical morphogen of Turing (1952) type, while in the latter the field is defined in terms of an order parameter which describes some aspect of molecular or cellular order relevant to the initiation of chondrogenesis. In yet another model which uses visco-elastic field equations to describe the properties of the limb field, there is a very interesting coupling between the cross-sectional eccentricity of the limb bud and the stability of solutions of differing periodicity (Oster *et al.*, 1984). As the wave of chondrogenesis proceeds in proximo-distal order, the mechanical forces which are involved in the condensation of the cells to form the chondrogenic sites cause a flattening of the limb, and this in turn leads to a bifurcation so that one solution ceases to be stable and another becomes stable. In between these solutions of different periodicity across the antero-posterior axis there is a domain of instability, no solution being stable. This is interpreted to constitute the joint, the region of separation between the skeletal elements. This type of model is extremely interesting, and its application to a number of different morphogenetic processes such as gastrulation, neurulation, and dermal placode formation has been described by Odell *et al.* (1981). Like the models considered above, it imposes important constraints on possible morphologies which have been considered in a paper by Oster *et al.* (1980).

All these specific models of pattern formation in limbs are, in a sense that can be made precise in abstract terms, contained within a higher-level generative theory, known as the polar co-ordinate model (French *et al.*, 1976; Bryant *et al.*, 1981). This has been extremely influential in stimulating the search for invariants and regularities, expressed in the form of abstract generative rules of limb formation not simply in tetrapod vertebrates but also in insects. It has been extended and elaborated in a variety of very interesting ways, a recent paper by Holder (1983a) illustrating its richness in explaining observed constraints such as the absence of tetrapod limbs with more phalanges on the outer digits (e.g., I and

Figure 6. The limb pattern generated by the 'energy' profile of the dotted curve in Figure 5

IV of Figure 1) than on the inner one, (I, II, and III), whereas the converse inequality is common (see the extra phalanx on digit III). A detailed empirical analysis of tetrapod limb morphology providing evidence for developmental constraints of this type can be found in Holder (1983b). Recent work by Alberch and Gale (1983) shows that digit reduction in amphibia can be achieved by a phenocopying process which mimics evolutionary reduction of the type described in Figure 3. Differences in the pattern of chrondrogenic sites in all these models arise as a result of the stabilization of different series of field solutions in the developing limb field. Figure 5 (dotted curve) shows the 'energy' gradient which gives rise to the pattern of elements shown in Figure 6, which represents the earliest stage of chondrogenesis in a four-digit limb of the type shown in Figure 1. Clearly the model aims to describe only the most basic aspects of the skeletal pattern and ignores all secondary effects such as the differential growth of elements, fusions, and asymmetries across the antero-posterior axis. These are regarded as effects superimposed upon the primary spatial array. Other patterns resulting from changes in the 'energy' profile are shown in Figure 7, corresponding roughly to tetrapod limbs with 3, 2, and 1 digit. What emerges quite clearly from this level of analysis is that all the limb patterns share certain basic features by virtue of the

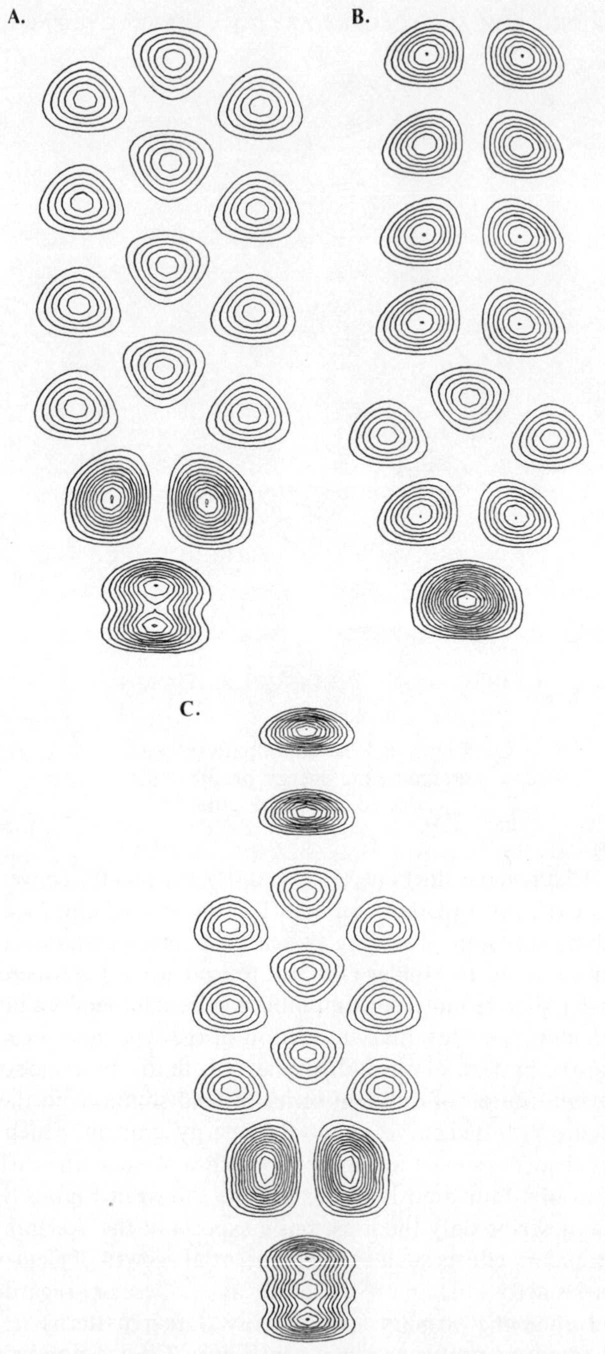

Figure 7. Limb patterns generated by modifications of the proximo–distal 'energy' profile of the limb bud, corresponding to limbs with three, two, and one digit

constraints on the generative process, so that they belong to a class which constitutes a 'typical form'. We thus arrive at a justification of the pre-Darwinian definition of homology, equivalence of forms under transformation, understood now in generative terms.

We can now ask more precisely what may be the nature of the transformations between the different limb patterns. What we see from the patterns shown in Figures 4, 6, and 7 is that, because of the constraints on the generative process, a transformation from one type of limb to another does not involve the loss or gain of individual elements, but a change of whole aspects of the pattern. Thus if we compare the five-digit pattern of Figure 4 and the four-digit pattern of Figure 6, we see that there is no way of numbering or labelling the phalanges, say, in such a way that we can describe which digit has been lost or gained in going from one to the other. A field solution with five-fold periodicity is simply a different solution to one with four-fold periodicity. They are transformations of one another under changes in some parameters, an 'energy' level, a diffusion constant, or a viscoelastic parameter. And the actual patterns are created anew in each generation. Thus from this viewpoint, tetrapod limbs are united in their structure not by virtue of descent from a common ancestor, with functional variants showing loss or gain of identifiable elements in the ancestral limb form. They are all members of a logical class of structure united by common generative principles. What actually happens in the historical lineage of these forms tells us something about adaptation to external contingencies but nothing about internal organizational principles. It is now necessary to recognize that biological process conforms not only to extrinsic functional stability criteria of the type expressed in the concept of fitness, but also to intrinsic principles of order or organization arising, in the case of morphology, from the spatial ordering properties of generative fields. Organisms are not aggregates of elements, whether molecules, cells, organs, skeletal or other components, whose random variation results in an unconstrained variety of forms. They are self-organizing wholes governed by laws describing spatial and temporal organization such that processes of biological change involve constrained transformation, whether ontogenetic or phylogenetic. Evolution and development then emerge as aspects of this generative process over different time-scales and constrained by different categories of parametric change.

4.5. THE GENERATIVE PARADIGM IN BIOLOGY

In order to pursue the above propositions further, it is necessary to get a clearer idea of the nature of the morphogenic fields which are being proposed as the source of developmental and evolutionary potential, and what factors are involved in the selection or stabilization of specific field solutions from the potential set, resulting in specific morphology. Changes in these factors will then result in changes of form, i.e., in the constrained transformations which explain both the unity and the diversity of organismic morphology. That morphogenetic fields are indeed domains of potential, being capable of generating a great diversity of forms which belong to a general type or class of structure, has been amply demonstrated in developmental biology, wherein systematic explorations have been made of the range of morphologies arising from defined classes of perturbation to different fields. In relation to the limb field, surgical manipulations involving axial re-orientations of presumptive limb tissue (Harrison, 1921), transplantation to different positions in the body (Swett, 1927; Slack, 1980), adding tissue with special 'organizing' properties to specific parts of the limb field (Wolpert and Hornbruch, 1981; Cameron and Fallon, 1977), severing the tip of the limb-bud, rotating the tip through 180° and replacing it on the stump (Maden and Goodwin, 1980; Maden, 1981), etc., have resulted in a great variety of structures, all recognizably of limb type but differing from one another in axial organization and in the number and spatial relations of skeletal and other elements.

Regenerative processes, in which whole structures are produced from parts, are also characterized by field behaviour of this kind. For example, amphibia such as newts and axolotls (classified as tailed amphibia or urodeles) are capable of regenerating their limbs from stumps remaining after accidental loss or surgery, and the systematic study of this process has provided much useful detail about the epigenetic principles of limb formation. The tissue which forms on the stump and gives rise to the missing part of the limb is called the blastema. It has field properties of the type described earlier, i.e. the capacity to reorganise itself after a variety of disturbances such as deletions, additions or translocations of blastemal tissue and then to produce the limb parts which were lost. Thus the blastema, like the embryonic limb domain, is a region of organized potential which exhibits characteristic field behaviour. And it, too, gives rise to a great variety of limb forms in response to surgical manipulations such as rotation of the blastema through 180° and replacement on the stump (Wallace and Watson, 1978; Maden and Turner, 1978). This particular operation results in what are known as supernumerary limbs, additional to the primary which grows from the rotated blastema. The structure of some of these supernumeraries in the axlotl, *Ambystoma mexicanum*, has been described by Maden (1982), examples of which are shown in Figure 8. This selection shows limbs with digit number varying from one to six simply to illustrate the variety observed, which is much greater than this.

The point that emerges from these results is that the limb field of a single species (i.e., of specific genotype) is capable of generating a variety of basic skeletal patterns (not, of course, the secondary modifications) which is greater than that observed in the full range of tetrapod limbs. In this sense the limb field has the potential for generating a greater range of limb forms than that which happens to have been produced by chance and the contingencies operating during evolution. In theoretical terms, the variety arises from the existence of a

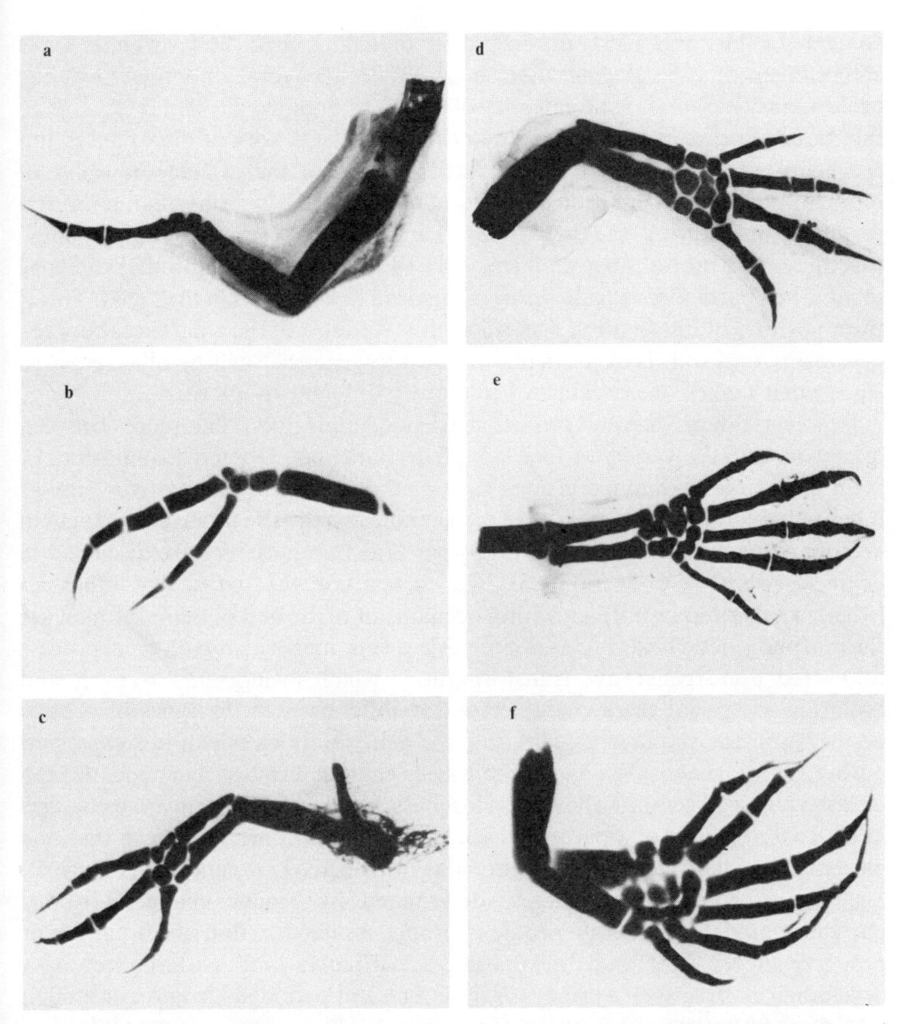

Figure 8. Cartilege patterns observed in supernumerary limbs after 180° ipsilateral rotations of blastemas in the axolotl, *Ambystoma mexicanum*, showing (a) one, (b) two, (c) three, and (f) six digit limbs together with the normal four-digit forelimb (d), and five-digit hind limb (e)

large number of solutions of the limb field equations, defining the potential or possible set of forms which can be generated. Particular solutions are selected and stabilized as a result of the initial and boundary conditions which act on the field, specifying relevant parameter values. Both internal (e.g., genetic) and external factors (such as surgical manipulations) can influence these conditions and hence the particular solution which is expressed, so that these factors have the same causal status in relation to morphogenesis. And both act within the context of an organized process, the morphogenetic field, which is an embodiment of general generative causes. Within a biology dominated by the evolutionary paradigm, one category of causes of organismic form is elevated to the status of primary cause, and a description of the history of these causes and their effects as seen in evolution is then regarded as the primary objective of biology. Weismann's dualistic description of organisms in terms of an immortal germ plasm carrying the hereditary determinants of specific morphology, together with a mortal somatoplasm, isolated for special attention that category of inherited particulars which is now recognized to be carried in the DNA. This is the repository of information which accounts for most of the differences between organisms, and so it is of particular interest to geneticists whose objective is to study what makes one organism hereditarily different to another.

However, inheritance and evolution are not simply about differences between organisms and the variety of their adaptive characters. Observed similarities of form over large taxonomic groups such as the tetrapods, arising from similar inherited generative processes, must also be considered. The necessary changes in Weismann's scheme to take account of generative regularities are discussed in detail elsewhere (Goodwin, 1983). The recognition that particular influences (whether inherited or not) act within domains of organized potential, which are the morphogenetic fields of developing organisms, means that biological process is centred upon generative transformations which include development and evolution as special cases characterized by differences in the types of change occurring in the parameters selecting the field solutions which are expressed during embryogenesis. We have already seen that, in terms of the model of limb morphogenesis discussed above, phylogenetic transformations involve changes in the spatial profile of a parameter such as that shown in Figure 5 so that one pattern of chrondrogenic sites (Figure 4) is transformed into another (Figure 6). Such transformations are largely determined by genetic changes. In this particular model the 'energy' profile is simply assumed to describe a particular property of the limb bud, generating a particular limb pattern, but in a developing embryo such a profile (or the equivalent parameter in another model) would itself be generated as a part of the process of limb ontogenesis. And this is an integral development from a previous ontogenetic process involving the patterning of the body axis and the emergence of the fore- and hind-limb fields from it. Thus transformations resulting in the different tetrapod limb patterns are likely to involve aspects of the generative mechanism which are operative in other morphogenetic processes so that a syndrome of changes occurs of which a transformation in limb pattern is but one aspect. Furthermore, realistic models of morphogenetic fields necessarily involve non-linearities in the equations so that the parameter space describing different stable domains (the 'points of organic stability', to use Bateson's (1894) phrase) is necessarily 'cellularized', as discussed by Oster *et al.* (1981), with the consequence that different morphologies are discretely separated from one another. This may be interpreted to imply that the process of speciation involves the generation of separate and distinct forms *ab initio* so that species are natural kinds (see Webster (1983) for a discussion of this). A further implication of this view is that species do not arise as a result of natural selection operating on a continuum of forms to give speciation, but come into existence by virtue of the spectral (quantized) nature of the morphological stability domains of the generative process. Of course, natural selection can act to eliminate forms which are unstable in relation to particular *external* environments, but this is a secondary process which does not tell us anything significant about the generative origins of species. These origins lie in the generative process which is structured by the intrinsic dynamics of the living state. As Oster *et al.* (1980) have put it:

> '... we view the evolutionary diversification of form as the product of regulation of developmental dynamics... In order to understand evolution we must understand developmental mechanics' (Oster *et al.*, 1980, p. 231).

If we exclude changes in genetic parameters by considering ontogenetic time-scales, then the generative process consists of those regular changes of shape and form which occur in embryos during their development. In the limb-bud the chondrogeneic sites are layed down in proximo-distal order. We have seen that fixing a spatially varying parameter such as the 'energy' profile, which is assumed to be genetically regulated, results in the selection of a particular sequence of field solutions and so generates a particular pattern. The regularity of the developmental process within a given species is due to the relatively high degree of genetic determination of such parameters. But once again we may note the causal fallacy inherent in the assertion that genes determine form, that a polydactylous gene causes six toes in a cat, for example. Of no mathematical theory can it be said that a knowledge of the parameters alone results in (or 'causes') a solution. What this ignores or forgets is that you have also to know the equations. It is these which embody the organizing principles, the order, the regularity of whatever is being described, hence defining a domain of constraint. And the discovery of what these may be is no trivial matter, but central to the whole enquiry. Would anyone dismiss the derivation of the equations of mechanics, of electromagnetism, of relativity, of quantum mechanics, as trivial affairs in the development of science? There is no doubt that in developing organisms the relevant equations are field equations, but exactly what form these will take for an adequate description of morphogenetic processes is very much a research problem at the moment. The particular model discussed in this essay is so crude as to be primarily of illustrative value; but its very crudity assists in the achievement of the degree of abstraction required for the present discussion. For with it, or indeed with any similar type of model mentioned, it is possible to see that biological process is centred upon transformations which are not arbitrary, but constrained; that the biological realm is a domain of order and intelligibility, structuring the effects of historical contingencies and randomness. And this is so because organisms are the embodiments of a particular type of order, that which is distinctive to the living state. I have focussed in this chapter on spatial order because morphology is such a significant aspect of biology, and is so central to taxonomy. But the same arguments can be applied to the study of temporal order and behaviour, and indeed this is implicit in the whole discussion since we must be concerned in biology with four dimensional order, with process.

This brings us finally to questions relating to the nature of what I have called generative transformations, the process wherein domains of distributed potential, which are the morphogenic field of developing organisms, give rise to actualized patterns of localized structure, which is overt organismic morphology. This is the process of biological creation, the real or the generative origin of specific biological forms. Each organism carries within it the potential of creating a great variety of forms, for each morphogenetic field is described by equations with many solutions which define the set of morphological possibilities. We have seen the experimental meaning of this in relation to tetrapod limbs in that manipulation of the particular conditions of development in limb buds or blastemas result in a great variety of limb forms, Figure 8 showing a small sample of the latter. This potential is not normally apparent, for unperturbed morphogenesis follows a regular sequence of spatial transformations characteristic of the species, a trajectory through the space of solutions of the morphogenetic field equations which Waddington (1957) called a creode. These creodes show a high degree of stability to the types of perturbations which normally occur during embryogenesis, such as variations in temperature, and there is an important genetic component to this stability which Waddington (1957) called genetic canalization. However, it is crucial to remember that genes are not sufficient causes of biological form any more than last year's elliptical passage of the earth around the sun is a sufficient cause of this year's elliptical trajectory. In both situations particular conditions, either of composition (gene products) or of position and velocity, select particular forms of space—time order from the set of possibilities defined by the laws governing fields, whether morphogenetic or gravitational. The actualization of specific morphological and behavioural patterns in organisms by the action of particular genes and environments on the space–time order of the developing organism described by the laws of organization of the living state is the biological process of creation. The exploration of the potential set of forms defined by these laws, by changes in genes and in the environment, is the process of evolution; while the generation of individual entities of specific form from this set is development. A biology based upon a generative paradigm focusses on these processes of biological creation as the central and distinctive features of the living condition, and sees the actual history of organisms (their contingent evolution) as intelligible only in relation to the logic of the creative process.

4.6. REFERENCES

Alberch, P., and Gale, E. A. (1983) Size dependence during the development of the amphibian foot. Colchicine-induced digital loss and reduction, *J. Embryol. exp. Morph.* (in press).

Bateson, W. (1894) *Materials for the Study of Variation.* Cambridge University Press, Cambridge.

Bryant, S. V., French, V., and Bryant, P. J. (1981) Distal regeneration and symmetry, *Science*, **212**, 993–1002.

Cameron, J., and Fallon, J. F. (1977) Evidence for polarizing zone in the limb buds of *Xenopus laevis*, *Devel. Biol.*, **55**, 320–330.

Darwin, C. (1859) *The Origin of Species*, 1st Edn, Penguin, Harmondsworth.

Eldredge, N. (1979) Alternative approaches to evolutionary theory, *Bull. Carnegie Mus. Mat. Hist.*, **13**, 7–19.

French, V., Bryant, P. J., and Bryant, S. V. (1976) Pattern regulation in epimorphic fields, *Science*, **193**, 969–981.

Goodwin, B. C. (1983) A relational or field theory of reproduction and its evolutionary consequences, in Ho, M.-W., and Saunders, P. T. (eds.) *Beyond Neo-Darwinism*, Academic Press, London, (in press).

Goodwin, B. C., and Trainor, L. E. H. (1983) The ontogeny and phylogeny of the pentadactyl limb, in Goodwin, B. C., Holder, N. J., and Wylie, C. C. (eds.) *Development and Evolution*, Cambridge University Press, Cambridge, pp. 75–78.

Gould, S. J., and Eldredge, N. (1977) Punctuated equilibria: the tempo and mode of evolution reconsidered, *Paleobiology*, **3**, 115–151.

Harrison, R. G. (1918) Experiments on the development of the forelimb of *Amblystoma*, a self-differentiating equipotential system, *J. exp. Zool.*, **25**, 413–461.

Harrison, R. G. (1921) On relations of symmetry in transplanted limbs, *J. exp. Zool.*, **32**, 1–136.

Hinchliffe, J. R., and Johnson, D. R. (1980) *Development of the Vertebrate Limb.* Oxford University Press, Oxford.

Holder, N. (1983a) Pattern formation and vertebrate limb evolution, in Goodwin, B. C., Holder, N. J., and Wylie, C. C. (eds.) *Development and Evolution*, Cambridge University Press, Cambridge, pp. 399–425.

Holder, N. (1983b) Developmental constraints and the evolution of vertebrate digit patterns, *J. theoret. Biol.* (in press).

Kent, G. C. (1965) *Comparative Anatomy of the Vertebrates.* C. V. Mosby Co, St. Louis.

Maden, M. (1981) Experiments on Anuran limb buds and their significance for principles of vertebrate limb development, *J. Embryol. exp. Morph.*, **63**, 243–265.

Maden, M. (1982) The structure of 180° supernumerary limbs and a hypothesis of their formation, *Devel. Biol.*, **93**, 257–265.

Maden, M., and Goodwin, B. C. (1980) Experiments on developing limb buds of the axolotl, *Ambystoma mexicanum*, *J. Embryol. exp. Morphol.*, **57**, 177–187.

Maden, M., and Turner, R. N. (1978) Supernumerary limbs in the axolotl. *Nature*, **223**, 232–235.

Newman, S. A., and Frisch, H. L. (1979) Dynamics of skeletal pattern formation in the developing chick limb, *Science*, **205**, 662–668.

Odell, G., Oster, G. F., Burnside, B., and Alberch, P. (1981) The mechanical basis of morphogenesis, *Devel. Biol.*, **85**, 446–462.

Oster, G. F., Murray, J. D., and Harris, A. K. (1984) Mechanical aspects of mesenchymal morphogenesis, *J. Embryol. exp. Morphol.*, (in press).

Oster, G. F., Odell, G., and Alberch, P. (1980) Mechanics, Morphogenesis and Evolution, *Lectures on Mathematics in the Life Sciences*, **13**, 165–255.

Patterson, C. (1983) How does phylogeny differ from ontogeny? in Goodwin, B. C., Holder, N. J., and Wylie, C. C. (eds.) *Development and Evolution*, Cambridge University Press, Cambridge, pp. 1–31.

Slack, J. M. W. (1980) Regulation and potency in the forelimb rudiment of the axolotl, *J. Embryol. exp. Morphol.*, **57**, 203–217.

Swett, F. H. (1927) Differentiation of the amphibian limb, *J. exp. Zool.*, **47**, 385–439.

Turing, A. M. (1952) The chemical basis of morphogenesis, *Phil. Trans. R. Soc. B*, **237**, 37–72.

Waddington, C. H. (1957) *The Strategy of the Genes*. Allen and Unwin, London.

Wallace, H., and Watson, A. (1979) Duplicated axolotl regenerates, *J. Embryol. exp. Morph.*, **49**, 243–258.

Webster, G. C. (1983) The relations of natural forms, in Ho, M.-W., and Saunders, P. T. (eds.) *Beyond Neo-Darwinism*, Academic Press, London, pp. 193–217.

Webster, G. C., and Goodwin, B. C. (1982) The origin of species: a structuralist approach, *J. Soc. Biol. Struct.*, **5**, 15–47.

Weismann, A. (1883) reprinted in Hall, J. L. (ed.) (1964) *A Source Book in Animal Biology*. New York: Hafner.

Weismann, A. (1885) reprinted in Moore, J. A. (ed.) (1972) *Readings in Heredity and Development*. Oxford University Press, N.Y.

Wilby, O. K., and Ede, D. A. (1976) Computer simulation of vertebrate limb development, in Lindenmeyer, B., and Rozenberg, G. (eds.) *Automata. Languages, Development*, North-Holland, Amsterdam, pp. 143–161.

Wolpert, L. (1969) Positional information and the spatial pattern of cellular differentiation, *J. theoret. Biol.*, **25**, 1–47.

Wolpert, L. (1971) Positional information and pattern formation, *Curr. Top. Devel. Biol.*, **6**, 183–224.

Wolpert, L. (1982) Pattern formation and change, in *Evolution and Development*, Dahlem Workshop Report 22, Springer-Verlag, pp. 169–188.

Wolpert, L., and Hornbruch, A. (1981) Positional signalling along the anteroposterior axis of the chick wing. The effect of multiple polarizing wing grafts. *J. Embryol. exp. Morphol.*, **63**, 145–159.

Evolutionary Theory: Paths into the Future
Edited by J. W. Pollard

Chapter 5

The complexity of organisms

PETER T. SAUNDERS
Department of Mathematics,
Queen Elizabeth College,
University of London, Campden Hill Road,
London W8 7AH,
UK

and

MAE-WAN HO
Biology Discipline,
Open University,
Milton Keynes,
Buckinghamshire MK7 6AA,
UK

5.1. INTRODUCTION

By comparison with the objects that we study in physics, living organisms are highly complex. Indeed, it is their complexity as much as anything else which suggests that there is a fundamental difference between animate and inanimate matter, and that the former requires something more by way of explanation than does the latter. There seems to be general agreement on this point; the disputes are about what form the additional explanation should take.

The simplest hypothesis is that of special creation. Each species of plant or animal was created out of nothing by some divine being. For those who hold this view, the great complexity of organisms presents no difficulties whatsoever; the creator is capable of producing anything with equal facility. Much the same is true of vitalistic explanations, for as Woodger (1929) observed, they are invoked precisely where the 'materialist mechanist' gets into difficulties, and no limits are placed or even implied on their powers.

At present, the most widely accepted way of explaining organisms in scientific terms is the so-called 'synthetic' or 'neo-Darwinist' theory of evolution by natural selection. But while this theory does provide a possible explanation for adaptation, it tells us little if anything about complexity, except insofar as we are prepared to accept that this has arisen through the accumulation of small changes over long periods of time. Complex structures are not really explained; they are only cited as examples of the great power of natural selection. In this respect neo-Darwinism is like the non-scientific theories we mentioned above: all three treat complexity as a property of organisms which is to be explained in terms of some all-powerful force. Once this force has been identified, very little remains to be said.

We see no reason to limit scientific inquiry in this way. The complexity of organisms is a property which, no less than any other, is a fit object for study. A crucial point is that while organisms are certainly very complex, they still evolve and develop through natural processes. Many features of organisms can be accounted for not by the selective advantage they may confer but by the limitations caused by the problem of producing a complex structure without a deity or entelechy as a guide. This idea is encapsulated in the principle of minimum increase in complexity (Saunders and Ho, 1981), but before we can introduce this we must first consider carefully what we mean by complexity. As is so often the case, an important step towards finding an answer is making sure that we know precisely what the question is.

5.2. COMPLEXITY

While complexity is an important characteristic of organisms, there are also inanimate objects which many would describe as complex. Their complexity is, however, of a different kind, and it arises in a somewhat different way. As Thom (1968) has remarked, when the sea and the weather erode a cliff face we may know the forces and the ways in which they act in great detail, but we are seldom able to say in advance what the final shape will be. In contrast, when an egg is fertilized we can predict the end result with considerable confidence, even though our understanding of the processes by which this will be achieved may be very scanty indeed. Inorganic objects are produced in nature by the blind action of physical and chemical forces. Organisms are produced by the self same forces, but not blindly. They contain hereditary material, and they develop under its influence. There is thus in the production of form in biology an element of control which is absent in physics. And because in this discussion we are concerned with the sort of complexity which is peculiar to animate matter, for our purposes it will be appropriate to choose a definition which has to do with instructions and control.

Our first inclination, therefore, might be to define the complexity of an organism to be the number of instructions required to specify its development. This is in effect the way that mathematicians have chosen to define the complexity of a sequence of numbers, but it is not suitable here because it ignores a consideration which is absent in mathematics but important for us. For a mathematician all digits are equivalent, in the sense that there is no *a priori* reason why one should appear in a sequence more often than any other. Any differences in frequency in the sequence are due solely to the algorithm which generates it.

This is not so in our case, and we have to make some allowance for this. We want to include in our definition the idea that something is more complex if it contains many elements which are themselves relatively unlikely to occur. This is necessary if only because we are going to remain intentionally vague about what precisely we mean by 'instructions', and the problems that this might cause are greatly diminished if we do not have to distinguish between the same number of instructions with different probabilities at one level of description, and different numbers of equally likely instructions at another.

What we require, therefore, is not the length of the list of instructions, but the information content (Shannon and Weaver, 1949). We want to weight each instruction according to its probability of occurrence in a manner that will satisfy the point we have just made and which will also have the property that the extra complexity gained by adding two instructions should be the sum of the amounts gained by adding the two instructions separately (and not the product, as we would expect with probabilities). This implies that our definition must be

$$-\Sigma \log_2 p_i$$

where p_i is the probability of the ith instruction. We conform to the usual convention in using logarithms to the base 2; in the case of equiprobability this means that the information content is just the number of yes/no choices. (The reader who has some familiarity with information theory or statistical mechanics may have been expecting the formula $-\Sigma p_i \log_2 p_i$ instead of the above, but that is the average information content per instruction, not the total.)

The first authors to propose a definition of the complexity of organisms along these lines were Dancoff and Quastler (1953). They took as their instructions the specification of the type, position and orientation of every atom or molecule in the organism. This does not seem to have been very productive. They did produce estimates of the information content of an adult human, which they found to be about 2×10^{28} with atoms as the fundamental unit or 5×10^{25} with molecules. Since there does not appear to be any particular reason for preferring one of these to the other, it is difficult to see how either value can be especially significant (cf. Apter and Wolpert, 1965).

It is easily seen where Dancoff and Quastler went wrong. What they were discussing were the instructions which would be required to tell someone how to construct a replica of a particular human being; what they evaluated was therefore what we might call the 'apparent complexity'. Our real concern, however, is with the 'intrinsic complexity', by which we mean the information content of the instructions which the epigenetic system actually uses.

A convenient measure of this would appear to be the information content of the genome, taking the frequencies of codons or amino acids as the p_i (Gatlin, 1972). Unfortunately, while this can be used to throw some light on genomic evolution, it is not adequate for our purposes. The difficulty is that the connection between the DNA and the phenotype is just not close enough. It is generally accepted that a high proportion of the DNA is not even expressed. Differences in DNA are also insufficiently correlated with phenotypic differences: sibling species of *Drosophila* or of mammals show genetic differences which are just as great as those between man and chimpanzee (King and Wilson, 1975), yet the anatomical and physiological differences between the latter two are great enough that they are placed in different genera. By comparison with laboratory stocks, the 'wild type' of an organism is typically more diverse genetically but less diverse in phenotype. And what is more, not all the instructions are in the DNA, for we have to take into account effects such as cytoplasmic or cortical inheritance (see Ho, 1984).

But while we can easily see why the information content of the genome is not a suitable definition of the complexity of an organism, it is hard to think of another way of measuring it directly. Rather than accept a definition which we find insufficiently realistic, we prefer to weaken our requirement in another direction, and not to insist that we must be able to produce a number which we can call the 'complexity' of any given organism. Accordingly, we have proposed the following (Saunders and Ho, 1981): Let C be the complexity of an organism at any stage of its evolution. Suppose that some change occurs, and that the probability of this particular change was p. Then the increase in complexity will be $\delta C = \log_2(1/p)$.

Our definition differs from the others that have been previously put forward in that it is a differential relation. It allows us to measure (in principle, at least) the change in complexity at any stage in the evolution of an organism. Like many other differential relations, however, it cannot in general be integrated. We cannot compare the complexities of two arbitrarily chosen organisms, nor can we compute the complexity of any one organism without tracing its entire evolutionary history.

As we shall see, this is enough to allow us to make use of our definition. And in fact we lose very little by not being able to assign a numerical value to the complexity of each organism. Whether or not an oak tree is more complex than a sparrow is not something we really wanted to discuss anyway. The major concern of those who have produced numbers seems to have been to discuss the apparent paradox that the adult contains so much more information than the egg (See, for example, Dancoff and Quastler, 1953; Elsasser, 1958; Raven, 1961).

Now this is not really a problem, for, as Apter and Wolpert (1965) and others have pointed out, an egg is a set of instructions, not a scale model. In the same vein, Riedl (1975) has suggested that we should distinguish between what he calls the 'law content' and the 'redundancy content'. The law content of two identical cells is the same as that of either on its own. Only the law content of the adult (together with a relatively few extra instructions to organize the repeats) has to be contained in the egg. By estimating the average number of instances of events in a human being to be of the order of 10^{19}, Riedl claims to reconcile Dancoff and Quastler's estimate of the information content of the germ plasm, 10^6, with that of the adult, roughly 10^{25}.

While the use of numbers may make the argument easier to follow, however, it does not contribute anything of substance, and the numerical agreement is not really significant. The reason for this is that repetitions are by no means the only way in which complexity (in the Dancoff and Quastler sense) can be increased; they are not even the most important. Many years ago, for example, D'Arcy Thompson (1917) remarked that the apparently highly complex form of a medusoid can be generated from remarkably little information: the instruction to allow a drop of fusel oil to fall into paraffin. So it is not just repeats of forms but the very forms themselves that do not have to be specified in detail in the genome. The extra information, the analogue of what Riedl calls the redundancy content, may be thought of as residing in the laws of physics and chemistry. Since there is no way in which we can assign a numerical value to the amount of information these laws contain, we cannot perform a calculation of the kind that Riedl has carried out for the simpler case. The concept of complexity as we have defined it still allows us to discuss and resolve the problem of the egg and the adult; having quantitative estimates of the complexities would not add anything to our understanding.

5.3. THE PRINCIPLE OF MINIMUM INCREASE IN COMPLEXITY

According to neo-Darwinism, evolution consists of the natural selection of random variations. A trait is explained by an account of the selective advantage that it gives, or might give, or might have given to the individuals that happen to possess it. It is not considered necessary to say anything about how the trait arose in the first place. As Dobzhansky *et al.* confidently assert:

> 'If the appropriate genetic variants to face an environmental challenge are not already present in the population, they are likely to arise soon by mutation' (Dobzhansky *et al.*, 1977, p. 72).

Nature has only to choose the best from a large and readily available supply of alternatives, and the task of the evolutionist is to study how this choice is made. If anything at all is said about complexity, it is to point to the role of natural selection as a generator of the improbable.

Facing up to the problem of complexity suggests a different approach. Organisms are not produced by mysterious forces; they arise through natural processes. Obviously these processes are capable of producing entities of great complexity—if they were not there would be no organisms. They are not, however, omnipotent. They cannot produce absolutely anything, and of those things they can produce some arise with very much higher probability than others. Nor is the probability necessarily related to the apparent complexity of the object in question; there can be little doubt that the shape of the medusoid is relatively easy to produce, once we notice that it occurs in inanimate matter as well.

We cannot, of course, neglect natural selection altogether. A variant must be viable, and if it competes directly with some alternative form it must have some selective advantage as well. But *natural selection cannot act on a variant which does not exist*. Whether or not a particular feature appears in evolution largely depends not on its relative fitness once it occurs, but on the probability that it ever occurs at all. If we denote this probability by p, then the occurrence of the character increases the complexity by $\delta C = \log_2 (1/p)$. Larger values of p imply smaller values of δC. Consequently we expect that evolution should proceed in accordance with a principle of minimum increase in complexity.

We can apply the principle without calculating any complexities or even being able to provide numerical estimates of the various probabilities. Instead, our task is to try to determine what sorts of features are comparatively easily produced by the means available to organisms. It is these that we expect to observe. Thus the demonstration that a certain feature is easy to produce becomes an important part of explaining its appearance during evolution.

It may seem odd that we do not actually have to evaluate anything, but the sort of reasoning that is involved is not at all novel. Consider what happens when some organization with which we deal changes from manual to computer accounting. There is usually a gain in efficiency, but this is not always apparent to an outsider. On the other hand, the customer is generally aware of certain other changes which are not necessarily improvements in themselves but only characteristic of computer-based systems. For example, almost everyone and everything acquires a code number, often a long one. Procedures tend to become standardized, with fewer allowances for individual circumstances. The statements that are sent out are often much more detailed, including more items that require some extra computation. Even the mistakes that occur are different: an error of five pence is less likely than before, an error of £1 million more so.

The point is that we can predict all this without knowing anything at all about the particular programs that were written nor even which machine was used. After we have experienced a couple of such changeovers we know what to expect. And if we know a bit about how digital computers work and have a general understanding of the principles of programming, we will be able to explain why these things happen and perhaps even anticipate others, still without any specific information. We do not have to see the program that the records office uses to be able to relate the long identification number on a driving licence to the fact that computers are very good at calculation and less efficient at looking up information.

The same idea can be used in evolution. The more we can learn about how organisms are produced, i.e., the more we discover about developmental biology in the widest possible sense, the more we will be able to account for what we observe today without a constant appeal to natural selection. Nor is this just a pious hope for the future. We now show how our present understanding is easily enough to enable us to make a start on the problem.

5.4. REPEATED STRUCTURES

Once we begin to look for ways in which organisms can and do conform to the principle of minimum increase in complexity, it is not hard to find them. The most obvious is the frequent use of sequences of structures which are identical or nearly so. It is, at the very least, open to doubt that the repetition of a single form should so often be the optimal solution to an environmental challenge, but it is certainly readily accomplished. There are many ways in which such a sequence can be brought about. Sometimes, as in the tapeworm, it is a matter of repeated execution of the same instructions, very much like the loops which are the stock in trade of the computer programmer. The same effect can also result from the partitioning of a region into similar sub-regions, as in segmentation in insects and somite formation in vertebrates. A number of different biochemical processes have been suggested which can account for this; many of them, following Turing (1952), are based on the interaction between reaction and diffusion.

A similar argument applies to other serial homologies: different features within the same organisms which have the same basic structure. A striking example can be seen in the crayfish, in which 19 kinds of appendages are all variations on a single design (cf. Dobzhansky *et al.*, 1977). Again, it does not appear likely that each of these is better suited to its function than any conceivable alternative would be. On the other hand, using the same bauplan for all appendages does keep the complexity to a minimum.

There are also many instances of the same feature in organisms which are not generally seen as homologies. For example, the construction of different parts of the body involves the same kinds of cell types, thus apparently relying on a standard set of 'sub-routines' which are called upon as required. One observable consequence of this is that the 'sub-routines' will often be correctly 'executed' even when the main 'program' has been 'corrupted': a malformed limb will still generally be properly covered with skin. (See Bonner (1965) for an interesting discussion of sub-routines in development.)

Repetition is a general phenomenon, occurring at all levels of biological organization. Gene duplication has given rise to many multigene families such as the ribosomal RNAs and histones in eukaryotes, which are repeated from several hundred to tens of thousands of times. This fulfils the need for a very large supply of the gene products during the initial phase of embryonic development. Other multigene familes include a more limited number of duplicated genes which undergo divergence both in base sequences and in function. The best known examples are the α- and β-haemoglobins, different genes of which are expressed during different developmental stages (see Proudfoot *et al.*, 1980). In addition, many 'families' of repeated stretches of DNA with no known function are interspersed throughout the eukaryotic genome (Doolittle and Sapienza, 1980), so much so that they have been called 'selfish DNA', as their sole function appears to be self duplication within the genome.

Duplication of genes may provide yet another relatively simple way in which repeated structures can arise. This is suggested by the bithorax complex (Lewis, 1978), a cluster of loosely-linked genes, the products of which are required for the formation of the thoracic and abdominal segments in *Drosophila*. The order of the genes in the cluster is roughly the same as the order of the segments each affects. This has led Lewis (1963) to postulate that the cluster arose from the duplication of a single ancestral gene which controlled the formation (and possibly also the repetition) of the archetypal segment, the mesothorax. The duplicated genes then diverged and took over control of different segments posterior to the mesothorax. Whether this is actually what happened should soon be revealed by DNA sequencing. (A restriction map of the complex recently obtained by Bender *et al.* (1983) gives no indication that there are large repeating units of DNA corresponding to different segments of the fly.) It is in any case intriguing that the gene order here appears to be a close internal representation of the segment order. Elucidation of the precise relationship between the two should reveal a great deal about the relationship between genes and morphology in general (cf. Ho, 1984).

There are many instances of repeated or homologous structures in organisms, and their study has much to tell us about evolution. We shall not dwell on them here, however, not because we consider them unimportant but because they have been discussed by other authors, especially Riedl (1975). Instead, we shall move on to a question which has received far less attention. We can see why organisms should as far as possible be made up of 'standard parts'. But where do the standard parts themselves come from? Are they accidents of history, or has the principle of minimum increase in complexity something to say about these as well? We now show that the latter is the case.

5.5. FORM

When applied to the problem of the origin of the standard parts, i.e., to the problem of form, the principle must be applied in a somewhat different way from before. It cannot tell us directly what we may expect to observe; instead, it indicates how we should set about discovering this. It predicts that the forms which we observe in organisms will be those which are most easily produced, i.e., those which can be created by comparatively simple physico-chemical processes with as little intervention as possible from the hereditary information. This still leaves us with the problem of determining which these forms are, but while this task is a hard one it is by no means impossible. There are essentially two approaches we can adopt, one specific and already bearing fruit, the other less well developed but more general and hence potentially more powerful.

The first is a part of the sort of biological modelling with which we are all familiar. We make certain hypotheses about how a certain form is produced and try to construct a model which brings it about. An early advocate of this approach to biology was D'Arcy Thompson (1917), who pointed out that many of the shapes that we see in organisms arise through the action of comparatively simple physical forces. One example of this was the medusoid which we mentioned earlier, although in this case it is easier to be convinced that the shape is due almost entirely to autonomous physical forces than to show precisely how they act. On the other hand, we can account for the shapes of many simple organisms by supposing them to be determined by elastic forces in their external surfaces responding to internal pressure. This idea has recently been successfully applied in microbiology (Saunders and Trinci, 1979).

Progress is also being made on more difficult problems. A recent example is an extensive study of the folding of tissues (Odell *et al.*, 1981; Oster *et al.*, 1981). The basic model is quite simple. The authors begin with a single layer of cells of uniform size, joined together at their apical periphery so that the external surface is maintained. Beneath the apical surface, each cell contains a network of contractile fibres. If the fibres are stretched by a small amount, they will return to their original lengths, but stretching the network beyond a certain threshold amount triggers an active contraction which reduces the apical surface area to below its original value. Reducing the area of one cell will naturally tend to stretch the apical surfaces of its neighbours, and this may trigger the active contraction process in them as well. In this way, a wave of contraction can pass through a tissue, and the end result will be that the tissue will be deformed. What the final shape will be, and through what sequence of intermediate forms this will be achieved, depend on the details of the model. The authors mention in particular the dependence on the viscoelastic properties of the fibre bundle, the viscous resistance of the cells to dilations and contractions, the mechanical characteristics of the interfaces between cells, and the external forces acting on the tissue.

The model itself consists of a very large system of differential equations. These cannot be solved analytically, but the authors have studied them extensively on a computer. They have been able to generate numerical solutions which quite realistically simulate processes such as gastrulation, neurulation, and the formation of the ventral furrow.

This is clearly of considerable interest for developmental biologists. It is also important for evolution, because it extends our understanding of what can happen without detailed control at every stage. As Odell *et al.* point out:

> '...the crucial aspect of our model is that, once triggered, the morphogenetic process of invagination proceeds on its own, directed solely by the global balance of mechanical forces generated locally by each cell, and with no requirement for individually preprogrammed sequences of patterns of cell shape change' (Odell *et al.*, 1981, p. 450).

In other words, if the model is correct, gastrulation is a far less complex phenomenon than it appears.

All the same, this sort of modelling does have some limitations, at least from our point of view. It assumes one particular mechanism, so it really only gives us information about forms which are generated in very much the way that is suggested. It is certainly an advance to be able to show that certain forms are not, in our sense of the word, complex, but we do not know whether these constitute a special small subset of the set of all forms or whether there are many others which can be produced by equally plausible mechanisms. Nor would these necessarily have to be very different. Even within this model the number of possible combinations of assumptions about the various parameters is immense. Moreover, all the calculations that have so far been carried out have been essentially two-dimensional, yet they require vast amounts of computer time. Modelling a three-dimensional situation without strong symmetry properties seems out of the question, yet we would very much like to know whether the extra degree of freedom substantially increases the possibilities. Experience in other fields suggests that it might. So even a model which was nothing more than a slight modification or generalization of the above could well predict the appearance of quite different forms.

Our aim in studying evolution is to explain how the organisms that we observe today came to be. And part of what we have to do is to show why it is that it is these particular organisms that exist, and not some different collection. A theory of evolution that does not even address this problem will always be incomplete. Like neo-Darwinism it will lay itself open to the charge of being more a rationalization than an explanation, and it will certainly not be falsifiable (Saunders and Ho, 1982). This means that as evolutionists we demand a great deal of a theory of form. Not only do we want to show that certain forms and processes are simple, we also want to show that they are the most simple, which would explain why they are the ones that so often occur. Models of the kind we have been describing cannot establish this, though they can provide evidence in favour. It is rather like the situation in mathematics: working out some special cases may suggest that a certain theorem is true, but it does not constitute a proof.

Of course in morphology the special cases are themselves not at all easy to analyse, and we may be sure that developing a general theory is going to be even harder. It is not even certain that it is possible, at least not in the sense of deriving a simple list of naturally occurring archetypes. A start has, however, been made, with the appearance of catastrophe theory (Thom, 1972). Most readers will be familiar with this theory through its application to other problems (including, incidentally, some of the examples which lead us to suspect that what happens in three dimensions may be quite different from what happens in two) but it was originally designed as 'an outline of a general theory of models', with biology especially in mind. To enable the reader to see how catastrophe theory approaches the problem of morphogenesis we give a brief account here, but with the warning that it is necessarily over-simplified. Those who are interested to learn more on the subject might consult Saunders (1980), but the full flavour of Thom's ideas can be obtained only by reading his own book.

The underlying principle is the same as in the other models in this section, i.e., we assume that biological form arises not from some kind of detailed preprogramming in individual cells but rather through processes which involve whole regions of tissue. If we also assume that neighbouring cells influence each other and generally, though not quite always, develop in similar ways, then we may suppose that the processses can be modelled by differential equations.

Now it is well known that even if a system of differential equations involves only smooth well-behaved functions, and even if the boundary conditions are also smooth, the solutions can still develop discontinuities, which may manifest themselves as sharp frontiers separating distinct regions. A familiar example is the shock wave associated with supersonic flight. Behind the shock wave we hear the noise of the aircraft, first loud and then gradually dying away: ahead of it we hear nothing at all.

This suggests a relatively simple mechanism for morphogenesis during development. It is only necessary to arrange that some process occurs which can be described by a suitable system of differential equations and that the boundary conditions are right. The frontier, i.e. the form, will then appear without further intervention.

That form can be produced in this way is not a new idea. What is new is that the number of different forms that can be produced in this way can be shown to be remarkably small. No matter how many differential equations there are in the system (though they have to be members of a certain large class; not quite anything will do) the form will have to be one of those that can be seen in the geometry of the seven elementary catastrophes.

This is a very powerful result. We do not have to solve the equations; we do not even have to be able to write them down. We are freed from the concern that a slight alteration in one of the equations or in the boundary conditions might introduce some unexpected phenomenon. When we speak of 'different' forms, to be sure, we mean qualitatively different, for the theory tells us nothing about the precise details of size and shape, but even this is an advantage: the fundamental questions in biology are surely themselves more topological than geometric. Within a species there is often considerable variation in the exact dimensions, but the general plan is the same for all individuals. Moreover, there is evidence that topology is more fundamental than geometry both in ontogeny and in phylogeny. We may cite as evidence D'Arcy Thompson's (1917) famous demonstrations that certain organisms or parts of organisms which have quite different geometries are nevertheless topologically almost identical. And Wolpert (1982) and his co-workers have found that in the development of the chick limb it is the topology that is specified first, with the geometry being adjusted afterwards in an apparently separate process.

It is, however, important to recognize that this is still only a beginning. Thom himself argues that many processes (including the formation of the medusoid) are described only by what he calls 'generalized catastrophes', and these are not yet well understood. More importantly, however, what the seven elementary catastrophes pick out are not so much the basic forms that we observe in organisms as the building blocks out of which they themselves are made.

So while Thom's approach and the more conventional one do attack the problem of form from opposite directions, they have rather more in common than might appear. On the one hand, the use of catastrophe theory does demand a familiarity with the details of the process much greater than a bald statement of the idea would suggest; it is striking the extent to which Thom's own work relies on analysis beyond that which his own theory provides. Conversely, while the conventional method may, strictly speaking, be limited to particular situations, in practice it tells us more. Whether the models of Oster and Odell and their co-workers turn out to be correct or not, their conclusion that gastrulation is not very difficult to bring about is likely to survive. If one mechanism can be found that will do it, no doubt there are others.

Whatever the ultimate fate of these particular theories, they serve to demonstrate how quite complex forms can be generated in remarkably simple ways. They provide ample evidence that the problem of form, while difficult, is by no means totally intractable, and that it is well worth tackling. In trying to understand how the forms of organisms have been produced through evolution, this is surely a more profitable approach than the mere statement of one's faith in the power of natural selection to produce anything at all, providing only that it is useful.

5.6. THREE SPECIFIC APPLICATIONS

We have shown that the principle of minimum increase in complexity has an important role to play in the study of evolution. It provides, or at least demands that we provide, explanation of a kind which is missing from neo-Darwinism and without which our understanding of evolution remains incomplete. To bring the issue into sharper focus, we now discuss three specific evolutionary phenomena, Williston's Law, parallelism, and mimicry. In each case we show that the explanation provided by the principle is more satisfactory than that suggested by the synthetic theory.

5.6.1. Williston's Law

Many palaeontologists have remarked that when an organism possesses many similar parts, there is a tendency for the number of these parts to be reduced, and for those that remain to become more specialized. A good example of this phenomenon, which is generally known as 'Williston's Law', can be seen in the evolution from primitive Crustacea like the trilobite, with many segments and pairs of legs, to the crab, with far fewer of each and with one pair of legs modified into claws.

Now it is quite possible to construct a neo-Darwinist explanation of this trend. It is not difficult to see that specialization could bring some selective advantage, and it can be argued that a reduction in the number of parts should lower the energy requirements of the organism.

There are, however, some difficulties. For one thing it is not at all clear that the selective advantage in reducing the number of parts is real: we doubt that the energy requirements of cats with extra digits on their paws are significantly greater. On the other hand, if it is real, how did the extra parts evolve in the first place? And how are we to explain those cases in which the most recent evolutionary trend has been towards *more* components of a given type, not fewer? As Stebbins writes, commenting on his observation that increases in the number of florets in a flower are probably as common as decreases:

> 'What evidence is there that these differences in number are of adaptive significance and that both upward and downward trends have been guided by natural selection?' (Stebbins, 1967, p. 118).

The principle of minimum increase in complexity provides a much more satisfactory explanation. It is no longer decrease in number which is significant, but change. Suppose, for example, there is some alteration in the environment. There might be many ways in which a species could adapt to this, but if one of them consisted chiefly of a change in the number of some part, then this is the adaptation that we would expect actually to occur. Equally, such an easily produced change is more likely to occur even without selection, say by genetic drift. As a result, changes in number should be relatively common in evolution, as indeed they are. But this is because they are relatively easy to bring about, and not because decrease—or increase—is in itself inherently advantageous.

5.6.2. Parallel evolution

Parallel evolution, or parallelism, is the development of similar characters in different lineages of common ancestry. There are many examples; one of the most striking is the existence in Australia of marsupials which closely resemble placental mammals such as wolves, cats, squirrels, ground hogs, anteaters, moles and mice (cf. Simpson and Beck, 1965).

According to neo-Darwinists (e.g., Dobzhansky *et al.*, 1977, p. 266) the phenomenon is due to a combination of two effects, homology and analogy. Some of the resemblance between, say, a timber wolf and a Tasmanian wolf, is attributable to the fact that they are descended from a common ancestor. The cause of the remaining similarity, and especially of those traits which are found in the two wolves but not in other mammals of either kind, is that they occupy similar ecological niches. This explanation, however, depends on the assumption that the two species have been subjected to very much the same selection pressures in the same sequence, and there is no evidence for this, nor for a struggle for existence between them and other variants, with the wolf-type winning in both cases. In any case, there is no shortage of examples of parallelism in situations where conditions are believed to have been not at all the same. We may cite the trend in fossil reptiles to develop mammalian characters in apparently quite different environments. (Grassé, 1973; see also Zuckerkandl, 1976, for a discussion of this point.)

Now let us apply the principle of minimum increase in complexity to the problem. Suppose that there are a number of possible ways in which a given species can adapt to a certain environmental challenge, and suppose that each of these adaptations has a probability p_i of occurring. Naturally we do not know all the possible adaptations, still less the corresponding probabilities. If, however, we observe a particular adaptation in one lineage, then by a Bayesian argument we infer that this is the adaptation with the largest p_i, and so we expect to observe it in other lineages as well.

Sometimes similar characters appear not in different lineages but in the same line on different occasions, in which case the phenomenon is referred to as iterative evolution. For example, of the Globigerinidae, which are planktonic forams, only the ancestors of the *Globigerina* appear to have survived an extinction at the end of the Eocene. Yet most of the other forms which were in existence during the Eocene have later counterparts (Cifelli, 1969). Iterative evolution is of course very much the same as parallelism, and the same argument applies.

5.6.3. Mimicry

Birds find the taste of the monarch butterfly offensive, and so they tend to avoid it. The viceroy, on the other hand, is perfectly palatable to the same birds, but it so closely resembles the monarch that it too is seldom taken. This is an example of Batesian mimicry, a phenomenon which is often cited by neo-Darwinists as evidence of the great power of natural selection. A number of researchers have gone to considerable trouble to demonstrate that the effect is real and to determine its evolutionary origin, by which they mean the pattern of mutation and selection required to bring it about.

One question that does not appear to have been addressed, however, is why the viceroy has chosen to protect itself from predation in this particular way. The monarch has a very beautiful and complicated wing pattern which the viceroy has had to copy in detail. Since it is obviously possible for a butterfly to evolve a noxious taste—the monarch is by no means the only species to have done this—we are bound to wonder why the viceroy did not simply do the same.

We have here an unusually clearly posed evolutionary problem. The viceroy is faced with a simple environmental challenge, that of avoiding predation by birds. There are at least two adaptations available. (If there are more it does not affect the argument.) Which will it choose?

According to neo-Darwinism, it should be the adaptation with the greater selective value that will be preferred. But this must be the evolution of a bad taste, which would give the viceroy the same increase in fitness, a say, that the bad taste gives to the monarch in the absence of the viceroy. The advantage due to mimicry, in contrast, is density dependent. It has the value a only when the viceroy is comparatively rare, and decreases as the ratio of mimic to model increases. This is a well known result and it is in any case easy to see why it should be true: the more viceroys there are the more birds will learn to associate the colouration with a palatable prey, not a noxious one.

The principle of minimum increase in complexity, in contrast, implies that what matters is not so much relative selective advantage as which variation is more likely to occur in the first place. Now it has been known for a long time that the majority of butterfly wing patterns are variations on a single theme, the so-called nymphalid ground plan, and recently Nijhout (1978) has proposed a mechanism for the patterning which suggests that for one butterfly to mimic another requires nothing more than a few minor alterations in the rate constants of certain reactions. This is of course just the sort of change we might expect to result from a very few allele substitutions. The correct answer to our question, therefore, appears to be that the reason that it is mimicry rather than bad taste which has evolved is that the patterns on butterflies' wings are far less complex (in our sense of the word) than they look. And we would expect similar arguments to apply in other cases of mimicry as well.

5.7. CONCLUSION

The great complexity of organisms makes them especially difficult to explain. One way of coping with the problem is simply to suppose that organisms owe their structure to the workings of some all-powerful force; this disposes of the question of *how* they could have come into being and leaves only the issue of *why*. The only mode of explanation is therefore in terms of function or, as the neo-Darwinists put it, selective advantage.

The alternative is to accept that most of the explanation of organisms must be in terms of the same forces that mould the inorganic world. This leads to a much more ambitious and difficult research programme, and it is unlikely to yield answers so readily as the other, but ultimately it is more satisfactory. It brings the study of evolution into line with the rest of science.

Rejecting the hypothesis of an omnipotent force brings a new consideration into play. For if there are limitations on what can be done, these must be of great importance in evolution. What changes will occur will be largely determined not by selective advantage but by whether or not the required variation is readily produced by the means available. It is this idea that is expressed by the principle of minimum increase in complexity. Here we have shown how the principle can be used even with our present understanding of development (see also Saunders, 1984); we may be confident that as more is learned about this important subject, more applications of the principle will be found.

There are other ways in which direct consideration of complexity can provide explanations for features of organisms and their evolution. For example, as Simon (1973) and others have shown, complex systems in general—and hence organisms in particular—are likely to be hierarchically organized. The trend towards increased complexity cannot be explained in terms of natural selection alone; it is a consequence of the tendency of a complex system to permit the addition of components more readily than their removal (Saunders and Ho, 1976).

Throughout this chapter we have been dealing almost exclusively with only one level of biological organization, the so-called higher organisms. We recognize, however, that complexity exists at all levels in biology, and we would not wish to end without acknowledging that these too can be studied along the lines that we propose, and that a major problem is to discover to what extent anything we learn about one level can help to illuminate others.

A number of workers have already turned their attention to such issues. For example, Zuckerkandl (1976) remarks that the spontaneous organization of material systems has been shown to prevail in both prebiotic and cellular evolution, and he adds that it would be 'strange' if it should be a pervasive influence up to a certain limiting level but not beyond. Novák (1982) argues that self organization does indeed exist at all levels, and that it operates in very much the same way in each.

Not only are there parallels between levels, but each influences the others, sometimes directly. It is not always easy to distinguish between the two effects: phenotypic organization is, as we have seen, sometimes a reflection of that in the genome and sometimes not. Equally, while it is obvious that thermodynamic considerations are important in prebiotic evolution, they may also have significant consequences in later stages as well (Wicken, 1979, 1984; Gladyshev, 1982).

Finally, we may profitably consider parallels between the evolution of organisms and the behaviour of other complex systems. The best understood of these are thermodynamical systems, and we may use what we already know about them as a starting point for the investigation of more complicated systems. It may even be that some of what we learn about the latter will be of assistance to those who are still struggling with the problem of trying to understand the thermodynamics of systems which are far from equilibrium. Moreover, there are problems which are common to many fields. Discussion of whether irreversibility in physics can be deduced from time-symmetric laws is clearly relevant to the question of whether trends in evolution can be accounted for by natural selection which, because it is an optimization theory, also has no built-in arrow of time. In economics, the analogous issue is the extent to which equilibrium theories, which again are time-symmetric, can adequately describe the behaviour of economic systems in which irreversible phenomena clearly do occur. It is a sobering thought that any attempt to manage an economy solely on the basis of an equilibrium theory may be no more likely to succeed than a project to construct a perpetual motion machine—and for essentially the same reason.

Many years ago there was a struggle between vitalism and mechanism in biology. In the end, the latter won out, and today we are all mechanists. Yet something has been lost as a result, for the vitalists did perform a useful function in continually stressing that organisms have important properties which are not observed in simpler systems. In this they were perfectly correct; it was the explanation they offered that was unacceptable (Needham, 1936). We are now in a position to take on the task of finding a better explanation, and one aspect of this will be the study of complexity as a property in itself, rather than as an epiphenomenon. It will not be easy, but for the increasing number of biologists who are becoming dissatisfied with *Just So Stories* and the Panglossian paradigm it is an important part of the way forward.

5.8. REFERENCES

Apter, M. J., and Wolpert, L. (1965) Cybernetics and development, *J. Theor. Biol.*, **8**, 244–257.

Bender, W., Akam, M., Karch, F., Beachy, P. A., Peifer, M., Spierer, P., Lewis, E. B., and Hogness, D. S. (1983) Molecular genetics of the bithorax complex in *Drosophila melanogaster*, *Science*, **221**, 23–29.

Bonner, J. T. (1965) *The Molecular Biology of Development*, Oxford University Press, Oxford.

Cifelli, R. (1969) Radiation of cenozoic planktonic foraminifera, *System. Zool.*, **18**, 154–168.

Dancoff, S. M., and Quastler, H. (1953) The information content and error rate of living things, in Quastler, H. (ed.) *Information Theory in Biology*, University of Illinois Press, Urbana, pp. 263–273.

Dobzhansky, Th., Ayala, F. J., Stebbins, G. L., and Valentine, J. W. (1977) *Evolution*, Freeman, San Francisco.

Doolittle, W. F., and Sapienza, C. (1980) Selfish genes, the phenotype paradigm and genome evolution, *Nature*, **284**, 601–603.

Elsasser, W. M. (1958) *The Physical Foundation of Biology*, Pergamon, London.

Gatlin, L. (1972) *Information Theory and the Living System*, Columbia University Press, New York.

Gladyshev, G. P. (1982) Classical thermodynamics, tandemism and biological evolution, *J. Theor. Biol.*, **94**, 225–239.

Grassé, P. P. (1973) *L'Evolution du Vivant*, Albin Michel, Paris.

Ho, M. W. (1984) Environment and heredity in development and evolution, in Ho, M. W., and Saunders, P. T. (eds.) *Beyond Neo-Darwinism*, Academic Press, London, pp. 267–289.

King, M. C., and Wilson, A. C. (1975) Evolution at two levels in humans and chimpanzees, *Science*, **188**, 107–116.

Lewis, E. B. (1963) Genes and developmental pathways, *Amer. Zool.*, **3**, 33–56.

Lewis, E. B. (1978) A gene complex controlling segmentation in *Drosophila*, *Nature*, **276**, 565–570.

Needham, J. (1936) *Order and Life*, Cambridge University Press, Cambridge.

Nijhout, H. F. (1978) Wing pattern formation in Lepidoptera: A model, *J. Exp. Zool.*, **206**, 119–136.

Novák, V. J. A. (1982) *The Principle of Sociogenesis*, Academia, Prague.

Odell, G. M., Oster, G. F., Alberch, P., and Burnside, B. (1981) The mechanical basis of morphogenesis. 1. Epithelial folding and invagination, *Develop. Biol.*, **85**, 446–462.

Oster, G. F., Odell, G. M., and Alberch, P. (1981) Mechanics, morphogenesis and evolution, in Oster, G. F. (ed.) *Lectures on Mathematics in the Life Sciences: Some Mathematical Questions in Biology Vol. 13*, American Mathematical Society, Providence, pp. 165–255.

Proudfoot, N. J., Shander, M. H. M., Manley, J. L., Gefler, M. L., and Maniatis, T. (1980) Structure and *in vitro* transcription of human globin genes, *Science*, **209**, 1329–1336.

Raven, Ch. P. (1961) *Oogenesis: The Storage of Developmental Information*, Pergamon, London.

Riedl, R. (1975) *Die Ordnung des Lebendigen*, Paul Parey, Hamburg.

Saunders, P. T. (1980) *An Introduction to Catastrophe Theory*, Cambridge University Press, Cambridge.

Saunders, P. T. (1984) Development and evolution, in Ho, M. W., and Saunders, P. T. (eds) *Beyond Neo-Darwinism*, Academic Press, London, pp. 243–263.

Saunders, P. T., and Ho, M. W. (1976) On the increase in complexity in evolution, *J. Theor. Biol.*, **63**, 375–384.

Saunders, P. T., and Ho, M. W. (1981) On the increase in complexity in evolution. II. The relativity of complexity and the principle of minimum increase, *J. Theor. Biol.*, **90**, 515–530.

Saunders, P. T., and Ho, M. W. (1982) Is neo-Darwinism falsifiable?—and does it matter?, *Nature and System*, **4**, 179–196.

Saunders, P. T., and Trinci, A. P. J. (1979) Determination of tip shape in fungal hyphae, *J. Gen. Microbiol.*, **110**, 469–473.

Shannon, C. E., and Weaver, W. (1949) *The Mathematical Theory of Communication*, University of Illinois Press, Urbana.

Simon, H. A. (1973) The organization of complex systems, in Pattee, H. H. (ed.) *Hierarchy Theory*, George Braziller, New York, pp. 1–27.

Simpson, G. G., and Beck, W. S. (1965) *Life*, 2nd Edn, Harcourt Brace and World, New York.

Stebbins, G. L. (1967) Adaptive radiation and trends of evolution in higher plants, in Dobzhansky, T., Hecht, M. K., and Steere, W. K. (eds.) *Evolutionary Biology*, North Holland, Amsterdam, Vol. 1, pp. 101–142.

Thom, R. (1968) Une théorie dynamique de la morphogénèse, in Waddington, C. H. (ed.) *Towards a Theoretical Biology* 1. *Prolegomena*, Edinburgh University Press, Edinburgh.

Thom, R. (1972) *Stabilité Structurelle et Morphogénèse*, Benjamin, Reading. (English translation by Fowler, D. H. (1975) *Structural Stability and Morphogenesis*, Benjamin, Reading.)

Thompson, D'A. W. (1917) *On Growth and Form*, Cambridge University Press, Cambridge.

Turing, A. M. (1952) The chemical basis of morphogenesis, *Phil. Trans. R. Soc. Lond.*, **B**, **237**, 37–72.

Wicken, J. S. (1979) The generation of complexity in evolution: A thermodynamic and information-theoretical discussion, *J. Theor. Biol.*, **77**, 349–365.

Wicken, J. S. (1984) On the increase in complexity in evolution, in Ho, M. W., and Saunders, P. T. (eds.) *Beyond Neo-Darwinism*, Academic Press, London, pp. 89–112.

Wolpert, L. (1982) Pattern formation and change, in Bonner, J. T. (ed.) *Evolution and Development*, Springer, Berlin, pp. 169–188.

Woodger, J. H. (1929) *Biological Principles*, Kegan Paul, Trench, Trubner & Co., London.

Zuckerkandl, E. (1976) Programs of gene action and progressive evolution, in Goodman, M., and Tashian, R. E. (eds.) *Molecular Anthropology*, Plenum, New York, pp. 387–447.

Evolutionary Theory: Paths into the Future
Edited by J. W. Pollard

Chapter 6

Evolution as an entropic phenomenon

DANIEL R. BROOKS
Department of Zoology,
University of British Columbia,
Vancouver, British Columbia V6T 2A9,
Canada

and

E. O. WILEY
Museum of Natural History,
and Department of Systematics and Ecology,
University of Kansas,
Lawrence, Kansas 66045,
USA

6.1. INTRODUCTION

Evolutionary biologists have been faced with four general questions about the phenomena they investigate. (1) Why is there order rather than chaos in biological diversity? (2) Why does the order we observe take the form of a hierarchy of descent (i.e., a phylogenetic tree)? (3) Why do particular organisms look the way they do? (4) How has evolution continued to generate diversity? Around 1930 evolutionary biologists such as Fisher (1930), Wright (1931), and Haldane (1932) sought partial explanation for these questions through studying the mathematical implications of population genetics. Progress was made and we now understand much about the changes of genes in populations, and how particular phenotypes may be favoured in particular environments. Further, we are gaining partial understanding of the speciation process itself (Mayr, 1963; Eldredge and Gould, 1972; Bush, 1975; Endler, 1977; White, 1978; Wiley, 1981; Templeton, 1981). But population genetics and modes of speciation are not enough (Gould, 1980a) nor is 'macroevolution' (Stebbins and Ayala, 1981; Charlesworth *et al.*, 1982). The hierarchy of species we observe cannot be adequately explained by a study of reversible phenomena because hierarchies are the product of irreversible processes. Further, an understanding of how some species speciate cannot explain how two or more species with different histories, intrinsic properties, and ecologies might show similar historical patterns in time and space. Thus, if we are to reach a new level of understanding we must generate new theories that link species into hierarchies and ask questions not previously approached. In this chapter we shall outline a theory of non-equilibrium evolution (Wiley and Brooks, 1982, 1983) which links the singular history of organic diversity to causal laws of greater applicability than biological evolution itself. We suggest that this theory provides an explanatory framework for the existence of the hierarchy and for certain findings in developmental and molecular biology. Further, we suggest that many of its major predictions can be tested directly.

6.2. AN ALTERNATIVE THEORY

We suggest that an alternative to the neo-Darwinian paradigm should be based on four major principles if it is to lead to new insights.

6.2.1. The principle of irreversibility

The larger aspects of evolution, the origin of species and the production of diversity within a clade, are time-dependent and thus irreversible phenomena. Evolution is also an entropy producing process characterized by increasing complexity of the system as a whole. In general terms, the principle of irreversibility coupled with increasing complexity assert that evolution is a manifestation of the second law of thermodynamics operating in accordance with the general principles of nonequilibrium thermodynamics (see Prigogine, 1980). If so, then living systems such as organisms and species must be dissipative structures (see Allen, 1981; section 6.3.1). This first principle suggests that evolution is (a) hierarchy producing, (b) time-dependent, and (c) thermodynamically driven.

6.2.2. The principle of individuality

Entities which evolve exhibit spatio-temporal continuity (Hull, 1976) and some intrinsic boundary conditions. In short, they must be individualized (Ghiselin, 1966, 1974, 1980, 1981). In a complementary manner, Goldschmidt (1952) and Riedl (1978) have noted that an irreversible process operating on discrete (individualized) entities produces a hierarchy. In contrast, reversible processes operating on classes of entities (or characters) do not produce hierarchies. The results are as different as a genealogy and the periodic table. Popper (1964) was among those to recognize that something was amiss in evolutionary theory because the evolutionary hierarchy is a singular, not a universal, statement.

However, this does not necessarily mean that there are no causal laws governing evolution. If evolutionary units are individualized physical systems such as dissipative structures, then their behaviour can be partly explained in terms of causal mechanisms governing dissipative structures generally. The principle of individuality serves as a bridge principle (see Hempel, 1965) between the singular statement that phylogeny represents and the dynamic processes of dissipative structures in general.

6.2.3. The principle of intrinsic constraints

Intrinsic constraints are of two kinds. Developmental constraints operate during the ontogeny of an individual organism and preclude the realization of certain phenotypes in a population. Historical constraints operate to limit logically the scope of future change to those changes compatible with the evolved ontogenetic program. Thus developmental constraints operate on individual organisms and historical constraints operate on individual species. Some recent attempts to improve evolutionary theory (cf. Seilacher, 1972; Løvtrup, 1974; Riedl, 1977, 1978; Ho and Saunders, 1979; Gould and Lewontin, 1979; Gould, 1980a,b; Wicken, 1979, 1980; Alberch, 1980; Saunders and Ho, 1981; Jantsch, 1981) have stressed the need to consider and discover internal factors governing the production of evolutionary novelties upon which natural selection works. The converse of such mechanisms is just as important: what constraining effects do historically and developmentally determined characters of living systems have on the production of novelties independent of the effects of selection (Lauder, 1982)? If evolution is a manifestation of the second law of thermodynamics, we suggest that no special explanation is required to explain increases in variation. Rather, we must explain why not all conceivable variation is realized. Evolutionary biologists have laboured under the belief that they have had to explain why there are so many species. We suggest that given the detectable levels of genetic variation that have been observed, we must actually explain why there are so few species.

6.2.4. The principle of compensatory change

Evolutionary novelties must integrate with the total developmental programme if the organisms in which they occur are to develop and reproduce. The ontogenetic programme must be capable of compensating for changes that occur. The more radical the novelty, the more the organism must compensate. In simplest terms, the principle of compensatory change is a conservation principle. In closed systems, matter and energy must be strictly conserved. Open systems are not bound absolutely by this requirement, but still exhibit a tendency to conserve, the magnitude of which is directly proportional to the complexity and organization of the system. If information and cohesion are deterministic factors exhibiting real entropic behaviour, they must exhibit some conservative tendencies. We shall attempt to show that the principle of compensatory change is an integral part of non-equilibrium evolution because organisms and species are dissipative structures and thus subject to the same physical laws as other dissipative structures. In doing so, we hope to demonstrate that there is evidence for an internal mechanism for producing variation that is different from the usual 'copy mistake' mechanism termed point mutation.

6.3. OUTLINE OF NON-EQUILIBRIUM EVOLUTION

We have previously published a theory of evolution attempting to tie together ideas derived from systematics, developmental biology, information theory and biophysical systems analysis (Wiley and Brooks, 1982, 1983). We take this opportunity to present a synopsis of our theory, concentrating on the nature of species as evolving biological systems and the dynamics of speciation as an entropic phenomenon.

6.3.1. The nature of biological systems

An organism is an energy-using system. It must take up energy to perform work, and it must dissipate excess energy if it is to continue to live. An organism is also an information system, and this system will determine what kinds of work can be performed and how energy will be taken up and dissipated. Because organisms take up energy, dissipate energy in the form of work, and dissipate excess energy in the form of heat, they constitute open physical systems (Jantsch, 1981). Because they can do so and even alter their appearance without losing their identities, organisms are dissipative structures (Allen, 1981). Dissipative structures are characterized, in part, by the interaction of (relatively) stochastic and (relatively) deterministic factors. For an organism, energy flowing through an environment is a stochastic factor while integration of the information system represents the deterministic factor. Contrary to some thinking (e.g., Jantsch, 1981), energy does not appear to be the factor providing for the boundary limits (and thus individuality) of organisms. Rather, it is provided by the organism's information system.

We have suggested that population genetics fell short of a more complete theory of evolution because it does not provide for the case of the evolution of partially closed systems. Ghiselin (1966, 1974, 1980, 1981) has suggested that species are individuals and that recognition of this view is vital to advancing our understanding of the evolutionary process. We agree, for considering species as classes composed of individual organisms renders species immutable (Hull, 1976, 1978; Beatty, 1982) and thus timeless abstractions. Classes cannot be physical systems. If, however, species are considered individualized entities, we may see them as changeable physical structures susceptible to the causal laws of physics and biology. Viewed in this manner the properties of a species must include its parts (organisms) in a way analogous to viewing the parts of a molecule or the cells of an organism.

6.3.2. Information and cohesion

We have stated that organisms and species can be viewed as physical systems bounded by information and cohesion. Information systems can be viewed as being organized into structural and regulatory components, or non-canalized and canalized information (Wiley and Brooks, 1982, 1983). Because a consistent set of regulatory information can tolerate a certain degree of variation in structural information without altering the final product, canalized information appears more deterministic than non-canalized information.

Not all of the information inherent in an organism is expressed all the time. Indeed, some is not expressed at all. We have termed unambiguous information, which is expressed at some time during the life of an organism, *stored information*, taking the term from information theory. Ambiguous, or *potential*, information is not expressed in the organism although it may be passed on in typical fashion to offspring where it may be expressed. Examples of potential information include recessive alleles or alternate metabolic ontogenetic pathways which could be expressed given the correct cue, but are not. Alternate metabolic pathways comprise a set of *alternate dissipative pathways* (see Allen, 1981) for homeostasis. The presence of alternate homeostatic and developmental pathways is an important factor for testing the principle of compensatory change, as we shall discuss shortly.

Cohesion in multicellular organisms is a phenomenon which involves cell recognition and adhesion. The ability of cells to recognize and interact with each other is a function of their information systems, expressed largely by surface proteins.

Species, as historically distinct lineages, may also be thought of as information and cohesion systems. At this level of individuality stored information refers to that part of the collective information system expressed by all viable and fertile members of the species. It is exhibited by the common possession of an ontogenetic plan coupled with the structured information fixed for the species. Potential information to a species is variable stored information observed in its parts (organisms). As a simple example, polymorphism is a manifestation of potential information. Thus, members of species recognize each other as being parts of a whole because of the common sharing of a stored information system, and they vary because of variations tolerated by the stored system.

Cohesion at the species level refers to the network of reproductive ties among its parts. These may be termed *linkages*. Obviously, purely asexual species are not cohesive. The extent to which a sexually reproducing species is cohesive depends on gene flow between populations within the species. That is, it depends upon the number of potential and realized linkages between members of that species. A species that exhibits cohesion can tolerate changes in its information system through time while maintaining its historical continuity. A species exhibiting no cohesion retains only historical continuity and its parts (=populations in sexual species) are free to vary independently of other parts.

The reality of species as cohesion and/or information systems can be confirmed by observing that the organic world is composed of groups of organisms which share unique traits and, in the case of sexually reproducing species, which breed with members of their own group more frequently (or exclusively) than with members of other groups. The reality of species as physically and informationally bounded entities is confirmed by the observation that there are more than one species, for if species were not individuated systems we should expect only a single species.

6.3.3. Entropy

There are clearly two major schools of thought on the issue of the nature of entropy. Traditionally, entropy has been viewed as a manifestation of energy flows, but an alternative perspective (see, for example, Gatlin, 1972) views entropy as a manifestation of time. That is, entropy production of some kind is associated with any time-dependent process. We endorse this second, more general, view. For example, we would observe the 'negentropic energetic' behaviour of living organisms and not conclude that a negentropic principle was at work. Rather, we would assume that energy was not the determining factor in the time-dependent behaviour of living things. This would lead us to search for determining factors which showed entropic behaviour.

Our view also requires that organized macroscopic structure can arise spontaneously in the absence of local negentropic phenomena. Layzer (1975) and Frautschi (1982) discussed the manifestation of entropy in an expanding universe. If gravitational effects affect the expanding universe, actual entropy increase will lag behind the maximum possible allowed by the expansion. The expansion defines a 'causal region' within which entropy increases but at a progressively slower rate relative to the cosmological expansion. Below a certain rate of actual increase, organized macroscopic structures begin to appear. To us, the analogue of gravitational constraints are the historical, developmental and genetic constraints which operate to allow fewer combinations than the maximal number possible to occur. The more constraints there are, the more organized the organism, population, or species, all achieved without invoking negentropic behaviour. Just as Brillouin (1962) showed that there was no need for a Maxwellian Demon at the level of basic physics, and Layzer (1975) and Frautschi (1982) showed that none is necessary at the cosmological level, we assert that none is necessary to account for biological organization and diversity. That this is more than wishful thinking can be shown by the use of information theory.

As there are two major views of entropy, there are two major views of the relationships between entropy and information. To some, the statistical entropy of a system represents its information content. Gatlin (1972) called this the entropy (H) of the system; Brillouin (1962) called it the bound information (I_b) of the system; Karreman (1955) called it the information content (I); and Wicken (1979) associated it with complexity.

To others, the information content is expressed as the difference between maximum possible entropy and the actual entropy. Gatlin (1972) called this the information content (I); Brillouin (1962) called this the free information (I_f); and Wicken (1979) associated it with the organization of the system. It is this sense of information which is usually associated with negentropy. Information theory usually begins with the assumption of an alphabet (A) of finite length. Maximum possible entropy is $\log A$. Since the difference between $\log A$ and the actual entropy (H) is information, the only way to create information is to lower the entropy of the system. However, if the entropy were lowered to zero, the system would not be able to change (it would be a monotone). Thus, in evolutionary models following this view, evolution is viewed as a process of random increase in entropy (mutation) offset by the negentropic behaviour of natural selection.

We asked if it were possible to generate information without negentropic behaviour. The answer is yes, if the alphabet actually increases through time, and increases at a faster rate than the increase in entropy. This would be a process in accordance with Layzer's (1975) and Frautschi's (1982) cosmological models. A computer simulation of simple population dynamics shows that so long as either (a) new mutations occur before old mutations are completely homogenized throughout the population or (b) some mutations are never homogenized throughout the population, information and entropy will increase together (Brooks, Leblond and Cumming, in press).

One comment which bears stating at this point is: if most agree that biological evolution is an open-system, non-equilibrium, phenomenon, why is it that previous authors have thought a measure of information based on the assumption of a closed, equilibrium, alphabet would accurately reflect the system's behaviour? And, what does it mean that an information measure, based on an open alphabet, shows an increase in entropy and information simultaneously? We think it means that the alphabet of evolution is not closed. Gatlin (1972) assumed that the four DNA bases were the letters of the alphabet of evolution; we consider those bases to be more analogous to the dots and dashes of Morse code than to letters of an alphabet.

Species experience an increase in information complexity with each evolutionary novelty that appears. That is, the entropy level of information increases with increasing levels of potential information. Further, decreasing the cohesion of a species increases the entropy level of cohesion by increasing the disorganization of that species. We may consider the entropy level of information a measure of the *order* exhibited by a species and the entropy level of cohesion as a measure of the *realized linkage patterns* or *organization* (see Wicken, 1979) of that species (see also Ho and Saunders, 1979). Asexual species are highly ordered but completely disorganized. Sexual species may exhibit some disorder just so long as it is not sufficient to cause disorganization. In other words, it is possible to have linkage patterns between two phenotypes, *cross linkages*.

Prigogine *et al.* (1972) suggested that the behaviour of non-equilibrium systems could be summarized as follows:

$$dS = d_eS + d_iS \qquad d_iS > 0$$

where dS refers to the entropy production by the system and its surroundings, d_eS refers to entropy production resulting from energy flows through the system and d_iS refers to entropy production by irreversible processes within the system. For thermodynamic systems, d_eS and d_iS are energetically linked and for non-equilibrium systems the result of these energy flows is macroscopic structure. Emergence of macrostates is signified by an increase in statistical entropy. In typical non-equilibrium systems, these macrostates are open systems whose existence and structure is determined by the energy flows, or the thermodynamic entropy production, of the system. Biological systems are different, however. Organisms, populations and species are not totally open, but are bounded by finite information content and unique genealogies. In our first paper (Wiley and Brooks, 1982), we lumped both external and internal energy flows under d_eS. This was not, strictly speaking, correct notation because it left the impression that we did not recognize the energy connections between the organism and environment which make life a thermodynamic phenomenon. Our notation should have been the following:

$$dS = d_eS + d_iS$$

where

$$d_iS = dS_i + dS_c + dS_m$$

where dS_i refers to entropy production resulting from information changes, dS_c refers to entropy production resulting from cohesion changes, and dS_m refers to entropy production resulting from internal energy flows, or metabolism.

6.4. DYNAMICS OF EVOLUTION

The appearance of evolutionary novelties is necessary for evolution to progress. Novelties have two effects. First, they raise the entropy level of information of the species in which they occur. In other words, they make the species more *complex*, and the more complex a system, the higher its entropy (Wicken, 1980). This may be expressed as:

$$S_p \leq S_a < S_{a+p}$$

Table 1. Possible linkage patterns resulting from the reproductive interactions of two phenotypes (from Wiley and Brooks, 1982)

	New patterns established and maintained	New patterns not established or lost
Old patterns maintained	Case I	Case II
Old patterns lost	Case III	Case IV

where S_p is a measure of the entropy level of information for the primitive (plesiomorphic) homologue of the novelty (apomorphy) that appears. Second, the appearance of a novel phenotype allows for the possibility of new reproductive linkage patterns. The fate of old and new phenotypes (=old and new information systems) depends on what linkage patterns are realized. Basically there are four possibilities for the simplest case of two phenotypes, as shown in Table 1. Case I outcomes result when the new phenotype can establish a linkage pattern, either with the old phenotype, others of its class, or both. Case II occurs if the new epiphenotype is lost in each generation, as for example, lethal gene effects or hopeful monsters without mates. Case III might occur if the old phenotype was eliminated or if linkage patterns established between old and new phenotypes under Case I were disrupted by further changes in information or extrinsic events such as geographical subdivision. Case IV represents extinction.

6.4.1. Evolution within species

If a rise in complexity of a species is not of sufficient magnitude to bring about its disorganization, then the realized linkage patterns may take one of three forms shown in Table 2. Polymorphism (Case I) results when both phenotypes are found in the same deme. Geographic variation (Case I) allows complexity in the species as a whole to rise while preserving simplicity (and thus relatively low entropy values) in any one deme. Stable polymorphisms are established when the realized linkage patterns between phenotypes are equally probable (no selection against heterozygotes). Such a species would be relatively disordered but as highly organized as a monotypic species. If, however, hybrids between the two

Table 2. Possible outcomes for various linkage pattern changes through time and within a lineage (from Wiley and Brooks, 1982)

	New pattern established and maintained	New pattern not established or lost
Old patterns maintained	Polymorphism or geographic variation (I)	Elimination of novel phenotype (II)
Old pattern lost	Anagenesis or concerted evolution (III)	Not realized (IV)

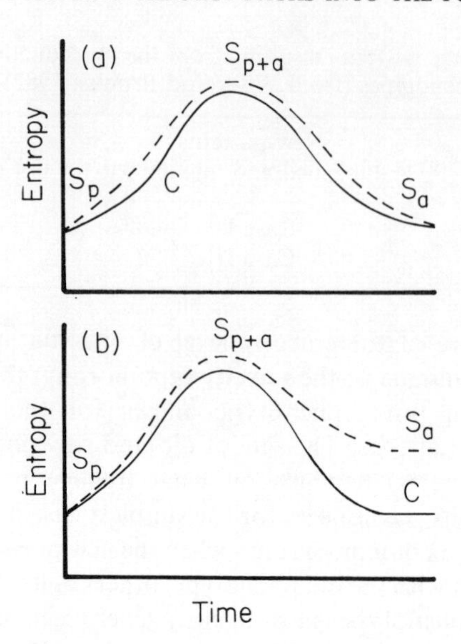

Figure 1. Evolution within a lineage. (a) Replacement of a plesiomorphic phenotype by an apomorphic phenotype by substitution. (b) Replacement of a plesiomorphic phenotype by an apomorphic phenotype by addition. (a) Results in no over-all rise in information complexity of the lineage information system whereas (b) results in such a rise. S_p = information complexity of the plesiomorphic information system; S_a = information complexity of the apomorphic information system; C = cohesion (from Wiley and Brooks, 1982)

phenotypes were selected against, the species would be disorganized in direct relation to the selection pressure. Species occupying more geographic space than the range of any one individual are inherently disorganized to a greater or lesser degree simply because there is no possibility that all linkage patterns are equally possible. This holds true even for monomorphic (highly ordered) species because gene flow is limited and probabilities of realized reproduction decrease with distance. In such a case, geographic variation may not cause a significant increase in disorganization, even when there is some selection against heterozygotes.

Case III is anagenesis (or concerted evolution (see Dover, 1982a,b)), the elimination of the old phenotype by the new phenotype. Figure 1 shows changes in entropy levels of this case. An increase in the entropy level of information and an increase in the entropy level of cohesion result from an increase in the frequency of the new phenotype in the species and a decrease in the realized number of linkages of the old phenotype. This is followed by a decrease in the entropy levels of both information and cohesion as the new phenotype becomes fixed. The species eventually settles down to an entropy level of cohesion that reflects its distribution and gene flow. Whether the relative value of information is at a higher level will depend on the complexity of the new phenotype. In the case of substitution, the relative values are the same because the new phenotype is no more complex than the old, only different (Figure 1a). In the case of addition, the relative values are different, with the new and more complex phenotype occupying a higher level of entropy (Figure 1b). These outcomes may be summarized thus:

substitution: $dS_l = S_a - S_p = 0$

addition: $dS_l = S_a - S_p \geq 0$

where dS_l is the change in entropy of the lineage l, S_p is the entropy level of the plesiomorphic phenotype and S_a is the entropy level of the apomorphic phenotype.

We have suggested that Case IV (Table 2) is not realized within species because change in information of sufficient magnitude to force complete disorganization and thus extinction would be eliminated under Case II (Wiley and Brooks, 1982).

6.4.2. Speciation

It is possible that the rise in complexity of a species is high enough to cause its disorganization (i.e. a loss of cohesion between phenotypes). The possible results in such a rise in complexity are shown in Table 3 and graphed in Figure 2. Two cases, II and IV, do not result in speciation. Case II is identical with Case II in Table 2. Case IV is extinction, a common phenomenon, but not one we would expect to result from information changes because aberrant individuals would be eliminated under Case II (Wiley and Brooks, 1982).

Case I is immediate, one generation, sympatric speciation with the species immediately reaching zero cohesion (indicated by the ZCL, or 'zero cohesion line', in Figure 2a). We can imagine such a situation with shifts from sexual to

Table 3. Possible outcomes for various linkage pattern changes through time during the diversification of organized lineages (from Wiley and Brooks, 1982)

	New patterns established and maintained	New patterns not established or lost
Old patterns maintained	Immediate sympatric speciation (I)	Elimination of novel phenotype (II)
Old patterns lost	Sympatric, parapatric or allopatric speciation (III)	Extinction (IV)

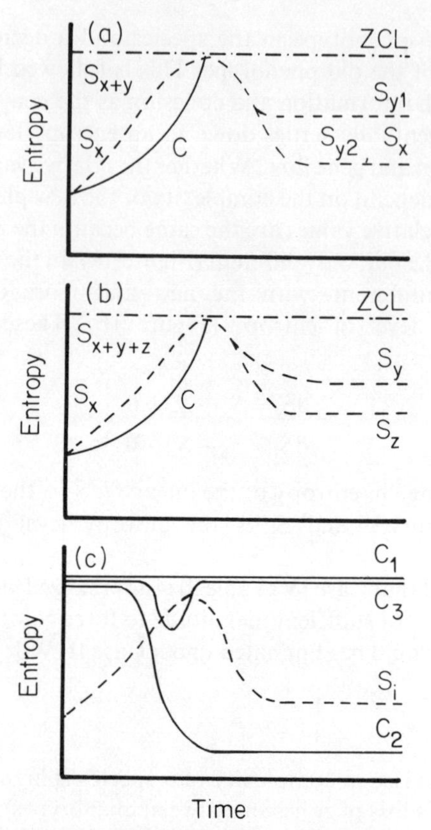

Figure 2. The dynamics of speciation. (a) Speciation with the persistence of the ancestral species. (b) Speciation with the extinction of the ancestral species. (c) Speciation in disorganized species or between species. S_x = entropy state of the plesiomorphic information system, S_y and S_z = entropy states of apomorphic information systems; zcl—zero cohesion line (—·—·—); —— = changes in cohesion (c); – – – = changes in information (from Wiley and Brooks, 1982)

asexual reproduction, as in the case of parthenogenetic shifts. The result is two species, the ancestor (S_x, Figure 2a) and a descendant (S_{y1} or S_{y2}, depending on substitution or addition of information, Figure 2(a). Case III sympatric and parapatric speciation is shown in Figure 2(a). In both cases information changes progressively affect reproductive linkages, forcing a rise in the entropy level of cohesion until the ZCL is reached. In either case, the shape of the curves would depend on the kind of information changes that occur. We may imagine two possibilities: (a) the establishment of a new phenotype whose hybrids with the ancestral phenotype are selected against, and (b) a sequential progression of phenotypes departing more and more from the ancestral phenotype as time goes on. The curves of possibility (a) result from the increase in the members of the new phenotype, with the species progressing to maximum polymorphism or polytypism and the occurrence of hybrids being progressively restricted to F_1 generation crosses. Thus the curve does not follow an intrinsic increase in information complexity *per se* but tracks the frequency of established phenotypes. Under possibility (b), there is a rise in information complexity as well as a rise in phenotype frequencies. Thus the curves track a rise in intrinsic information content of one or more phenotypes as well as the dynamics of their interactions. A measure of the intrinsic change in information complexity is the difference in entropy values between the results. Figure 2(a) graphs changes where the ancestral species survives a speciation event (cf. peripheral isolation, Wiley, 1981). There may be no change in the entropy of intrinsic information if the descendant species (S_{y2}) has substituted information. In such a case, change in entropy production is restricted to cohesion relationships. There may be a change if the descendant species (S_{y1}) has added information and a measure of this change in the distance between S_x and S_{y1} on the entropy scale. In such a case, entropy changes in the system involve both cohesion and information.

Speciation may also involve the extinction of the ancestor (Wiley, 1981). One possible outcome of this is shown in Figure 2(b), with both descendants, S_z and S_y, showing intrinsic increases in information.

Entropy changes in sympatric and parapatric speciation are the easiest to model because they involve the direct interplay of information and cohesion. Allopatric speciation differs because cohesion linkages are disrupted by extrinsic events which split the ancestral species. This allows the two (or more) allopatric populations to vary without having to integrate their information systems. The effect is a functional lowering of the ZCL with speciation going to completion as a within population phenomenon, similar in each of the separate populations to the curve shown in Figure 1. The main difference between allopatric speciation and the sympatric and parapatric modes is that in the allopatric mode information changes may lag behind cohesion changes between populations, whereas in the sympatric and parapatric modes cohesion changes lag behind information changes because information changes are causing cohesion changes. Thus the allopatric speciation model involves the extrinsic (geographic) disorganization of a species. The cohesion of the species, between populations, is already at the ZCL. Any population which experiences a rise in information complexity of sufficient magnitude to establish a new canalized information system becomes a new species. Information changes within the affected population are labelled 'S_i' in Figure 2(c).

There are other modes of speciation involving disorganized species (Table 4,

Table 4. Possible outcomes for various linkage pattern changes through time involving disorganized species and different species (from Wiley and Brooks, 1982)

	New patterns established and maintained	New patterns not established or lost
Old patterns maintained	Speciation via hybridization, asexual speciation (I)	Hybridization not leading to speciation (II)
Old patterns lost	Reduction speciation	Extinction

Figure 2(c)). Asexual speciation is determined entirely by changes in information systems (Figure 2(c), line labelled C_1). Speciation via hybridization involves the temporary establishment of reproductive linkages between two species (Figure 2(c), line labelled C_3). The theoretical possibility of reductive speciation involves the permanent fusion of two species (Figure 2(c), line labelled C_2).

The dynamics of speciation, dS, is always greater than zero, indicating that speciation is an entropic phenomenon. This is summarized by the following equations where dS is the change in entropy of a monophyletic group, and S_x, S_y, and S_z are the entropy levels for species X, Y, and Z (from Wiley and Brooks, 1982).

1. Persistence of the ancestral species (this includes the special case of stasipatric speciation):

$$dS = S_x + S_y - S_x > 0$$

2. Extinction of the ancestral species:

$$dS = S_y + S_z - S_x > 0$$

3. Asexual speciation:

$$dS = S_x + S_y - S_x > 0$$

4. Reduction speciation:

$$dS = S_z + \infty + \infty > 0$$

5. Speciation via hybridization:

$$dS = S_x + S_y + S_z > 0$$

6.4.3. Thresholds and mechanisms of change

We may define a threshold as the point at which a potentially reversible genetic system becomes an irreversible system (cf. anagenesis) or systems (cf. speciation). In an evolutionary sense a threshold is reached when the outcome of evolutionary events becomes unpredictable. At this point there is a shift between events where the rules of equilibrium analysis apply (cf. population genetics) to the rules where non-equilibrium analyses apply (cf. phylogenetic analysis). The individual organisms which are involved in this shift probably do not all spring from a common source; rather they form a class of like phenotypes which belong to the same ancestral species. Transforming this class of phenotypes into a new individual, a new species, involves something analogous to a *phase transition*. The point at which this occurs is the threshold. There are two ways this might happen. An *active phase transition* would involve mechanisms such as sexual selection, active mate choice. *Passive phase transitions* are the result of isolation. The individuals do not 'know' they are different since they never meet individuals of the ancestral phenotype, or the ancestral phenotype is eliminated by selection, chance or concerted evolution without reproductive choices being exercised by either phenotype. This may be accomplished by any of the mechanisms listed below. Obviously, the longer daughter populations remain isolated, or the greater the rise in information complexity, or the stronger the selective forces, the greater the chances that a passive transition over the threshold will occur. If an active phase transition occurs in concert with passive isolation, the threshold will be reached sooner.

In reference to our graphs (Figure 2a–c), we may define the speciation threshold as the point at which the entropy curves of information and cohesion intersect, i.e., where two or more information systems in a sexually reproducing species can no longer integrate to form long-term reproductive linkages (which does not mean that they can no longer hybridize, only that hybridization does not result in total lineage fusion). Where this threshold lies for any one speciation event depends on the canalized potential information which accumulates in the ancestral species. This, in turn, depends on the ancestral information system of the ancestor itself. Thus, every speciation event has a unique component bound by the history of the species involved. Changes in some characters may produce instant speciation in one event and minor polymorphism in another. Thus it is no surprise that many of the same characters found to vary within some species belong to the same classes of characters which distinguish other species. This does not necessarily mean that there is a smooth transition between 'microevolution' and 'macroevolution' as Rensch (1959) suggested. Rather, such observations can be just as easily explained by the fact that the same character shift may have different effects depending on the boundary conditions of the ancestral information programme. The types of characters correlated with speciation range widely (cf. Stebbins, 1950, 1971; Mayr, 1963; Grant, 1971; White, 1973, 1978; Rachootin and Thompson, 1981; Dover and Brown, 1981; Dover, 1982a,b). A larger question asks how canalized potential information originates and how non-canalized information is canalized given the historical constraints present (Goldschmidt, 1940, Ohno, 1970; Britten and Davidson, 1971; Løvtrup, 1974; Van Valen, 1974; Waddington, 1975; Riedl, 1978).

We do think we know something about how canalized potential information is processed. The fate of a new phenotype is determined by one of four processes which may be conveniently termed *proximal mechanisms of change*, to distinguish them from the ultimate mechanisms responsible for the production of the characteristics themselves. These are listed below.

6.4.3.1. Canalizing selection

Any evolutionary novelty must be minimally accommodated with the rest of the information system of an individual for that individual to grow and reproduce.

6.4.3.2. Directional selection

A new phenotype will replace an ancestral phenotype if it has an advantage in relative reproductive success. Directional selection is an attractive mechanism because it works in both large and small populations. It has been abused (in our opinion) by some evolutionary theorists (cf. Dobzhansky *et al.*, 1977) because it has been tied to the concept of adaptation. If speciation is primarily an adaptive phenomenon, then we should expect always to see positive correlations between the morphologies or behaviour species exhibit and the environments they inhabit. In some cases there is a correlation: it is no mistake that polar bears have white pelts. In some cases there does not seem to be a correlation, and indeed, no perceptible difference in environment.

6.4.3.3. Drift

A new phenotype may replace the ancestral phenotype by chance (Wright, 1931). This mechanism is attractive because it can explain cases of speciation which result in no correlation between environment and morphology and cases where the new species is maladapted compared to the ancestor. Its major drawback as a general mechanism is that it works most effectively only in populations with small effective breeding sizes.

6.4.3.4. Concerted evolution

A new phenotype may replace an old phenotype because of the wholesale conversion of the gene pool via *molecular drive* (see Dover, 1982b for review). This mechanism is attractive because (a) it is an irreversible phenomenon, (b) it works on populations of all sizes, (c) variance remains low so that the hopeful monsters are, indeed, hopeful, and (d) speciation does not have to be correlated with environmental differences. Its major drawbacks are (a) the generality of the mechanisms has not been established and (b) the link between the phenomenon on the molecular level and changes during ontogeny (which produces the characters we normally observe) has not been established. However, if molecular drive is established as a common phenomenon, it will take its place as a major evolutionary force. Further, it is possible, though highly speculative, that an analogous mechanism works on the ontogenetic level.

6.4.4. Extrinsic factors

We do not wish to leave the impression that extrinsic factors cannot affect the evolutionary process. Indeed, one way of producing entropy is to force a young equilibrium (i.e. one that has been recently reached) away from its equilibrium point by an extrinsic event (Hollinger and Zenzen, 1982) and thus force periodic systems into non-equilibrium behaviour.

One set of extrinsic factors is provided by geography and geological change. It follows that geologic and climatic changes which cause partial or complete disjunction will promote diversification by not allowing for the establishment of equilibrium conditions, by disrupting an equilibrium or by disrupting a constant departure from an equilibrium condition.

Another set of extrinsic factors is ecological. If a species is to survive it must alter or accommodate to its environment, including both abiotic factors and other information systems (other species). The interfacing of two or more information systems may or may not result in co-accommodation. We have discussed the outcomes of various ecological interactions elsewhere (Wiley and Brooks, 1982, 1983).

6.5. TESTING THE THEORY

There are three areas we need to address if the theory of non-equilibrium evolution is to be considered seriously. First, we must identify results which, if found, would falsify the theory. This amounts to identifying a class of potential falsifiers (Popper, 1968). Second, we need to establish whether evolution is associated with increasing entropy. Third, we need to demonstrate that some measure of entropy change can be quantified by a research programme. This amounts to establishing a research paradigm for the theory (Kuhn, 1970).

6.5.1. Potential falsifiers

The single most important potential falsifier of non-equilibrium evolution would be the demonstration that organisms, species, or their characters occur periodically, an attribute our theory disallows. This periodicity could be demonstrated in two ways. First, one might demonstrate that the attributes of organisms are better described by a non-hierarchical classification analogous to the periodic table rather than a hierarchical classification. This might result in a failure of our systematic methods to show genealogy but a success of our systematic methods to show periodicity in character evolution that is analogous to the periodicity observed in atomic structure. Or, it might result when one demonstrates that a systematic method based on a periodic theory explains character distributions better than a method based on a theory of hierarchy such as ours. Second, if we observed spontaneous withdrawals from character equilibrium followed by re-establishment of the equilibrium, our theory would be falsified because such withdrawals are characteristic of systems displaying Newtonian periodicity (Hollinger and Zenzen, 1982). (The problem with the second potential falsifier is that the period of equilibrium may exceed the age of the universe, or at least the age of our local system!)

6.5.2. Is evolution an entropic phenomenon?

Commonsense dictates that increases in complexity must be accompanied by decreases in entropy. This is an example of how commonsense can be wrong. Increases in information complexity represent *increases* in entropy (Karreman, 1955; Moroshowitz, 1968a,b,c; Gatlin, 1972; Denbigh, 1975; Wicken, 1979; Ho and Saunders, 1979; Saunders and Ho, 1981). Our manifestation of this phenomenon is the reproducibility of results. Hollinger and Zenzen (1982) state:

> '*Increasing entropy is a sign of assurance for reproducibility* . . . The clue comes to us from comparison of statistical entropy and the entropy of observed processes is that we should focus our attention on the reproducibility of the observed process. Because they are reproducible, as signalled by their increasing entropy if nothing else, they are candidates for statistical descriptions' (Hollinger and Zenzen, 1982, p. 32).

Reproducibility in physical systems means running the experiment over and over again. In biological systems reproducibility is represented in two ways. The first is ontogeny. If ontogeny is an entropic phenomenon it should be possible to predict the ontogenetic sequence of events given the initial boundary conditions. Of course, this is what we observe. Further, departures from the expected should be correlated with events beyond the initial conditions specified by the parent or parents of the organism. We also observe this phenomenon. The second is phylogeny. Since phylogeny is an historical phenomenon it cannot be repeated like a physical experiment. However, the reconstruction of the genealogy should be repeatable using different characters given that we have established the correct boundary conditions and given the limitations of the characters.

6.5.3. Entropy changes

Viewing evolution as an irreversible entropic process solves an apparent paradox: how to achieve order on the macroscopic level represented by phylogeny from the relative chaos at the microscopic level represented by phenomena occurring within species. A problem remains: if evolution is an entropic phenomenon, how do we measure entropy changes? There are two approaches, measuring the dynamics of entropy change and measuring the history of entropy change.

Measuring the dynamics of entropy change amounts to quantifying the dS_i and dS_c parts of our general equations. We are not in a position to accomplish this at present. To do so we would have to catch a speciation event 'in the act' and, further, we would have to know a great deal more about development and mating systems than we know now. We are encouraged, however, by data such as Williamson's (1981). These data show an increase in variance followed by a decrease during lineage differentiation which conforms to the rise and fall of complexity our theory predicts.

Measuring the history of entropy state changes is less satisfactory than a direct measure. Nevertheless, such measures would permit us to establish the historical reality of entropy change and they would allow us to test the proposition that evolution tends to follow the principle of minimum entropy increase (see Saunders and Ho, 1981; Wiley and Brooks, 1982).

The extent to which evolution has followed a minimum entropy production principle can be measured by treating information changes as messages about the history of a group. The more reliable (=organized) a message about the history of a group the lower the entropy of the message relative to its maximum possible value and the easier it will be to pick a data set which will accurately tell us about the history (the phylogeny). Gatlin (1972) has pointed out that low entropy messages are characterized by increasing divergence from independence of the parts of the message ($D_2 = \text{minimal}$) while maintaining an *a priori* degree of divergence from equiprobability as constant as possible ($D_1 = k$). Brooks (1981) has pointed out that phylogenetic methods according to Hennig (1966) are capable of giving a measure of these parameters for historical information changes.

We may quantify D_1 and D_2 in a way that relative entropy production of information can be measured heuristically for different tree topologies of the same species. There is a finite number of ways any group of species (or populations) may be genealogically related. The possible genealogical topologies of three species forming a clade are seven, three dichotomies, one trichotomy, and three topologies involving ancestral species (Figure 3). However, there is only one true phylogeny. The best of all possible worlds would be a series of data sets having only homologies because such a data set, when analysed correctly, would favour only one of the seven trees. If particular characters are independently derived in two species the effect would be a data set with some characters favouring the correct solution (the synapomorphies) and some characters favouring one or more incorrect solutions (the homoplasies). In the worst of all possible worlds we would have a maximum entropy message with all trees having equal support. The result would be convergence on equiprobability which is represented by a trichotomy. We can calculate the extent to which a

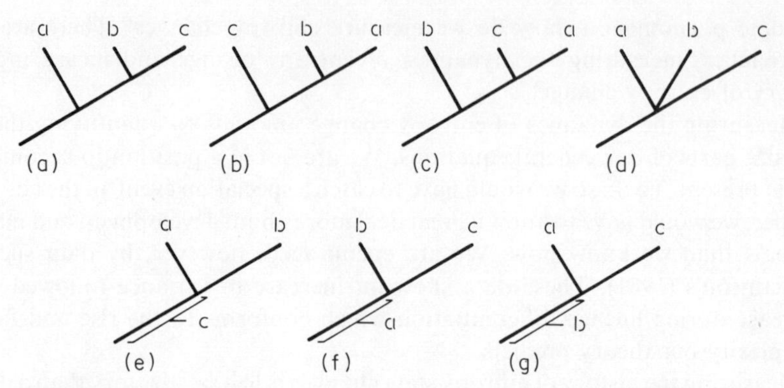

Figure 3. Seven possible phylogenetic trees for three species. (a)–(c) are the three dichotomous solutions given that no ancestor is present in the analysis; (d) would be the result if the characters analysed were converging on equiprobability or if an ancestor was present in the analysis; (e)–(f) are possible solutions to (d) given an ancestor in the analysis (redrawn from Wiley, 1981)

data set departs from minimum entropy production of characters by defining D_1 as:

$$D_1 = \frac{C_1}{\Sigma C_n}$$

where C_1 is the number of characters corroborating our 'best' tree and ΣC_n is the sum of characters corroborating the n dichotomous tree solutions for the taxa involved. Phylogeneticists (cf. Hennig, 1966; Eldredge and Cracraft, 1980; Nelson and Platnick, 1981; Wiley, 1980, 1981) and critics of phylogenetics (cf. Mayr, 1974) both agree that synapomorphies (shared derived characters) corroborate trees while symplesiomorphies and automorphies do not. Homoplasies (convergences, parallelisms, reversals) are recognized as such only because they do not fit properly on the 'best' tree. Thus homoplasies are characters which 'corroborate' trees other than the 'best' tree. The extent to which homoplasies are present will determine the extent to which D_1 is less than one. The 'best' tree can be defined by the D_2 criterion which is met by picking as 'best' the tree which maximizes the divergence from independence of characters. D_2 can be calculated for any one tree thus:

$$D_2 = \frac{H_n}{C_n + H_n}$$

where H_n is the number of homoplasies of the nth tree and C_n is the number of characters corroborating the nth tree.

We can see that $D_1 + D_2 = 1$ for all cases where $C_n > 0$. Further, when the data are converging on equiprobability D_1 approaches the reciprocal value of the number of possible dichotomous tree solutions for the problem at hand such that the value at equiprobability is $D_1 = 1/n$ where n equals the total number of dichotomous tree solutions for the taxa.

We may use slightly different measures of D_1 and D_2 to compare trees of different groups. To do so we must take into account the total historical entropy production of each group and not simply the extent to which the values are diverging from equiprobability. Specifically, we take into account autapomorphies (uniquely derived characters of terminal species, see Wiley, 1981). Such characters signal increases in information complexity without resulting increases in the order or disorder of the group. When we compare two different clades D_1 and D_2 may be defined as shown below.

$$D_1 = \frac{C_n}{C_n + H_n + A_n}$$

$$D_2 = \frac{H_n + A_n}{C_n + H_n + A_n}$$

where:

C_n = Synapomorphies corroborating the nth tree
H_n = homoplasies associated with the nth tree
A_n = autapomorphies for terminal species of the group.

and:

n = the 'best' tree of the alternatives as defined by our former D_1 criterion of minimum entropy configuration of information.

In this situation $D_1 + D_2 = 1$ even where $C_n = 0$ so long as at least one species has at least one autapomorphy. Thus an equiprobable answer which results from a lack of synapomorphies will produce a $D_1 + D_2 = 0$ result for within-group measurements but a $D_1 + D_2 = 1$ with $D_1 = 0$ for a between-group measurement.

The extent to which such estimates represent real values depends on sampling distributions. For a true quantitive measure we must assume that we have a random sample of changes that have occurred. Thus when comparing different clades it is important to use large samples of characters, or several different data sets derived from different classes of characters. In spite of the practical problems associated with sampling error (a ubiquitous problem for any quantitative measure), we feel that such measures are at least theoretically possible and serve to move our theory from the realm of qualitative thought to the realm of quantitative measure. For example, a statistical estimate of entropy changes during phylogenesis (Wiley and Brooks, 1983 adapted from Karreman, 1955) shows that historically constrained systems will minimize the amount of entropy increase through time. Thus, the principle of minimum entropy production is simply an expected outcome in systems having constraints on the range of possible changes inherent in their initial conditions for existence.

6.6. WHAT DOES THIS THEORY EXPLAIN?

A new theory is useful to the extent that it explains more phenomena than previous theories. We have previously suggested (Wiley and Brooks, 1982, 1983) that our theory explains three features of evolution not predicted by the current paradigm of neo-Darwinism: (a) why organisms are related in a hierarchial manner, (b) why allopatric speciation is to be expected *a priori* as the most common mode of speciation, and (c) how it is possible to have replicated biogeographic distributions among certain groups of organisms without involving dispersal explanations. We do not suggest that theories such as neo-Darwinism are incompatible with these features of evolution. But, there is a difference between not precluding certain phenomena from occurring and explaining them. In this section we shall address additional phenomena which may be explained by non-equilibrium evolution.

If individual organisms, species and/or populations act as individuals in an irreversible entropy-producing process they may be viewed as dissipative structures. Maintenance of a dissipative structure in the face of changing conditions involves the use of *alternate dissipative pathways* (Allen, 1981). If such a structure is perturbed beyond its ability to maintain itself it may either cease to function (die or become extinct), or fluctuate to a new ordered state(s) characterized by an interaction of new deterministic and stochastic factors. Prigogine (1980) referred to the latter as 'ordering through fluctuations'. The result is a historically unique hierarchy. A pertinent question to our theory is this: *If the evolutionary process produces more and more determinism as it proceeds, then why have not developmental programmes become so canalized that no mutations are successful?* In other words, why are species still variable?

The principle of compensatory change may provide a partial explanation. Processes occurring in a dissipative structure which increase the stochastic component of the system are accompanied by an increase in determinism in another part of the system if that system is to continue to maintain its individuality. The reverse is also true. If such a system is evolving in a manner that is making it more deterministic then it will generate its own variation to compensate for this increase in determinism. If it does not, evolution will shut down or be restricted to changes of a non-compensatory nature (and thus a non-regulatory nature). The new kind of variation that is generated may not be predictable *a priori* because any part of the system not affected by a new deterministic factor could compensate by becoming more stochastic.

Consider a regulatory gene situated on a particular chromosome. Following Kauffman (1983), this regulatory gene might influence structural genes proximal to it in the *cis* or *trans* position. If so, the effect of such a regulatory gene is an increase in determinism of the structural gene products. As Kauffman (1983) has shown, if the regulatory gene jumps to another chromosome or another position on the same chromosome it will exert a *cis* or *trans* influence on proximal structural genes. And, a series of such jumps will produce a canalized, highly ordered, system. However, what becomes of the structural genes left behind by such jumping? They would become more stochastic in their output.

Apparent examples of compensatory changes in living organisms have been known for a long time, but under a Darwinian paradigm such findings were anomalous. For example, Sturtevant (1915) posed the questions, 'Does crossing over in one chromosome have any effect upon the other chromosomes in the cell? Is there anything corresponding to interference taking place between different chromosomes?, four years later, Sturtevant (1919) discovered that the presence of an inversion on one chromosome of a homologous pair significantly loosened the linkage of genes on another pair. However, in the light of T. H. Morgan's findings supporting Mendel's law of independent assortment (see Morgan *et al.*, 1931, 1932, 1933; also Sutton, 1903; Carothers, 1913), little significance was attached to findings of the sort recorded by Sturtevant. Such 'interchromosomal effects' were persistent findings in genetics laboratories and, beginning in the late 1960s (see Lucchesi and Suzuki, 1968), interest was renewed in trying to explain the phenomena. Interchromosomal effects on recombination and such other phenomena as dosage compensation and hybrid dysgenesis (see Lucchesi, 1976, 1977; Suzuki, 1973; Sved, 1979; Yamamoto, 1979) have received intense scrutiny but an evolutionary significance has not been proposed for them. We consider such findings as evidence that the entire genome is an evolutionary entity which can tolerate only those changes for which it can compensate. Thus, the principle of compensatory changes is closely allied with the principle of intrinsic constraints.

At the organismic level, energy taken up by any individual can be dissipated into growth, differentiation, homeostasis, or reproduction. Consider a genetic programme which normally allocates a given amount of energy dissipation to differentiation. The instructions for taking up the energy responsible for differentiation precede the instructions for executing the transformation of energy into structure. If the differentiation programme is not followed completely, and a viable organism results, we would predict that it would exhibit enhanced attributes relating to one or more of the other dissipative pathways. This might explain why neotenic ambystomatid salamanders, such as individuals of *Ambystoma tigrinum*, grow much larger than their non-neotenic relatives within the species.

6.7. RESEARCH PROGRAMMES

The theory of non-equilibrium evolution asserts that character changes during descent are historically and developmentally constrained. Thus, characters should be good predictors of history, given that we can determine the boundary conditions of the system. The boundary conditions are determined by the information system of the ancestor of the taxa studied, Phylogenetic systematics

(Hennig, 1966; see Chapter 2) provides the research programme for recovering the history of descent because it is based on determining character transformations in reference to these historical boundary conditions. Such analyses provide the base line of character information necessary to pursue studies of relative entropy change, biogeography, and development.

Developmental biologists may provide data bearing on the validity of this theory from at least three different sources: (a) distinguishing between canalized and non-canalized information, (b) formulating a technique for measuring entropy changes during ontogeny, and (c) determining if there are any instances in which a state of decreased complexity (= lowered entropy) is achieved in a portion of an organism during ontogeny without a compensatory increase in complexity in another part. Much data exist supporting the notion that there are regulatory (canalized) and structural (non-canalized) portions of many genomes (Jacob and Monod, 1961; Britten and Davidson, 1969; Hedrick and McDonald, 1980). In addition, Maze and Scagel (1982) have suggested that plant morphogenesis is an entropic phenomenon which could potentially be mapped using information theoretic equations. Lewin (1982) has suggested that realization of such formulations is crucial to the theory.

Another promising research programme would involve a combination of phylogenetic analysis and developmental biology. For example, if ontogeny and phylogeny are linked through non-equilibrium evolution, we should expect that phylogenetic analysis applied to various cell types in a specialized cell lineage would provide a summary fate map indicating the various pathways of development. This would also indicate that ontogeny is an entropic phenomenon which minimizes entropy production, as suggested by Saunders and Ho (1981; see also Chapter 5). Prigogine and Wiame (1946) suggested a similar conclusion based on metabolic changes during an organism's life. Thus far, we know of only one attempt to map cell ontogeny in this manner (J. Caira, 1981, unpublished M.Sc. thesis, University of British Columbia); in that study the ontogeny of hemocytes of some insects was reconstructed using phylogenetic systematics and found to be in agreement with previously published developmental studies.

Hypotheses concerning the transformation of characters (i.e., transformation from primitive to derived) may provide hypotheses of character changes indicating directly the degree of developmental constraint in phylogenesis. These may be postulates about precursor stages, or 'building blocks', from ancestors which are necessary for the production of derivative characters seen in some species today (see Ho and Saunders, 1979; Saunders and Ho, 1981; Chapter 5). These hypotheses could be tested directly by interrupting the development of organisms representing species with the derivative trait, attempting to provoke expression of the primitive trait. Transformation series may also serve to pinpoint architectural constraints so pronounced that one might be able to predict the morphology of species not yet discovered (e.g., see Brooks, 1982b).

Some evolutionary views of ontogeny (e.g., Haeckel, 1866) explained the succession of stages in development as a succession of ancestral adult forms. Each episode of evolutionary change resulted in the formation of a new adult form, produced by natural selection, accreted onto the previous adult form. Natural selection could affect only the terminal stages. Later modifications of this view (see Maienschein, 1978 for a review) incorporated the notion that natural selection could affect all stages in ontogeny so that the true evolutionary sequence might be obscured. However, in neither case was the notion entertained that organized evolutionary divergence might result from non-terminal changes. Central to our theory is the confirmation that ontogenetic programmes are causal sequences, coherent units, and the units of evolutionary change. A combination of developmental biology and phylogenetics provides two types of research programmes for examining this question.

Developmental biology has provided much direct evidence that ontogeny is a causal sequence in which each derivative stage is a modification of its precursor stage (Løvtrup, 1974; Ho and Saunders, 1979; Saunders and Ho, 1981; Chapter 5). Many biologists have recognized that historical and developmental constraints are linked (e.g., Goldschmidt, 1940, 1952; Waddington, 1975; Seilacher, 1972; Riedl, 1977, 1978; Gould and Lewontin, 1979; Gould, 1980a,b; Lauder, 1982).

An approach to determining whether or not ontogeny is a causal sequence and is a unit of evolutionary change involves (a) searching for particular classes of character changes based on phylogenetic changes and (b) a type of congruence study. Our theory asserts that non-terminal ontogenetic changes may occur during the evolutionary process as well as terminal changes. We suggest that in fact such changes are not rare. For example, in Figure 4 the phylogenetic relationships of the major groups of parasitic flatworms are shown (from Brooks, 1982a). The Digenea (flukes) are distinguished as a real evolutionary group (clade, monophyletic group, see Hennig, 1966) in part by the presence of unique larval stages, sporocysts, rediae and cercariae, all non-terminal additions. Some groups have lost the redial stage. All of these findings represent non-terminal phenomena. However, by themselves, they are not sufficient grounds for viewing ontogenetic programmes in terms of non-equilibrium evolution.

If ontogenetic programmes are causal sequences and are the units of evolutionary change, we must find that non-terminal changes in ontogenetic programmes affect all subsequent stages in the programme. This second requirement would be met if it were discovered that cladograms based on data from different developmental stages (*semaphoronts* of Hennig, 1966) were all congruent, that is, they all predict the same tree. This would tell us that no matter what part of the ontogenetic programme were affected during a given evolutionary event, all character changes would still be correlated with the phylogenetic history of the group involved. This has been shown to be true for digenetic trematodes (Brooks, Glen and O'Grady, in preparation). Few other studies examining this question have been undertaken, although Hennig (1950, 1953,

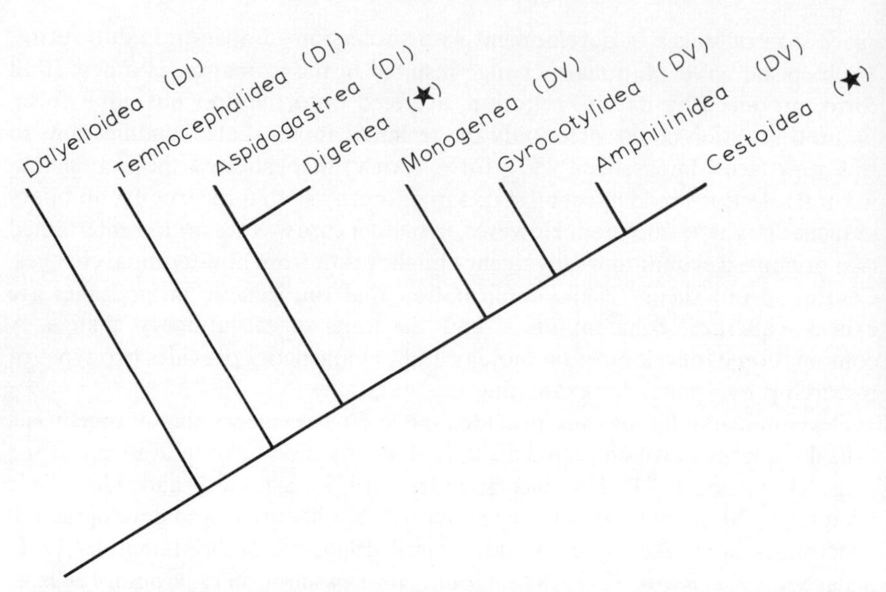

Figure 4. A phylogenetic hypothesis of the relationships among parasitic flatworms from Brooks (1982a) illustrating that non-terminal changes in ontogeny result in congruence with terminal changes in determining phylogenetic relationships. DI = direct life cycle with invertebrate host; DV = direct life cycle with vertebrate hosts; '*' = complex life cycle with both invertebrate and vertebrate hosts. For the full analysis see Brooks (1982a)

1969) and Howden (1982) examined phylogenetic classifications of various insect groups based on larval and adult characters, finding substantial agreement.

6.8. SUMMARY

We suggest that viewing evolution as an entropic phenomenon helps us approach the four general questions we posed in our introduction. There is organization in biological diversity because the process of modification with descent involves dissipative structures which fluctuate from one steady state to another. The organization we observe is hierarchical because the process involves the production of entropy in both information systems and cohesion systems. Particular organisms look the way they do because history constrains the future and because organisms must interact with their environments. Most species do not become static because placing constraints on one part of their information systems is compensated by a decrease of constraints on other parts of their information systems. Further, the history of change appears deterministic because variation is constrained by the deterministic factors in the information systems of organisms. Discovering the pattern of descent is necessary if we are to make further progress because the pattern tells us what we must explain.

ACKNOWLEDGEMENTS

This chapter was prepared during the tenure of Operating Grant No. A7696, Natural Sciences and Engineering Research Council of Canada to D. R. Brooks and NSF Grant DEB-8103532, National Science Foundation, USA to E. O. Wiley. Our thanks to both organizations for support. We thank Dr Bob Holt (University of Kansas) and Dr George Karreman (University of Pennsylvania) for discussing various aspects of measuring entropy.

6.9. REFERENCES

Alberch, P. (1980) Ontogenesis and morphological diversification. *Amer. Zool.*, **20**, 653–667.

Allen, P. M. (1981) The evolutionary paradigm of dissipative structures, in Jantsch, E. (ed.) *The Evolutionary Vision*. AAAS Selected Symposium No. 61. Westview Press, Boulder, Colo., pp. 25–72.

Beatty, J. (1982) Classes and cladists. *Syst. Zool.*, **31**, 25–34.

Birch, L. C., and Raven, P. R. (1967) Evolutionary history and population biology. *Nature*, **214**, 349–352.

Brillouin, L. (1962) *Science and Information Theory*. Academic Press, New York.

Britten, R. J., and Davidson, E. H. (1969) Gene regulation for higher cells: A theory, *Science*, **165**, 349–357.

Britten, R. J., and Davidson, E. H. (1971) Repetitive and non-repetitive DNA sequence and a speculation on the origins of evolutionary novelty, *Quart. Rev. Biol.*, **46**, 111–138.

Brooks, D. R. (1981) Classifications as languages of empirical comparative biology, in Funk, V. A., and Brooks, D. R. (eds.) *Advances in Cladistics: Proceedings of the First Meeting of the Willi Hennig Society*, New York Bot. Garden, N.Y., pp. 61–70.

Brooks, D. R. (1982a) Higher level classifications of parasitic platyhelminthes and fundamentals of cestode classification, in Mettrick, D. F., and Desser, S. S. (eds.) *Parasites—Their World and Ours*, Elsevier Biomedical, Amsterdam, pp. 189–193.

Brooks, D. R. (1982b) A simulations approach to discerning possible sister groups of *Dioecotaenia* Schmidt, 1969 (Cestoda: Tetraphyllidea: Dioecotaenidae). *Proc. Helminthol. Soc. Wash.*, **49**, 56–61.

Brooks, D. R., Glen, D. R., and O'Grady, R. T. (1984) Phylogenetic analysis of the *Digenea* Platyhelminthes: *Cercomeria*) with comments on their adaptive radiation. In preparation.

Brooks, D. R., Leblond, P. H., and Cummings, D. D. (1984) Information and entropy in a simple evolution model, *J. Theor. Biol.* (in press).

Bush, G. L. (1975) Modes of animal speciation, *Ann. Rev. Ecol. Syst.*, **6**, 339–364.

Carothers, E. E. (1913) The Mendelian ratio in relation to certain orthopteran chromosomes, *J. Morph.*, **24**, 487–511.

Charlesworth, B., Lande, R., and Slatkin, M. (1982) A neo-Darwinian commentary on macroevolution, **36**, 474–498.

Denbigh, K. G. (1975) A non-conserved function for organized systems, in Kubat, L., and Zemen, J. (eds.) *Entropy and Information in Science and Philosophy*, American Elsevier, New York, pp. 83–92.

Dobzhansky, Th., Ayala, F. J., Stebbins, G. L., and Valentine, J. W. (1977) *Evolution*, W. H. Freeman, San Francisco.

Dover, G. A. (1982a) A molecular drive through evolution, *BioScience*, **32**, 526–533.

Dover, G. A. (1982b) Molecular drive: A cohesive mode of species evolution, *Nature*, **299**, 111–117.

Dover, G. A., and Brown, S. M. (1981) The evolution of genomes in closely-related species, in Scudder, G. G. E., and Reveal, J. L. (eds.) *Evolution Today*, Hunt Inst. Bot. Documentation, Carnegie-Mellon Univ., Pittsburgh, pp. 337–349.

Eldredge, N., and Cracraft (1980) *Phylogenetic Patterns and the Evolutionary Process*, Columbia University Press, New York.

Eldredge, N., and Gould, S. T. (1972) Punctuated equilibria, an alternative to phyletic gradualism, in Schopf, T. (ed.) *Models in Paleobiology*, Freeman, Cooper and Co., San Francisco, pp. 82–115.

Endler, J. A. (1977) *Geographic Variation, Speciation and Clines*, Princeton Univ. Press, New Jersey.

Fisher, R. A. (1930) *The Genetical Theory of Natural Selection*, Clarendon Press, Oxford.

Frautschi, S. (1982) Entropy in an expanding universe, *Science*, **217**, 593–599.

Gatlin, L. (1972) *Information Theory and the Living System*, Columbia University Press, New York.

Ghiselin, M. T. (1966) On psychologism in the logic of taxonomic controversies, *Syst. Zool.*, **15**, 207–215.

Ghiselin, M. T. (1974) A radical solution to the species problem, *Syst. Zool.*, **23**, 536–544.

Ghiselin, M. T. (1980) Natural kinds and literary accomplishments, *The Michigan Quart. Rev.*, **19**, 73–88.

Ghiselin, M. T. (1981) Categories, life and thinking, *The Behavioral and Brain. Sci.*, **4**, 269–313 (with peer comments).

Goldschmidt, R. B. (1940) *The Material Basis of Evolution*. Yale Univ. Press, New Haven, Connecticut.

Goldschmidt, R. B. (1952) Evolution as viewed by one geneticist, *Amer. Sci.*, **40**, 84–98.

Gould, S. J. (1980a) Is a new and general theory of evolution emerging? *Paleobiol.*, **6**, 119–130.

Gould, S. J. (1980b) The evolutionary biology of constraint, *Daedalus*, **109**, 39–52.

Gould, S. J., and Lewontin, R. C. (1979) The spandrels of San Marco and the Panglossian paradigm: a critique of the adaptationist programme, *Proc. R. Soc. London B*, **205**, 581–598.

Grant, V. (1971) *Plant Speciation*, Columbia Univ. Press, New York.

Haeckel, E. (1866) *Generelle Morphologie der Organismen*, G. Reimer, Berlin.

Haldane, J. B. S. (1932) *The Causes of Evolution*, Harper, New York.

Hedrick, P. W., and McDonald, J. F. (1980) Regulatory gene adaptation: an evolutionary model, *Heredity*, **45**, 85–99.

Hempel, C. G. (1965) *Philosophy of Natural Science*, Prentice-Hall, New Jersey.

Hennig, W. (1950) *Die Larvenformen der Dipteran*. Eine Uberricht uber die bisher bekannten Jugendstudien der Zweiflugeligen Insekten. Verlag, Berlin.

Hennig, W. (1953) Kritische Bemerkungen zum phylogenetischen System der Insekten, *Beitr. Ent.*, **3**, 1–85.

Hennig, W. (1966) *Phylogenetic Systematics*, Univ. Illinois Press, Urbana.

Hennig, W. (1969) *Die Stammesgeschichte der Insekten*. W. Kramer, Frankfurt am Main.

Ho, M. W., and Saunders, P. T. (1979) Beyond neo-Darwinism—an epigenetic approach to evolution, *J. Theor. Biol.*, **78**, 575–591.

Hollinger, H. B., and Zenzen, M. J. (1982) An interpretation of macroscopic irreversibility within the Newtonian framework, *Philo. Sci.*, **49**, 309–354.

Howden, H. F. (1982) Larval and adult characters of *Frickius* Germain, its relationship to the Geotrupini, and a phylogeny of some major taxa in the Scarabaeoidea (Insecta: Coleoptera), *Can. J. Zool.*, **60**, 2713–2724.

Hull, D. L. (1976) Are species really individuals? *Syst. Zool.*, **25**, 174–191.

Hull, D. L. (1978) A matter of individuality, *Philo. Sci.*, **45**, 335–360.

Jacob, F., and Monod, J. (1961) Genetic regulatory mechanisms in the synthesis of proteins, *J. Mol. Biol.*, **2**, 318–356.

Jantsch, E. (ed.) (1981) *The Evolutionary Vision*. AAAS Selected Symposium No. 61. Westview Press, Boulder, Colorado.

Karreman, G. (1955) Topological information content and chemical reactions, *Bull. Math. Biophysics*, **17**, 279–285.

Kauffman, S. (1983) The evolution of metazoan gene regulation, in Polking, J. (ed.) *Non-linear Problems in Science*, Rice University Studies, Texas.

Kuhn, T. S. (1970) *The Structure of Scientific Revolutions*. 2nd Edn., University of Chicago Press, Chicago.

Lauder, G. V. (1982) Historical biology and the problem of design, *J. Theor. Biol.*, **97**, 57–68.

Layzer, D. (1975) The arrow of time, *Sci. Amer.*, **223**, 56–69.

Lewin, R. S. (1982) A downward slope to greater diversity, *Science*, **217**, 1239–1240.

Løvtrup, S. (1974) *Epigenetics, a Treatise on Theoretical Biology*, Wiley–Interscience, New York.

Lucchesi, J. C. (1976) Interchromosomal effects, in Ashburner, M., and Novitski, E. (eds.) *The Genetics and Biology of Drosophila melanogaster*, Academic Press, New York. Vol. 10, pp. 315–329.

Lucchesi, J. C. (1977) Dosage compensation: transcription-level regulation of x-linked genes in *Drosophila*, *Amer. Zool.*, **17**, 685–693.

Lucchesi, J. C., and Suzuki, D. T. (1968) The interchromosomal control of recombination, *Ann. Rev. Genetics*, **2**, 53–86.

Maienschein, J. (1978) Cell lineages, ancestral reminiscence and the biogenetic law, *J. Hist. Biol.*, **11**, 129–158.

Mayr, E. (1963) *Animal Species and Evolution*, Belknap Press, Cambridge, Massachusetts.

Mayr, E. (1974) Cladistic analysis or cladistic classification? *Zeit. Zool. Syst. Evolut.-Forsch.*, **12**, 95–128.

Maze, J., and Scagel, R. K. (1982) Morphogenesis of the spikelets and inflorescence of *Andropogon gerardii* Vit. (Gramineae) and the relationship between form, information theory and thermodynamics, *Can. J. Bot.*, **60**, 806–817.

Morgan, T. H., Bridges, C. B., and Schultz, J. (1931) Constitution of the germinal material in relation to heredity, *Carnegie Inst., Washington, D. C., Yearbook*, **30**, 408–415.

Morgan, T. H., Bridges, C. B., and Schultz, J. (1932) Constitution of the germinal material in relation to heredity, *Carnegie Inst., Washington, D.C., Yearbook*, **31**, 303–307.

Morgan, T. H., Bridges, C. B., and Schultz, J. (1933) Constitution of the germinal material in relation to heredity, *Carnegie Inst., Washington, D. C., Yearbook*, **32**, 298–302.

Moroshowitz, A. (1968a) Entropy and the complexity of graphs: I. An index to the relative complexity of a graph, *Bull. Math. Biophysics*, **30**, 175–204.

Moroshowitz, A. (1968b) Entropy and the complexity of graphs: II. The information content of digraphs and finite graphs, *Bull. Math. Biophysics*, **30**, 225–270.

Moroshowitz, A. (1968c) Entropy and the complexity of graphs: III. Graphs with prescribed information content, *Bull. Math. Biophysics*, **30**, 387–414.

Nelson, G., and Platnick, N. (1981) *Systematics and Biogeography*, Columbia University Press, New York.

Ohno, S. (1970) *Evolution by Gene Duplication*, Springer Verlag, Berlin.
Popper, K. R. (1964) *The Poverty of Historicism*, Harper and Row, New York.
Popper, K. (1968) *The Logic of Scientific Discovery*, Harper and Row, New York.
Prigogine, I. (1961) *Introduction to Thermodynamics of Irreversible Processes*, Interscience Publ., New York.
Prigogine, I. (1980) *From Being to Becoming*, W. H. Freeman, San Francisco.
Prigogine, I., Nicolis, G., and Babloyantz, A. (1972) Thermodynamics of evolution, *Physics Today*, **25**(11), 23–28; **25**(12), 38–44.
Prigogine, I., and Wiame, J. M. (1946) Biologie et thermodynamique des phénomènes irréversibles, *Experimentia*, **2**, 451–453.
Rachootin, S. P., and Thompson, K. S. (1981) Epigenetics, paleontology and evolution, in Scudder, G. G. E., and Reveal, J. L. (eds.) *Evolution Today*, Hunt Inst. Bot. Doc., Carnegie-Mellon Univ., Pittsburgh, pp. 181–193.
Rensch, B. (1959) *Evolution above the Species Level*, Columbia University Press, New York.
Riedl, R. (1977) A systems-analytical approach to macro-evolutionary phenomena, *Q. Rev. Biol.*, **52**, 351–370.
Riedl, R. (1978) *Order in Living Organisms*, Wiley–Interscience, New York.
Saunders, P. T., and Ho, M. W. (1981) On the increase in complexity in evolution. II. The relativity of complexity and the principle of minimum increase, *J. Theor. Biol.*, **90**, 515–530.
Seilacher, A. (1972) Divariate patterns in pelecypod shells, *Lethaia*, **5**, 325–343.
Stebbins, G. L. (1950) *Variation and Evolution in Plants*, Columbia Univ. Press, New York.
Stebbins, G. L. (1971) *Chromosomal Evolution in Higher Plants*, Addison-Wesley, Reading Massachusetts.
Stebbins, G. L., and Ayala, F. J. (1981) Is a new evolutionary synthesis necessary? *Science*, **213**, 967–971.
Sturtevant, A. H. (1915) The behavior of the chromosomes as studied through linkage, *Zeitsch. Induktive Abstammung Verebungslehre*, **13**, 234–287.
Sturtevant, A. H. (1919) Contributions to the genetics of *Drosophila melanogester*. III. Inherited linkage variations in the second chromosome, *Carnegie Inst., Washington, D.C., Publ.*, **278**, 305–341.
Sutton, W. S. (1903) The chromosomes in heredity, *Biol. Bull.*, **4**, 231–251.
Suzuki, D. T. (1973) Genetic analysis of crossing-over and its relation to chromosome structure and function in *Drosophila melanogaster*, in Bogard, R. (ed.) *Genetic Lectures*, Oregon State University Press, Corvallis. Vol. 3, pp. 7–32.
Sved, J. A. (1979) The 'hybrid dysgenesis' syndrome in *Drosophila melanogaster*, *BioScience*, **29**, 659–664.
Templeton, A. R. (1981) Mechanisms of speciation—a population genetic approach, *Ann. Rev. Ecol. Syst.*, **12**, 33–48.
Van Valen, L. (1974) A natural model for the origin of some higher taxa, *J. Herpetol.*, **8**, 109–121.
Waddington, C. H. (1975) *The Evolution of an Evolutionist*, Cornell University Press, Ithaca, N.Y.
White, M. J . D. (1973) *Animal Cytology and evolution*, Cambridge University Press, Cambridge.
White, M. J. D. (1978) *Modes of Speciation*, W. H. Freeman and Co., San Francisco.
Wicken, J. S. (1979) The generation of complexity in evolution: A thermodynamic and information-theoretical discussion, *J. Theor. Biol.*, **77**, 349–365.
Wicken, J. S. (1980) A thermodynamic theory of evolution, *J. Theor. Biol.*, **87**, 9–23.

Wiley, E. O. (1978) The evolutionary species concept reconsidered, *Syst. Zool.*, **27**, 17–26.
Wiley, E. O. (1979) Cladograms and phylogenetic trees, *Syst. Zool.*, **28**, 88–92.
Wiley, E. O. (1980) Phylogenetic systematics and vicariance biogeography, *Syst. Bot.*, **5**, 194–220.
Wiley, E. O. (1981) *Phylogenetics. The Theory and Practice of Phylogenetic Systematics.* Wiley–Interscience, New York.
Wiley, E. O., and Brooks, D. R. (1982) Victims of history—a nonequilibrium approach to evolution, *Syst. Zool.*, **31**, 1–24.
Wiley, E. O., and Brooks, D. R. (1983) Nonequilibrium thermodynamics and evolution: a response to Løvtrup, *Syst. Zool.*, **32**, 209–219.
Williamson, P. G. (1981) Palaeontological documentation of speciation in Cenozoic molluscs from Turkana Basin, *Nature*, **293**, 437–443.
Wright, S. (1931) Evolution in Mendelian populations, *Genetics*, **16**, 97–159.
Yamamoto, M. (1979) Interchromosomal effects of heterochromatic deletions on recombination in *Drosophila melanogaster*, *Genetics*, **93**, 437–448.

Evolutionary Theory: Paths into the Future
Edited by J. W. Pollard

Chapter 7

Movable genetic elements and evolution

HOWARD M. TEMIN AND WILLIAM ENGELS
Department of Oncology,
McArdle Laboratory and Department of Genetics,
University of Wisconsin,
Madison, Wisconsin 53706,
USA

7.1. INTRODUCTION

The structure of the eukaryotic genome is more complex and contains more mechanisms for variation than previously perceived. Recombinant DNA technology, DNA sequencing, and other new techniques have given us much new and surprising information. Even many classical genes, that is, controlling sequences and coding sequences, have been found to be complex and composed of exons and introns.

We shall first very briefly describe these complexities as an introduction to the concept of movable genetic elements.

At a DNA sequence level, the genome can be separated into non-transcribed and transcribed sequences. In non-transcribed sequences, near the boundaries with transcribed sequences there may be sequences that control transcription of RNA. Transcribed RNA can be processed and/or become a stable RNA. Stable RNA can be messenger RNA or not translated. Messenger RNA is composed of coding and non-coding sequences.

The genome can undergo variation by base pair mutations (that is, transitions, transversions, small deletions, and small duplications), formation of larger tandem duplications, chromosome rearrangements, translocations, duplications, or deletions, and transpositions. Transposition represents a form of genetic variation in which sequences are duplicated in the genome at a location separated from the original sequences. Since this process can be repeated, transposed DNA can be present in one to several hundred thousand copies per haploid genome. (In theory, the parent of a transposed sequence could be deleted and so only one copy of a transposed sequence remain per haploid genome. It would be difficult, but probably not impossible, to distinguish such DNA from

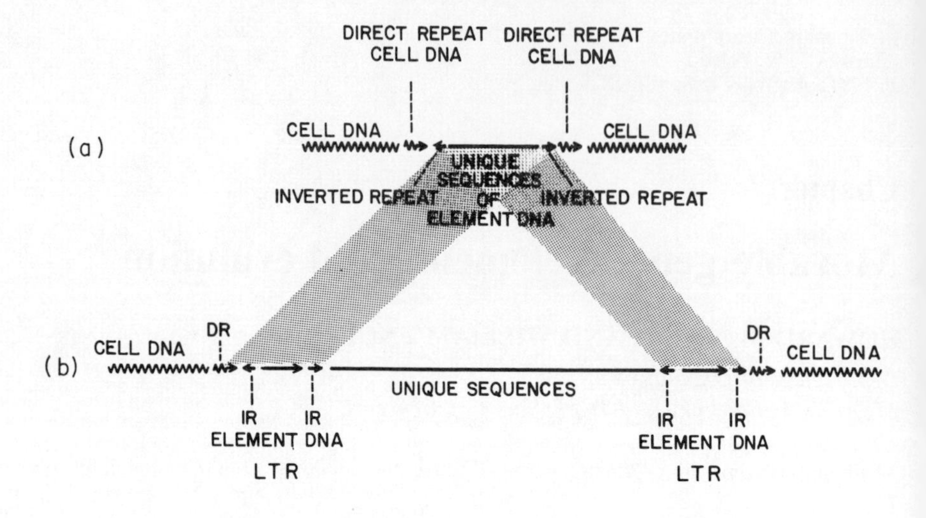

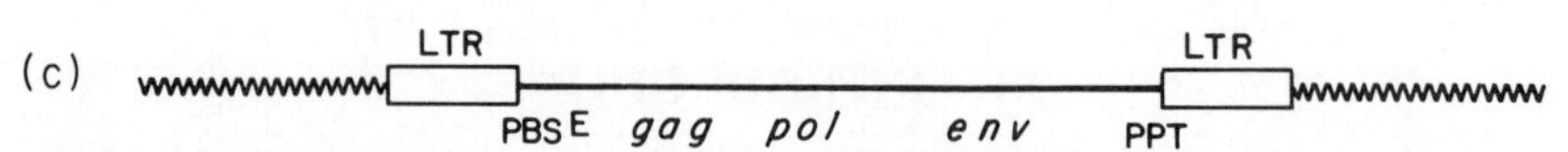

Figure 1. Structures of movable genetic elements.

Symbols: zig-zag line, cellular DNA; straight line, element DNA; arrows in the same direction and the same size (DR), direct repeats; arrows in the opposite direction and the same size (IR), inverted repeats; LTR, long terminal repeat; PBS, primer binding site; E, encapsidation sequences; *gag*, *pol*, *env*, genes for virion proteins; PPT, polypurine tract.

(a) Insertion sequence or transposon with no large terminal repeats.
(b) Transposon with large terminal repeats.
(c) Retrovirus provirus

TIME O ··········· G T C G [T C A T C] G T C G A T ·······

TIME I ··· G T C G [T C A T C] **TG** · · · · · · **CA** [T C A T C] G T C G A T ·····

CELL DNA DR ELEMENT DNA DR CELL DNA

Figure 2. Formation of direct repeat of cell DNA upon integration of movable genetic element or retrovirus provirus.

One strand of cell DNA is shown as well as the terminal dinucleotides of the element (in bold type). The boxed sequence at time 0 is that duplicated upon integration to form a direct repeat (DR) unique DNA that had never been transposed.) This reiterated DNA differs from tandemly duplicated DNA by being dispersed in the genome.

The nucleotide sequences of small circular viruses do not contain such repeated sequences. In fact, such nucleotide sequences often have multiple meaning, that is, they contain overlapping genes. The sequences of small and medium sized linear viruses also do not contain repeats, with the exception sometimes of small repeats at their ends.

Larger viruses and bacteria have some repeated nucleotide sequences which are not at their ends. These repeated sequences can be tandem, that is, next to each other, or dispersed. The dispersed repeated nucleotide sequences in larger viruses and bacteria often provide a means of amplification of essential genes, for example, ribosomal genes in bacteria. In addition, there are dispersed repeated DNA sequences in bacteria that (a) do not contain essential genes, (b) are present at different locations in the cell genome in different cells, and (c) may be entirely absent from individual cells. Such dispersed, non-essential DNA sequences are movable genetic elements.

In bacteria, movable genetic elements are of two major types: insertion sequences and transposons (Calos and Miller, 1980; Kleckner, 1977). Insertion sequences are small (usually less than 1 kbp) and code for no proteins or only code for proteins involved in their transposition (movement to a new place in the bacterial DNA). By contrast, most transposons are formed of two insertion sequences flanking some additional coding sequences. These additional coding sequences can code for proteins involved in transposition or in other functions, for example, antibiotic resistance. There are also derivatives of these transposons that have only one insertion sequence or even, apparently, no insertion sequences flanking the coding sequences (Machida *et al.*, 1982). Thus, there has been apparent evolution of movable genetic elements from insertion sequences to transposons (with two flanking insertion sequences) to transposons with one insertion sequence to transposons with no insertion sequences.

Some bacterial viruses, λ and Mu, have some characteristics similar to movable genetic elements (Campbell, 1979). Thus, there may have been evolution in prokaryotes from cellular movable genetic elements to certain viruses. In eukaryotes a similar phenomenon has apparently occurred with retroviruses, a family of RNA viruses of vertebrates (see below).

All insertion sequences and transposons have inverted repeats at their ends and make a direct repeat of cell DNA upon transposition (see Figures 1 and 2). These two properties define movable genetic elements and can be used to distinguish them from other possible types of dispersed repeated DNA.

There is much repeated DNA in most species, up to almost 80% of the genome in some species of eukaryote. This repeated DNA is found both as tandem repeats and dispersed throughout the genome (see Singer, 1982; Bouchard, 1982 for reviews). In this article we shall discuss the dispersed repeated sequences that appear to be movable genetic elements. These elements appear to be most (or all) of the dispersed repeated sequences in mammals, *Drosophila*, and yeast. However, there may be other kinds of dispersed repeated DNA in addition in other organisms, but critical evidence is not yet available to us. Our guess is that there will not be a completely different form of element.

Movable genetic elements affect cells and organisms by increasing variation, by duplicating coding sequences, by inserting positive and negative control sequences, and by enabling some sequences to spread both vertically and horizontally through and between populations. In addition, they alter the overall structure of the genome DNA and produce gene products that interact with cellular processes. Thus, they provide a mechanism that could have large effects on evolution.

7.2. OVERVIEW OF MOVABLE GENETIC ELEMENTS

Movable genetic elements were first recognized by McClintock in maize by their genetic effects which were inconsistent with previous models of mutation. These effects have been summarized as: creating somatic instability in chromosomes; inserting into coding or regulating sequences; adding new control sequences; and changing gene activity at later times (Shapiro and Cordell, 1982).

Later a new class of polar mutations was recognized in bacteria. Genetic analysis indicated they had similar properties to McClintock's elements.

Nucleic acid hybridization techniques then led to the recognition of multiple families of repetitive DNA sequences. Finally, analysis by DNA cloning and nucleotide sequencing has enabled a fuller classification of the types of dispersed repetitive elements in the genome. The different types of elements differ in their relationship to coding sequences (more usual genes), in their sequence organization, in their frequency, and in their mechanism of transposition.

First, there are functional dispersed duplications of coding sequences. These repeated sequences may be coding the same product as the original gene or have evolved to code for different products. Ribosomal RNA genes in *E. Coli* are examples of the first type of gene; actin genes of the second. Second, there are non-functional dispersed duplications of coding sequences—the various pseudogenes. The α-globin pseudogene of the mouse is an example of this type. Third, there are short (less than 500 bp) sequences present in high copy number in some eukaryotic genomes and coding for RNA and pseudogenes of these genes. The RNA can be found as a discrete molecule of the size of the repeated sequence, for example, small nuclear RNAs, or as part of larger molecules, for example, Alu sequences in heterogeneous nuclear RNA. Fourth, there are larger dispersed repeated sequences that code for proteins, for example, *copia* and P factor in *Drosophila* and TYl in yeast, and are analogous to bacterial insertion sequences and transposons and their defective derivatives.

It is likely that only the fourth type of dispersed element codes for proteins involved in transposition. The other types of repeated sequences apparently transposed using proteins coded for by single copy cellular genes or by this fourth type of dispersed repeated element.

There appear to be at least two distinct modes of transposition of these elements differing in whether there is an RNA or a DNA intermediate. The evidence for a DNA intermediate in transposition is most compelling for some bacterial transposons that form a cointegrate structure (two DNA copies of the element) as an intermediate in transposition, for example Tn3 and $\gamma\delta$ (see Galas and Chandler, 1981 for a recent discussion of mechanisms). The evidence for an RNA intermediate in transposition is conclusive for retroviruses and substantial for cDNA pseudogenes (see below).

In transposition with either DNA or RNA intermediates, a short unique region of the recipient DNA is duplicated around the transposed element (Figure 2). Presumably, this duplication is formed by a nuclease making staggered single strand cuts in the recipient DNA, like some restriction endonucleases or λ phage *int* protein, followed by ligation of the movable element, and DNA synthesis to fill in both single-stranded ends. The staggered cuts indicate possible evolutionary relationship of cellular movable elements and λ phage (as mentioned above).

All dispersed repeated elements of the type that code for proteins also have an inverted repeat at the ends of their sequence (Figure 1). Thus, the direct repeat in cell DNA is attached to the inverted repeat of element DNA. Presumably the inverted repeat serves to mark the end of the element for recognition by a protein, presumably another nuclease activity.

These descriptions are primarily relevant to the cell genome and the repeated sequences in it. It is also possible to hypothesize an evolution of the movable genetic elements themselves. This evolution is driven by selection of variant repeated sequences better able to replicate relative to other sequences in the genome and able to move from one position in the genome to another. It can be hypothesized that originally such variant sequences did not have their own coding sequences and transposed using products of other genes. However, once the variant sequences were able to code for proteins involved in their transposition, evolution would have been more rapid. A later stage in this evolution may be represented by some viruses that can be described as movable genetic elements with an extracellular phase. For example, Mu-phage and retroviruses are viruses that have integration with host DNA as an essential part of their life cycle. Thus they can be said to alternate between being a virus and being a cellular movable genetic element. Alternatively, they might remain in the genome and thus return to being a cellular movable genetic element.

Other viruses that do not have to integrate with host DNA in their normal replication may represent still further evolution of movable genetic elements (Temin, 1976; 1980; Campbell, 1979). λ-Phage is one example. Another is hepatitis B virus. For example, hepatitis B viruses are DNA viruses that replicate through an RNA intermediate and do not integrate in their replicative cycle

(Summers and Mason, 1982). Cauliflower mosaic virus may replicate in a similar fashion (Pfeiffer and Hohn, 1983). Therefore, the latter two can be considered *para*-retroviruses.

7.3. SOME MOVABLE GENETIC ELEMENTS

7.3.1. Structure of *Drosophila* transposons

In *Drosophila melanogaster* it is estimated that 10%–20% of the genome comprises moderately repetitive sequences. These sequences are mostly present as five to 1,000 single copies scattered throughout the genome in positions that vary widely among individuals. These repeated elements are highly heterogeneous in size, structure, and copy number, but those that have been studied in detail fall into four major categories. A brief description of the structure of each category is given below; see Rubin (1982) for a more detailed review.

7.3.1.1. Description of elements

7.3.1.1.1. Copia-like elements

The *copia* sequence is 5.2 kbp long with direct repeats of 276 bp at its termini. Each long terminal repeat (LTR) is itself flanked by a short inverted 17 bp repeat of imperfect homology (Finnegan *et al.*, 1978; Georgiev *et al.*, 1981; Levis *et al.*, 1980). Any element whose structure is analogous to *copia* but whose sequence is not necessarily homologous to *copia* is known as a 'copia-like' transposon (Rubin *et al.*, 1981). These elements might also be called 'retrovirus-like' because of the strong structural and behavioural similarities (Shiba and Saigo, 1983, Kugimiya *et al.*, 1983).

The copia-like sequences fall neatly into non-homologous 'families' within which there is very little nucleotide sequence variation. Seven such families including *copia* itself have been described in detail, and there are probably many others. They range from 5 kbp to 8.5 kbp and the copy number is usually between ten and 100 per haploid genome. Each family has a characteristic average copy number, but the actual numbers of copies vary greatly from one individual to another. Upon insertion, each copia-like transposon produces a repeat of three to five bases of the genomic DNA at the site of integration, the length of this repeat being characteristic of the family. There is a single observation of the LTR of a copia-like element ('gypsy', Modolell *et al.*, 1983) being separated from the rest of the element following an excision event, but otherwise the copia-like elements have been seen only as complete units in the genome.

7.3.1.1.2. Foldback elements

Another class of transposons called *foldback* (FB) elements has size and distribution characteristics of the same order as the copia-like elements, but their structure is very different (Potter *et al.*, 1980; Truett *et al.*, 1981). FB elements have long terminal repeats, but they are in reverse orientation rather than direct, and in most cases these repeats compose the entire element. Furthermore, the FB repeats themselves contain several long series of short direct repeats ranging in size from 10 bp to 31 bp with some unique DNA interspersed among the repeats. The copia-like LTRs, in contrast, are entirely unique DNA. In some cases, this inversely repeated region is much longer on one side of an FB element than on the other. Insertion of an FB element results in duplication of 9 genomic bp at the integration site. To date, only one homologous family of FB elements has been described, but there is much more variability within this family than in any of the copia-like families.

7.3.1.1.3. P elements

P elements (reviewed by Engels, 1984) occur at 30–50 scattered and variable genomic positions in most *D. melanogaster* strains recently derived from wild populations (P strains), but they are completely missing in most long-established laboratory stocks (M strains) (Kidwell *et al.*, 1977; Engels, 1981a; Bingham *et al.*, 1982). (Genomes with an intermediate number of P elements can, of course, be constructed by crosses between P and M strains, but as we will discuss below this genome is apparently not in a stable condition.) O'Hare and Rubin (1983) find that approximately one-third of the P elements in one strain tested have a sequence of 2,907 bp with 31 bp inverted terminal repeats. The rest of the P elements range from 500 bp to 2,000 bp and have sequences similar to the 2,907 elements except for an internal deletion. These deletions are highly variable in length and position, but they never overlap the 31 bp repeat. Their breakpoints sometimes coincide with duplications or triplications of a few bp in the 2,907 bp sequence. P element insertion events yield an eight bp duplication of genomic DNA at the integration site.

7.3.1.1.4. Clusters of smaller elements

Wensink *et al.* (1979) observed 17 families of smaller transposable elements each shorter than 1 kb. These elements are present in copy numbers exceeding 100 at scattered locations, frequently in heterochromatin. Unlike the elements described above, these elements tend to occur in 'scrambled clusters' with different combinations of individual units arranged in various permutations. The clusters studied by Wensink *et al.* (1979) had an average of 13 elements each, but the structure of individual unit elements was not determined.

7.3.1.2. Transcription

The *copia* element, so named for the copiousness of its cellular RNA transcripts, accounts for several per cent of all the polyadenylated RNA in some cell culture lines, and other copia-like elements are also transcriptionally active (Finnegan *et al.*, 1978). Some of the transcripts have been shown to be translatable messages, but the vast majority appears to be stable, full-length copies of the elements (Flavell *et al.*, 1980; Georgiev *et al.*, 1981; Schwartz *et al.*, 1982). In this respect, copia-like elements resemble retrovirus proviruses (see below).

Several species of P-factor transcripts have been found by R. Karess (personal communication), but none is long enough to represent the entire element. Comparisons of different *Drosophila* crosses showed that the presence/absence of these RNAs does not coincide with the P factors' active/repressed states (see below) as would be expected if these transcripts were related to P factor activity. One way to reconcile these observations is to note that the observed RNA species come from whole embryos and thus primarily from somatic cells whereas the P factor's transpositional activity is primarily in the germline.

7.3.1.3. Transposition

The ability of copia-like, FB, and cluster elements to transpose can be inferred from their dispersed positions and also from the variability of these positions in different genomes. Some details of the transposition events have been obtained by comparing the flanking DNA sequences in genomes with and without the elements (Rubin *et al.*, 1981). However, direct studies of the transposition of these elements has been difficult because of the apparently low rate of such events.

P elements, however, can be considerably more active in transposition than the other kinds of elements. They have both an inactive state where the transposition rate is of the same order as that of the other classes of elements and also an active state where they transpose at much higher frequencies. In their inactive state they can normally be observed only by molecular methods, but in their active state they cause a syndrome of germline abnormalities including frequent production of mutations and chromosome rearrangements. These traits are collectively known as 'hybrid dysgenesis' because they were first observed in the hybrids from crosses of P-strain males with M-strain females (Kidwell *et al.*, 1977). Subsequent work suggested that there is an element-encoded repressor involved with this regulation and also an element-encoded transposase present in the active state.

7.3.1.3.1. The P element repressor

The inheritance of the repressed state, known as the P cytotype, is unusual; it has characteristics of both Mendelian and cytoplasmic transmission (Engels, 1979a). Cytotype is easily monitored by observing the mutation rate of sn^w, a P-element insertion at the *singed* bristle locus. When P transposition is active, this mutation reverts or mutates to other allelic forms at rates of up to 60% per generation (Engels, 1979b). Studies of sn^w mutability and other dysgenic traits have shown that the Mendelian components of cytotype are dispersed, polygenic, and

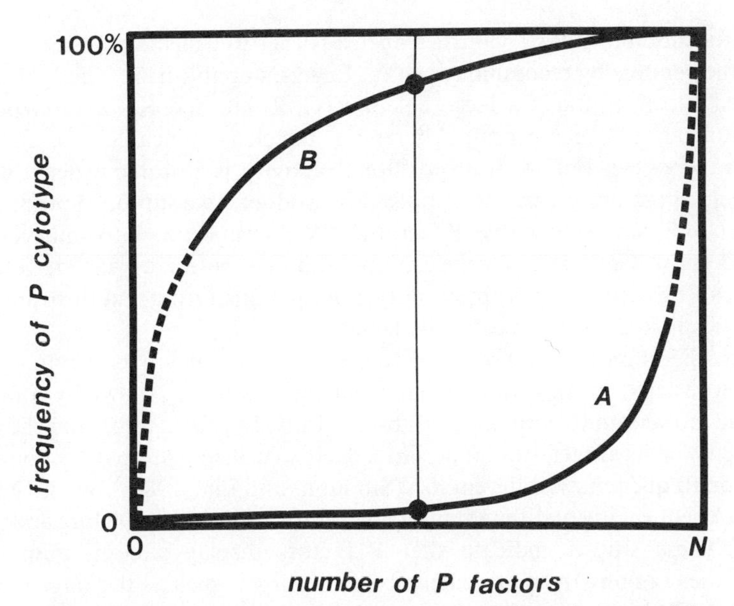

Figure 3. Rules of inheritance of cytotype for P factor in *Drosophila melanogaster*.

Frequency of the P (suppressed) cytotype is shown as a function of the number of P factors in the genome (Engels, 1979a). *N* is the total number of P factors in a given P strain (diploid) genome that are capable of influencing cytotype. These represent some subset of the 60–100 elements normally present. Curve A applies to individuals whose mother had the M (non-suppressed) cytotype and curve B is for those with P cytotype mothers. The two reciprocal types of first generation hybrids between P and M strains are indicated by dots. Hypothetical regions of the curves for which adequate data are not available are shown as dashed lines

presumably carried by the P elements themselves. The extrachromosomal components are identified by an 'inertia' of the repressed state through the female line over several generations (Engels, 1979a,b, 1981b; Kidwell, 1981; Engels and Preston, 1981b). Thus, as illustrated in Figure 3, the cytotype of a given individual appears to depend not only on the number of P elements in his own genome, but also on the cytotype of his mother. One possibility is that the P cytotype represents the presence of a P-encoded repressor, whereas the active state (M cytotype) is the absence of that repressor. If so, the female-line inertia indicates that the production of this repressor is under a complex type of control.

7.3.1.3.2. The P transposase

The existence of a P-encoded transposase is indicated by the ability of P factors to act as 'helper' elements to allow the transposition of defective P elements.

Thus, the mutability of *sn*^{w} vanishes when all other P elements have been removed from the genome by recombination (W. Engels, unpublished). The P element(s) inserted at *sn*^{w} contains a large deletion (G. Rubin and K. O'Hare, personal communication) which presumably renders the element incapable of making its own transposase, but its transpositional activity is restored when there are complete P factors present to supply this product. In addition, Spradling and Rubin (1982) have shown that P factor DNA after injection into embryos of the M cytotype will integrate into the genome, and that these newly inserted elements are capable of activating *sn*^{w} mutability, thus demonstrating that the transposase activity is linked to the injected P element.

Two other important aspects of P factor transposition are its strong tendency to be more active in the germline than in somatic tissues, as shown by comparing somatic mosaicism to germline mutability of *sn*^{w} (Engels, 1979b), and the partial specificity of its target sites. The latter has been demonstrated by comparing insertion frequencies at different loci (Simmons and Lim, 1980; Green, 1977) and by DNA sequencing of different insertions at the *white* locus (O'Hare and Rubin, 1983). These studies indicate that P factors display a continuum of site specificities ranging from extremely frequent targets such as the *singed* locus to loci where P elements rarely if ever insert. The *white* locus is intermediate in this continuum, but three of four independent insertions there occurred at the same nucleotide pair.

7.3.1.4. Distributions within and between species and higher taxa

The transposable elements of Drosophila vary in both genomic positions and in total numbers between individuals of a species. This variability is most pronounced for P factors, where individuals from P strains can have many copies of the element and individuals from M strains have none. With copia-like elements the variability is not as extreme, but it is still sufficiently great that in a sample of 20 flies taken by Montgomery *et al.* (1983) from a natural population, each element position was unique or nearly so.

Comparisons between species also reveal great variability. Preliminary data (Dowsett and Young, 1982; Martin *et al.*, 1983), indicate that mobile sequences are not well preserved even between closely related Drosophila species. By testing a wide range of species for hybridization to a set of copia-like and clustered elements, Martin *et al.* (1983) found that some elements are present over a wide taxonomic expanse, whereas others are limited to a small group of sibling species. Even among sibling species there was considerable variability in the number and distribution of each family. For P elements, studies by J. Brookfield, E. Montgomery, and C. Langley and others by J. Powell (personal communication) indicate that the elements are limited to *D. melanogaster*, at least among closely related species. This result might mean that P elements appeared relatively recently (following the divergence from the sibling species, *simulans* and *mauritiana*), perhaps through a rare invasion from a more distant species.

For the most part, these studies showed that the presence of a given family of transposable elements followed the accepted taxonomic lines, suggesting that vertical rather than horizontal transmission is the rule. The latter might occur only at the point of the initial introduction of the element. One interpretation consistent with the data is that most elements appeared only once in the evolution of the genus, but, in a few cases, were lost in one or more of the lineages. In these cases of apparent element loss, it is not possible to determine if the element actually became extinct in a given lineage, or if it merely diverged in sequence so much that it could no longer be detected by hybridization.

7.3.2. Retroviruses and retrovirus-like elements

In many respects, retroviruses, even though they are animal viruses and thus not themselves considered to be movable genetic elements, may be the most significant elements in a consideration of movable genetic elements and evolution. Retroviruses are RNA-containing enveloped animal viruses with a virion DNA polymerase (see Weiss *et al.*, 1982). They replicate through a DNA intermediate integrated with cellular DNA and can integrate in germ-line DNA. Since they are viruses, consideration of their role in evolution involves all of the normal considerations of parasitism, mutualism, and commensualism. In this case, however, the parasitism is at the level of the genome itself. But since retroviruses have numerous structural and functional homologies with true cellular movable genetic elements and, thus, are evolutionarily related to them, all movable genetic elements can be considered protoviruses and, thus, quasi-independent and semi-parasitic genetic elements.

Retrovirus DNA consists of two long terminal direct repeats (LTR) containing transcriptional control sequences surrounding coding sequences for virion proteins and some sequences involved in DNA synthesis from the viral RNA template and in encapsidation. The entire viral DNA has smaller inverted terminal repeats. This arrangement is the same as that of numerous bacterial transposons and numerous eukaryotic movable genetic elements, for example, TYl, *copia*, etc. (Figure 1B and 1C). Also similar is the small direct repeat of cell DNA resulting from integration (Figures 1 and 2).

In addition to these organizational homologies, there is direct terminal sequence homology between an avian retrovirus, spleen necrosis virus, and yeast and *Drosophila* movable genetic elements, TYl, sigma and *copia* (Del Rey *et al.*, 1982).

Retroviruses, like orthodox cellular movable genetic elements, cause genetic changes in the host either as a result of the insertion of the coding sequences in the unique portion of the DNA or the control sequences in the long terminal repeat. The latter can result in activation or inactivation of cellular genes. By analogy with *Drosophila* where many spontaneous mutations are the result of insertion of movable genetic elements, *notch*, *bithorax*, and *white*, at least one spontaneous mutation, *dilute*, in the mouse has been shown to be the result of retrovirus insertion (Jenkins *et al.*, 1981).

There is a complete series of retrovirus-related nucleotide sequences in eukaryotic DNA including (1) retroviruses that integrate in somatic cells, (2) retroviruses that have infected germ-line cells, (3) cellular elements whose RNA forms non-infectious retrovirus-like particles, (4) cellular elements whose RNA can be packaged by retroviruses, (5) cellular elements with sequence homology to retrovirus coding sequences, (6) cellular elements with homology to retrovirus LTRS, (7) cellular movable genetic elements that transpose through RNA intermediates, and (8) cellular elements that were transposed through a retrovirus-like process.

This series has been interpreted as demonstrating both ascending and descending evolutionary sequences. However, it appears to us that this series represents first evolution of retroviruses from cellular movable genetic elements followed in some cases by further evolution of endogenous (germline) retroviruses and repeated elements from exogenous retroviruses and then formation of other exogenous retroviruses from endogenous retroviruses.

7.3.4.1. *Retroviruses that integrate in somatic cells*

These retroviruses are termed exogenous and behave in many respects like any virus of vertebrates (see Weiss *et al.*, 1982). There are three genes in non-defective retroviruses, *gag*, *pol*, *env*, each of which gives rise to several polypeptides (Figure 1C). All the polypeptides are found in the virus particle and at present, only the product(s) of the *pol* gene are thought to be involved in viral nucleic acid replication. In addition, variant exogenous retroviruses are found that transduce modified cell genes. These cell genes are modified in the retrovirus and are termed oncogenes because they cause cancer. The oncogenes usually substitute for viral coding sequences, resulting in a replication defective virus.

Essentially all of the *cis*-acting sequences of retroviruses (including endogenous ones) are found near the ends of the genome in the LTRs and nearby sequences. Only one or two splice acceptors and possible ribosome binding sites occur in the coding sequences.

7.3.4.2. *Retroviruses that apparently infected germline cells*

These retroviruses are termed endogenous and are found in the DNA of many vertebrates (see Coffin, 1982). It has been estimated that as much as 0.05% of the total mouse genome consists of endogenous-retrovirus-related information (Callahan and Todaro, 1978).

Endogenous retroviruses are defined by their ability to form infectious retrovirus particles which then behave like exogenous viruses or by being related to such endogenous retroviruses (defective proviruses of endogenous retroviruses).

It seems useful to consider two types of endogenous viruses. The first type is represented by AKR or ecotropic murine leukemia virus, mouse mammary tumour virus, and Rous-associated virus-O (endogenous avian leukosis virus). Exogenous viruses formed from these endogenous viruses (proviruses) can infect somatic cells of some individuals of the species in whose genome they reside. Furthermore, there is good evidence from the existence of individual members of a species without any of these endogenous viruses and from acquisition of new proviruses in laboratory populations for recent germline infections by exogenous viruses to form these endogenous viruses (see Buckler *et al.*, 1982).

Such endogenous viruses might affect the response of the host to infection with a related exogenous virus (see Crittenden *et al.*, 1982). In laboratory situations this type of endogenous virus can yield cancers. Jaenisch has developed an experimental system of infection of mouse blastocysts with Moloney murine leukemia virus which results in some germline integration (Jaenisch *et al.*, 1981). However, most endogenous viruses do not cause cancer or any other observable pathologies.

Other endogenous retroviruses, the second type, have been found to be related to exogenous or endogenous retroviruses isolated from completely different species or even orders of vertebrates. This relationship supports the hypothesis of the formation of these endogenous retroviruses by germline infection by an exogenous retrovirus a long time ago, even millions of years ago. In some cases, especially studied by Benviniste and Todaro, the related endogenous and exogenous viruses are found at present in distantly related species—for example, an endogenous retrovirus of cats is related to an endogenous retrovirus of baboons, another cat virus is related on one of rats, and an endogenous virus of pigs is related to one of mice (see Todaro, 1980).

Relevant here is that at present this second type of endogenous virus cannot infect cells of the species in whose DNA they reside. This fact indicates that, if there was an original and ancient germline infection, a change in cellular receptors to virus infection has since occurred. This resistance to infection makes the copy number (five to 50) of these endogenous viruses difficult to understand. The amplification of endogenous proviruses could have been the result of multiple infections and/or intracellular duplications (see Tereba and Astrin, 1982). Multiple infections are prevented by superinfection resistance when a related retrovirus is being produced (see Tooze, 1973). Therefore, multiple germline infections requires the existence of germline repression which prevents superinfection resistance and allows superinfections. Janeisch has recently reported such a repression phenomenon with exogenous virus infection of early embryonic cells and teratocarcinoma cells (Stewart *et al.*, 1982).

7.3.4.3. *Cellular elements whose RNA forms non-infectious retrovirus-like particles*

These elements are called intracisternal A-particle genes and are found in mice and other mammals. However, while they are found in the DNA of many species, they are reported to be absent in other related species, for example, they are present in Syrian hamsters (Suzuki *et al.*, 1982), but absent from Chinese hamsters (Leuders and Kuff, 1981).

Intracisternal A-particle genes have the structure of a retrovirus provirus and code for proteins that can encapsidate their RNA in a non-infectious intracellular particle. The intracisternal A-particle genes are reiterated almost one thousand times. Therefore, either at one time the intracisternal A-particles were infectious and infected germline cells repeatedly, that is, they were an unusually successful endogenous virus (see above), or the intracisternal A-particle genes transpose, perhaps through their intracellular RNA containing particle. Evidence for the latter possibility has recently appeared (Hawley *et al.*, 1982; Shen-ong and Cole, 1982). If this hypothesis is correct, different extents of intracellular germline amplification in related species could explain the different distribution of these genes in related species.

7.3.4.4. *Cellular elements whose RNA can be packaged by exogenous retroviruses*

These elements are termed VL30 genes and have been described in the DNA of several species of mice and rats. Although no nucleotide sequence homology has been reported between these elements and exogenous retroviruses, the VL30 DNA has the structure of a retrovirus provirus and some of the sequences in VL30 RNA are recognized by some retrovirus proteins leading to encapsidation of VL30 RNA and potential reverse transcription and integration (Scolnick *et al.*, 1979). VL30 genes like intracisternal A-particle genes are reiterated almost 1,000-fold, indicating either multiple germline infections with selection against concomitant infection by the helper retrovirus or, more likely, some mechanism of intracellular duplication. Against the hypothesis of recent germline infection is the stability of VL30 genes in mouse strains during times when murine leukemia virus infection of the germline can be shown to occur (Courtney *et al.*, 1982).

The VL30 genes could have arisen by recombination of cell sequences with retrovirus control sequences, by convergent evolution of cell sequences and retrovirus sequences, or by descent of VL30 sequences from an ancestral retrovirus. However, the distribution of VL30 DNA sequences is broader than that of endogenous retroviruses, requiring in the ancestral retrovirus hypothesis conservation of the homology to retrovirus encapsidation sequences for millions of years in the absence of apparent selection (Courtney *et al.*, 1982).

A recent report of stimulation of VL30 RNA synthesis after cell stimulation by a polypeptide growth factor may indicate a physiological role of this RNA or, more likely, accidental activation as a result of chromosomal location of the VL30 genes (Foster *et al.*, 1982).

7.3.4.5. *Cellular elements with sequence homology to retrovirus coding sequences*

Cloned retrovirus DNAs have been used as hybridization probes with cellular DNAs. Under high stringency conditions, DNA of type one endogenous retroviruses hybridizes to proviruses of endogenous retroviruses (2 above). However, in some cases it has been found that further related DNA sequences can be found by hybridization under conditions of low stringency. The distribution of these related sequences is wider in terms of species than that of most endogenous viruses and indicates that these sequences do not have the same origin as endogenous retroviruses (Callahan *et al.*, 1982).

These retrovirus-related sequences hybridize to DNA of retrovirus coding regions (*gag-pol*). These sequences are reiterated more than most endogenous retroviruses, but less than VL30 or intracisternal A-particle genes. They appear to be incapable of producing infectious virus even when they have a provirus-like structure. Their over-all distribution indicates possible evolutionary descent from an ancestral virus, but this hypothesis requires deletions of the sequences in some species (Bonner *et al.*, 1982).

Similarly, use in mice of DNA of some of the second type of endogenous viruses as a hybridization probe has led to discovery of another widespread class of repeated DNA sequences called C-I. In many cases these sequences can give rise to infectious virus, but their widespread distribution and stability indicate that they are the result of even more ancient events than postulated for other endogenous retroviruses (Callahan *et al.*, 1979; Phillips *et al.*, 1982).

7.3.4.6. *Cellular elements with sequence and/or structural homology to retrovirus LTRs*

Two reports have appeared about insertions found in DNA clones of satellite DNA (Streek, 1982; Brown and Huang, 1982). (Satellite DNA is formed of multiple copies of tandemly repeated DNA. Satellite DNA can form several percent of the DNA of a cell and appears to be fundamentally different from the dispersed repetitive DNA which is the subject of this article.) In the DNA clones are sequences with either nucleic acid sequence homology or organizational similarity to retrovirus LTRs. In one hypothesis, the LTR-related sequences were passengers in the amplification of the satellite DNA which occurred subsequent to the integration of the LTR-related sequence. The mechanism of amplification was unequal crossing over and unrelated to transposition of movable genetic elements. In an alternative hypothesis, the LTRs could have led to the amplification.

The LTR-related sequences could be precursors to retrovirus LTRs or be the result of unequal crossing over in a provirus deleting all proviral sequences except one LTR. Homology of LTRs of otherwise unrelated repeated genetic elements (see below) might support the former hypothesis.

7.3.4.7. Cellular movable genetic elements that transpose through an RNA intermediate

Cellular movable genetic elements of bacteria seem to use a DNA form of transposition involving replicative recombination (see above). However, since the retrovirus provirus, which clearly is the result of an RNA form of transposition, has the same final DNA structure as bacterial cellular movable genetic elements, *a priori*, an RNA form of transposition cannot be excluded for cellular movable genetic elements.

Evidence in favour of such RNA-mediated transposition is the organization of transcription in some cellular movable genetic elements (*copia*, TYl), the presence of sequences complementary to tRNA 3′ to the 5′ LTR (analogous to retrovirus primer binding site), existence of unintegrated element DNA, and concordance of LTR sequences (see references in Temin, 1982; also Flavell and Ish-Horowicz, 1981, 1983; Scherer *et al.*, 1982).

Direct nucleotide sequence homologies between some exogenous retroviruses and LTRs of cellular movable or repeated genetic elements indicate an evolutionary relationship between them and supports this hypothesis (Del Rey *et al.*, 1982; Kugimiya *et al.*, 1983). For example, almost all eukaryotic movable genetic elements and retroviruses are bounded by $\begin{matrix}\text{TG...CA}\\\text{AC...GT}\end{matrix}$ in an inverted repeat (Temin, 1980). It may be significant that the binding sites for resolving cointegrates of bacterial transposons of $\gamma\delta$ and Tn3 are bounded by $\begin{matrix}\text{TGT...ACA}\\\text{ACA...TGT}\end{matrix}$ (Grindley *et al.*, 1982).

7.3.4.8. Cellular elements that apparently were transposed through a retrovirus-like process involving an RNA intermediate

These elements are called cDNA or processed pseudogenes (or genes). They have been widely described in vertebrate DNA and represent copies of unique protein coding genes—α-globin, β-tubulin, immunoglobulin ε, dihydrofolate reductase, etc.—or shorter DNA sequences—Alu family of primates and rodents, R family of rodents, and genes for small nuclear RNA genes (Vanin *et al.*, 1980; Nishioka *et al.*, 1980; Wilde *et al.*, 1982; Battey *et al.*, 1982; Chen *et al.*, 1982; Schmid and Jelinek, 1982; Monson *et al.*, 1982; Gebhard *et al.*, 1972; Denison and Weiner, 1982; Hammarström *et al.*, 1982). These DNA sequences are characterized as being non-tandem or dispersed duplications (or more reiterations, there are 500,000 copies of the Alu sequences), surrounded by a different direct repeat of cell DNA at each site (a mark of transposition), and loss of intervening sequences and/or a 3′ poly(A) stretch (mark of an RNA intermediate) (see Figure 4). The F and G elements of *Drosophila* have some of these same properties (DiNocera and Dawid, 1983).

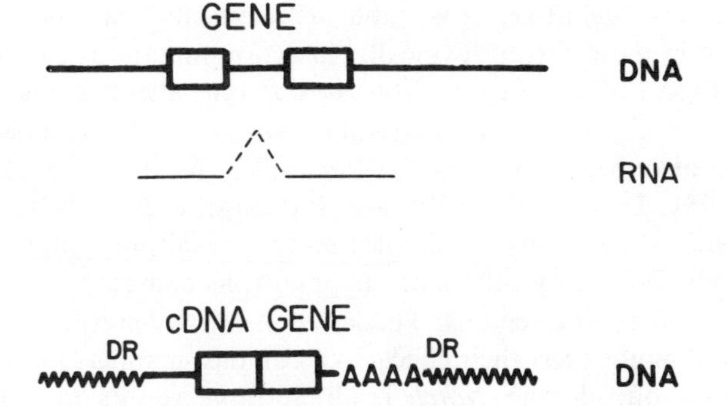

Figure 4. Formation of cDNA or processed gene. The open boxes represent exons; the dashed line, intervening sequences which are spliced out; direct repeat; AAA, poly(A) tract

The prevalence of cDNA genes indicates active reverse transcriptase and integrase (a nuclease making staggered cuts that result in the direct repeats) activities in germline cells. Models have been proposed to provide primers for DNA synthesis from a cellular RNA (see Temin, 1982). These reverse transcriptase and integrase activities are definitely not coded for by the sequences so transposed. The enzyme activities could be from retroviruses that are actively produced in germline cells (such viruses have not been described) (see Stewart *et al.*, 1982), from cellular genes for these activities, or from movable genetic elements that transpose through an RNA intermediate (see 7.3.4.7. above). In any event, the existence of these cDNA genes indicates continued presence of DNA to RNA to DNA information transfer in germline cells.

Of course, there also are elements that do not apparently transpose through an RNA intermediate, for example, the P factor of *Drosophila* as indicated by the apparent absence of full length transcripts and some genes for small nucleus RNAs (Denison and Weiner, 1982).

7.4. BIOLOGICAL EFFECTS OF MOVABLE GENETIC ELEMENTS

7.4.1. Genetic effects

Thus, there are in normal eukaryotic genomes many elements capable of causing mutations by their insertion. The presence of so many movable genetic elements in the cell genome provides an enormous pool of potential genetic variability that goes far beyond usual chromosomal changes and base pair mutations.

The most conspicuous effect of transposable elements on the host genome are the mutations produced by insertion at genetic loci. Molecular studies of a large number of spontaneous mutations at the *bithorax*, *white*, and *Notch* loci indicate that many, or even the majority of these mutations are actually insertions.

Involvement of *copia*-like, FB, and P elements has been specifically demonstrated in these events. Typically, insertion mutations can revert to wild type or undergo subsequent mutations to different allelic forms. Some of the reversions are precise excision events (Rubin *et al.*, 1981), but reversion by partial excision has also been observed for *copia*-like (Modellel *et al.*, 1983), FB (Bingham, 1981; Levis *et al.*, 1982), and P (Searles *et al.*, 1983) elements.

Some insertion mutations are simple amorphs resulting from interruption of coding regions, but many others are hypomorphs caused by insertions up to several kbp from coding regions. The latter insertions presumably affect gene regulation and might exert their effects in a complex manner. For example, one *copia* insertion outside the *Notch* reading frame results in a temperature-dependent phenotype (M. Young, personal communication), and a P factor insertion at the *held-up wings* locus cause no departure from wild type except when the element is involved in a chromosome rearrangement (Engels and Preston, 1981a). Many independent P insertions at the *singed* locus have been observed; these have a continuous range of phenotypes between wild type and the most extreme *singed* effect. Some but not all of these insertions also cause female sterility—a previously identified pleiotropic effect of some *singed* alleles. Preliminary results (W. Engels and C. Preston, unpublished) indicate that these phenotypic differences result from different kinds of partially deleted P elements. The most striking example of complex control is that exerted by *gypsy*, a *copia*-like element which has been found at nine widely separated loci resulting in a mutant phenotype in each case. However, these mutant phenotypes occur only in flies that are not homozygous for *su*(*Hw*), a recessive suppressor on the third chromosome (Modolell *et al.*, 1983). Thus, *gypsy* enables this suppressor gene to exert synchronous control over many loci in many genomic positions.

Transposable elements also cause large scale chromosome rearrangements. For example when P factors are in their active state, they can produce cytologically visible rearrangements at rates up to 10% per chromosome arm per generation. These include complex rearrangements as well as simple inversions (Berg *et al.*, 1980). The breakpoints of these rearrangements occur primarily at pre-existing P element sites, and they are sometimes associated with visible or lethal mutations (Engels and Preston, 1981a; Simmons and Lim, 1980). In some instances, the P factors remain at the breakpoints after the rearrangement has taken place. In those cases, the rearrangements can be observed to revert to the standard sequence at high frequencies, restoring the function of any mutated loci at the breakpoints (W. Engels and C. Preston, unpublished).

In addition, because LTRs often contain promoters and other transcriptional enhancing sequences, movable genetic elements can cause insertional activation. These have been described in yeast for TYl and in chickens for avian leukosis and reticuloendotheliosis viruses (see McKnight *et al.*, 1981; Hayward *et al.*, 1981; Payne *et al.*, 1982). In the latter case, insertion of a retrovirus provirus near (almost always 5′) to a cellular gene, called *c-myc*, leads to increased transcription of the *c-myc* gene. This increased transcription is correlated with formation of a B-cell lymphoma in the infected chickens.

7.4.2. Evolution of mobile DNA

This topic has been discussed recently by Orgel *et al.* (1980), Dover and Doolittle (1980), Jain (1980), Campbell (1981), and Bouchard (1982).

As discussed above, sequences with the ability to increase their copy number in the genome with respect to other sequences have the potential to spread through sexually replicating populations without help from, or even opposed to, natural selection at the organismal level. Originally these over-replicating sequences would be merely origins of replication. At the start of DNA replication, an origin of replication has replicated more than other DNA sequences. Evolution of cellular movable genetic elements would then be the result of co-evolution of the genome as a whole and 'over-replicating' DNA sequences.

In this perspective, cellular mechanisms that control initiation of cellular DNA synthesis are another form of this co-evolution. A later form of such co-evolution is the development of cellular mechanisms that control the frequency of transposition by cellular movable genetic elements. As mentioned above, some elements such as FB and P factors in Drosophila tend to transpose more frequently in the germline than in the soma. There is no apparent benefit to the element to transpose in somatic cells if we assume that they are parasitic (J. Crow, personal communication).

Mutations of replication origins to very frequent replication would lead to localized over-replication such as seen in induction of SV40 replication by T antigen (Conrad *et al.*, 1982). With no efficient excision and ligation mechanisms, such replication is lethal. However, if there are appropriate nucleases and ligases, the over-replicated DNA can be excised. (Cytoplasmic circular molecules of DNA with Alu-related sequences may represent a product of such a process (Krolewski *et al.*, 1982).)

Further mutations could lead to sequences that were transposed by replicative recombination rather than excised. Chance insertion of these transposed sequences next to genes coding for nucleases and/or ligases would lead to formation of insertion sequences.

Further evolution from insertion sequences to transposons in bacteria has been described (Ross *et al.*, 1979). Evolution of movable genetic elements using RNA intermediates and of intervening sequences could have been another part of this process.

Under circumstances where DNA transposition could not occur, an RNA intermediate for transposition would have an advantage. DNA sequences without a nearby promoter could be copied as part of a larger transcript. Excision of these sequences could yield a spliced transcript and a possible transposition intermediate. Thus, the maturase would be equivalent to an excisase and derived from a transposase (Lazowska *et al.*, 1980; Borst and Grivell, 1981; Kruger *et al.*, 1982). (Of course for such transposition reverse transcription must exist (Temin, 1982).) Once such a mechanism existed, insertion of certain types of movable genetic sequences into coding sequences would not necessarily be lethal. However, if the excision mechanism was not 100% efficient, there would be selection for insertion of the movable genetic sequences between functional domains of proteins.

This hypothesis is consistent with the distribution of intervening sequences in coding sequences, often between functional domains, but different in evolutionarily related coding sequences (Leicht *et al.*, 1982; Zakut *et al.*, 1982). (It also might explain the common terminal dinucleotides of intervening sequences (Mount, 1982).)

Presence of a nearby promoter would lead to another kind of transposition with an RNA intermediate. Polymerase III transcripts could be directly reverse transcribed and integrated (providing suitable primers were available), since the promoter is internal to the transcript (Sakonju *et al.*, 1980). Polymerase II transcripts that were directly reverse transcribed and integrated would no longer have a polymerase II promoter and thus, would be unable to transpose further. However, a mode of reverse transcription such as in retrovirus replication leads to resynthesis of promoter sequences 5′ to the RNA transcript. Thus, further transposition is possible even for polymerase II transcripts.

Finally, since each of these movable genetic elements would be restricted to a single lineage, there would be strong selection for an element that could infect other lineages (see Shimotohno and Temin, 1981). Such an element would be a virus. A virus of this type would be able to infect other lineages and integrate in their DNA. Several possibilities then arise. The proviral DNA could be able to form further virus particles which could lead to further infection in that or other lineages. The proviral DNA could in addition or instead be able to transpose as a purely intracellular movable genetic element. In this case it would be a cellular movable genetic element. Alternatively, the proviral DNA could be defective for virus formation and transposition and only passively amplified by other mechanisms. *cis* mechanisms would involve linkage to an amplified sequence, that is, another movable genetic element or tandemly repeated sequences. *trans* mechanisms would use proteins produced by other genes such as in formation of cDNA pseudogenes.

It is also possible that a purely cellular movable genetic element with such a history might again produce an infectious virus repeating the whole cycle.

It is entirely conceivable that many cycles of this type have occurred during evolution of eukaryotes so that there is now no real distinction between cellular movable genetic elements and many endogenous retroviruses.

At every stage in this evolution there would be other genomic mutations to modify the behaviour of the movable element so that it is not deleterious to the host. As is discussed in the next section, the whole system may further evolve so there are strong positive effects on fitness of the host organism (also see Jain, 1980).

This evolution of transposons might require development of some special theories. In the short run, the transposition rate of most such elements is low enough that they can be treated as standard Mendelian factors. However, for long-term equilibria and evolutionary predictions, their deviations from Mendelian behaviour will prove to be important. For the simplest kind of model in which self-regulation, partial site-specificity, and heterogeneity within families are ignored, the equilibrium between selection and transposition is given by classical results for meiotic drive (Crow, 1979; Hickey, 1982). Charlesworth and Charlesworth (1983), and Langley *et al.* (1983) have recently devised equilibrium conditions for a class of specific models of transposable element spread. However, a more realistic description of the behaviour of transposable elements taking into account such things as cytotype and interactions with the host genome will require considerably more complex models. A likely possibility is that the evolution of transposons will require the use of models designed for co-evolution of host–parasite systems (Anderson and May, 1982; Roughgarden, 1979).

7.4.3. Possible evolutionary roles of mobile DNA

What roles might mobile elements play in the evolution of the genome as a whole? As a null hypothesis we can suppose that they are merely rather mild parasites and largely irrelevant to all but their own evolution. To date, there are no conclusive arguments against this possibility. However, we can at least consider several ways that these elements might influence their hosts' evolutionary pathways. (Rose and Doolittle (1983) have also discussed this question.)

The earliest suggestion is that of McClintock (reviewed by Federoff, 1982) who considered transposable elements to be 'controlling elements' for single-copy genes. In *Drosophila* where the positions of transposons vary widely between individual genomes this kind of role seems unlikely. However, Davidson and Posakony (1982) point out that in some cases, especially where elements have strong developmental specificity in their transcription, the possibility of dispersed sequences acting as controlling elements cannot be excluded.

The most direct way that transposable elements might influence evolution is by the variability they generate through transposition into host genes. There is some evidence in *Drosophila* that insertions represent a major fraction of the total spontaneous mutation rate. This indication comes mainly from visible loci—*biothorax*, *Notch*, and *white*—where the effects are most conspicuous. It is not known whether these results will prove to be common for loci in general. However, even if insertions represent only a minority of spontaneous mutations, they might have disproportionate significance to evolution because they are qualitatively different from base substitutions. For example, the evolution of a system in which several genes at different loci are under the control of a single locus might involve, at least as an intermediate step, a transposable element with the special properties of *gypsy* as described by Modellel *et al.* (1983).

Gross rearrangements and duplications induced by transposons might also be important in opening new avenues for natural selection as discussed by Ohno (1970). In *Drosophila*, P factors (Engels and Preston, 1981a and unpublished) and other transposable elements (e.g., Laverty and Lim, 1982 and references therein) have been shown to catalyse cytologically visible duplication events. Furthermore, there is growing evidence that extremely rapid changes in total DNA content can be brought about by the amplification of transposable elements. Bouchard (1982) points out that most cases in which two closely related species differ greatly in DNA content the difference occurs primarily in the moderately repetitive and dispersed portion of the genome. This difference usually is the result of a set of several dispersed repetitive sequences being in low copy number (but not unique) in one of the species and in much higher copy number in the other. Subsequent divergence of the two species might then involve the gradual conversion of moderately repetitive DNA to unique sequence differences by nucleotide substitution.

Another possibility is that transposons can serve as vehicles by which single-copy genes might be transferred from one species to another. Rubin and Spradling (1982) have demonstrated the usefulness of transposable elements for gene transfer under laboratory conditions, and analogous processes, possibly involving viruses, could occur in nature. Even if such events are rare, they might have far-reaching effects. Four candidates for horizontal gene transfer between species are reviewed by Lewin (1982).

Transposable elements might contribute to the speciation process by acting as reproductive barriers between sub-populations. This possibility is indicated most strongly by the P elements which, by virtue of their distribution, have dichotomized *D. melanogaster* into two sub-categories (P strains and M strains) which are to some extent mutually incompatible because of the dysgenic traits produced upon hybridization. This mutual incompatibility is a direct outcome of the P factors' ability to produce a cellular state (the P cytotype) in which its own transpositional activity is repressed. It is reasonable to suppose that most other transposable elements will also have some kind of self-regulation, and thus they might also serve as reproductive barriers in a similar fashion.

7.4.3.1. *Theoretical questions*

Most of population genetics theory deals with the logical consequences of Mendelian heredity and the processes of mutation and recombination. Since the vast majority of genetic loci remain subject to the laws of Mendel, this theory will remain useful regardless of new information concerning transposable elements.

Furthermore, the theory of phenotypic evolution and quantitative characters will also be largely unaffected. Most of this theory simply bypasses the actual details of how genetic variability is generated and, therefore, applies equally well to variability generated by insertion mutations as to that from base substitutions provided the phenotype itself is polygenic. On the other hand, mutations of major effect such as those likely to result from the activity of transposable elements might prove to be important to evolution. If so, then quantitative genetic theory will decrease in value irrespective of whether such mutations come from transpositions or from more conventional events.

The major effect this new information about movable genetic elements is likely to have on existing theory is a change of emphasis. For example, much of experimental and theoretical population genetics in the last two decades has been directed toward describing and explaining protein variability as seen by differences in electrophoretic mobility and also the amino acid substitutions seen by interspecific comparisons. In particular, the question of how much of this variability and substitution rate is selectively neutral has received considerable attention (Lewontin, 1974). However, if a major part of evolution is independent of the process of base substitution then this kind of variability will be less important than previously thought.

Instead, greater emphasis will be placed on population variability and species differences which involve DNA rearrangements of up to several kbp especially in non-coding regions. Further observations of this kind of variability might well require different kinds of theoretical considerations.

7.5. SUMMARY

Most cells have a significant number of reiterated DNA sequences. Many of these sequences are dispersed in the genome and appear fundamentally different from tandemly repeated DNA sequences. The dispersed DNA sequences have a variety of structures, but all appear to represent some form of movable genetic element. In addition, several lines of evidence indicate a unified evolutionary lineage for some cellular movable genetic elements and some viruses.

Movable genetic elements have been shown to exert many different kinds of effects on organisms. It is an attractive hypothesis that movable genetic elements are responsible for major genetic effects important in evolution. Such effects could come about through the elements' demonstrated capacity to bring about massive genomic changes in evolutionarily short times. However, it is just as possible that the movable genetic elements merely provide another form of genetic variation whose effects are primarily mildly deleterious.

ACKNOWLEDGEMENTS

We thank J. Crow and K. Wilhelmsen for useful suggestions. The work from Dr Temin's laboratory is supported by grants CA-07175 and CA-22443 from the

National Cancer Institute. The work from Dr Engels' laboratory is supported by grants GM-30948 from the National Institutes of Health and PCM8104332 from the National Science Foundation. Dr Temin is an American Cancer Society Research Professor. This is paper number 2629 from the Laboratory of Genetics.

7.6. REFERENCES

Anderson, R. M., and May, R. M. (eds.) (1982) *Population Biology of Infectious Diseases*, Springer-Verlag, Berlin.

Battey, J., Max, E. E., McBride, W. O., Swan, D., and Leder, P. (1982) A processed human immunoglobulin ε gene has moved to chromosome 9, *Proc. Natl. Acad. Sci. USA*, **79**, 5956–5960.

Berg, R., Engels, W. R., and Kreber, R. A. (1980) Site-specific X-chromosome rearrangements from hybrid dysgenesis in *Drosophila melanogaster*, *Science*, **210**, 427–429.

Bingham, P. M. (1981) A novel dominant mutant allele at the *white* locus of *Drosophila melanogaster* in mutable, *Cold. Spring Harbor Symp. Quant. Biol.*, **45**, 519–527.

Bingham, P. M., Kidwell, M. G., and Rubin, G. M. (1982) The molecular basis of P-M hybrid dysgenesis: The role of the P element, a P strain-specific transposon family, *Cell*, **29**, 995–1004.

Bonner, T. I., Birkenmeier, E. H., Gonda, M. A., Mark, G. E., Searfoss, G. H., and Todaro, G. J. (1982) Molecular cloning of a family of retroviral sequences found in chimpanzee but not human DNA, *J. Virol.*, **43**, 914–924.

Borst, P., and Grivell, L. A. (1981) One gene's intron is another gene's exon, *Nature*, **289**, 439–440.

Bouchard, R. A. (1982) Moderately repetitive DNA in evolution, *Int. Review of Cytology*, **76**, 113–193.

Brown, A., and Huang, R. C. C. (1982) Mouse *Eco*RI satellite DNA contains a sequence homologous to the long terminal repeat of the intracisternal A particle gene, *Proc. Natl. Acad. Sci. USA*, **79**, 6123–6127.

Buckler, C. E., Staal, S. P., Rowe, W. P., and Martin, M. A. (1982) Variation in the number of copies and in the genomic organization of ecotropic murine leukemia virus proviral sequences in sublines of AKR mice, *J. Virol.*, **43**, 629–640.

Callahan, R., Drohan, W., Gallahan, D., D'Hoostelaere, L., and Potter, M. (1982) Novel class of mouse mammary tumor virus-related DNA sequences found in all species of *Mus*, including mice lacking the virus proviral genome, *Proc. Natl. Acad. Sci. USA*, **79**, 4113–4117.

Callahan, R., Meade, C., and Todaro, G. J. (1979) Isolation of an endogenous type C virus related to the infectious primate type C viruses from the Asian rodent *Vandeleuria oleracea*, *J. Virol.*, **30**, 124–131.

Callahan, R., and Todaro, G. J. (1978) Four major endogenous retrovirus classes each genetically transmitted in various species of *Mus*, in Morse, H. C. (ed.) *Origins of Inbred Mice*, Academic Press, New York, pp. 689–713.

Calos, M. P., and Miller, J. H. (1980) Transposable elements, *Cell*, **20**, 579–595.

Campbell, A. (1979) Structure of complex operons, in Goldberger, R. F. (ed.) *Biological Regulation and Development*, Plenum, New York, Vol. 1, pp. 19–55.

Campbell, A. (1981) Evolutionary significance of accessory DNA elements in bacteria, *Ann. Rev. Microbiol.*, **35**, 55–83.

Charlesworth, B. and Charlesworth, D. (1983) The population dynamics of transposable elements. *Genetical Research*, **42**(1), 1–27.

Chen, M. J., Shimada, T., Moulton, A. D., Harrison, M., and Nienhuis, A. W. (1982) Intronless human dihydrofolate reductase genes are derived from processed RNA molecules, *Proc. Natl. Acad. Sci. USA*, **79**, 7435–7439.

Coffin, J. (1982) Endogenous viruses, in Weiss, R., Teich, N., Varmus, H., and Coffin, J. (eds.) *RNA Tumor Viruses*, Cold Spring Harbor Laboratory, Cold Spring Harbor, New York, pp. 261–368.

Conrad, S. E., Liu, C. P., and Botchan, M. R. (1982) Fragment spanning the SV40 replication origin is the only DNA sequence required in cis for viral excision, *Science*, **218**, 1223–1225.

Courtney, M. G., Elder, P. K., Steffen, D. L., and Getz, M. J. (1982) Evidence for an early evolutionary origin and locus polymorphism of mouse VL30 DNA sequences, *J. Virol.*, **43**, 511–518.

Crittenden, L. B., Fadley, A. M., and Smith, E. J. (1982) Effect of endogenous leukosis virus genes on response to infection with avian leukosis and reticuloendotheliosis virus, *Avian Diseases*, **26**, 279–294.

Crow, J. F. (1979) Genes that violate Mendel's rules, *Scientific American*, **240**, 134–146.

Davidson, E. H., and Posakony, J. W. (1982) Repetitive sequence transcripts in development, *Nature*, **297**, 663.

Del Ray, F. J., Donahue, T. F., and Fink, G. R. (1982) Sigma, a repetitive element found adjacent to tRNA genes of yeast, *Proc. Natl. Acad. Sci. USA*, **79**, 4138–4142.

Denison, R. A., and Weiner, A. M. (1982) Human Ul RNA pseudogenes may be generated by both DNA- and RNA-mediated mechanisms, *Mol. Cell. Biol.*, **2**, 815–828.

Di Nocera, P. P., and Dawid, I. B. (1983) Interdigitated arrangement of two oligo(A)-terminated DNA sequences in *Drosophila. Nucl. Acids Res.*, **11**, 5475–5482.

Dover, G., and Doolittle, W. F. (1980) Modes of genomic evolution, *Nature*, **288**, 646–647.

Dowsett, A. P., and Young, M. W. (1982) Differing levels of dispersed repetitive DNA among closely related species of Drosophila, *Proc. Natl. Acad. Sci. USA*, **79**, 4570–4574.

Engels, W. R. (1979a) Hybrid dysgenesis in *Drosophila melanogaster:* rules of inheritance of female sterility, *Genetic. Res., Camb.*, **33**, 219–236.

Engels, W. R. (1979b) Extrachromosomal control of mutability in *Drosophila melanogaster*, *Proc. Natl. Acad. Sci. USA*, **76**, 4011–4015.

Engels, W. R. (1981a) Hybrid dysgenesis in Drosophila and the stochastic loss hypothesis, *Cold Spring Harbor Symp. Quant. Biol.*, **45**, 561–565.

Engels, W. R. (1981b) Germline hypermutability in Drosophila and its relation to hybrid dysgenesis and cytotype, *Genetics*, **98**, 565–587.

Engels, W. R. (1984) The P family of transposable elements in *Drosophila*, *Ann. Rev. Genet.*, **17** (in press).

Engels, W. R., and Preston, C. R. (1981a) Identifying P factors in Drosophila by means of chromosome breakage hotspots, *Cell*, **26**, 421–428.

Engels, W. R., and Preston, C. R. (1981b) Characteristics of a 'neutral' strain in the P-M system of hybrid dysgenesis, *Dros. Inf. Serv.*, **56**, 36–37.

Federoff, N. (1983) Controlling elements in maize, in Shapiro, J. A. (ed.) *Mobile genetic elements*, Academic Press, New York, pp. 1–63.

Finnegan, D. J., Rubin, G. M., Young, M. W., and Hogness, D. S. (1978) Repeated gene families in *Drosophila melanogaster*, *Cold Spring Harbor Symp. Quant. Biol.*, **42**, 1053–1063.

Flavell, A. J., Ruby, S. W., Toole, J. J., Roberts, B. E., and Rubin, G. M. (1980)

Translation and developmental regulation of RNA encoded by the eukaryotic transposable element *copia*, *Proc. Natl. Acad. Sci. USA*, **77**, 7107–7111.

Flavell, A. J., and Ish-Horowicz, D. (1981) Extrachromosomal circular copies of the eukaryotic transposable element *copia* in cultured Drosophila cells, *Nature*, **292**, 591–595.

Flavell, A. J., and Ish-Horowicz, D. (1983) The origin of extrachromosomal circular *copia* elements, *Cell*, **34**, 415–419.

Foster, D. N., Schmidt, L. J., Hodgson, C. P., Moses, H. L., and Getz, M. J. (1982) Polyadenylated RNA complementary to a mouse retrovirus-like multigene family is rapidly and specifically induced by epidermal growth factor stimulation of quiescent cells, *Proc. Natl. Acad. Sci. USA*, **79**, 7317–7321.

Galas, D. J., and Chandler, M. (1981) On the molecular mechanisms of transposition, *Proc. Natl. Acad. Sci. USA*, **78**, 4858–4862.

Gebhard, W., Meitinger, T., Höchtl, J., and Zachau, H. G. (1972) A new family of interspersed repetitive DNA sequences in the mouse genome, *J. Mol. Biol.*, **157**, 453–471.

Georgiev, G. P., Ilyin, Y. V., Chmeliavskaite, V. G., Ryskov, A. P., Kramerov, D. A., Skyrabin, K. G., Krayev, A. S., Lukanidin, E. M., and Grigoryan, M. S. (1981) Mobile dispersed genetic elements and other middle repetitive DNA sequences in the genome of Drosophila and mouse: Transcription and biological significance, *Cold Spring Harbor Symp. Quant. Biol.*, **45**, 641–654.

Green, M. M. (1977) Genetic instability in *Drosophila melanogaster*: *de novo* induction of putative insertion mutations, *Proc. Natl. Acad. Sci. USA*, **74**, 3490–3493.

Grindley, N. D. F., Lauth, M. R., Wells, R. G., Wityk, R. J., Salvo, J. S., and Reed, R. R. (1982) Transposon-mediated site-specific recombination: Identification of the three binding sites for resolvase at the *res* sites of $\gamma\delta$ and Tn3, *Cell*, **30**, 19–27.

Hammarström, K., Westin, G., and Pettersson, U. (1982) A pseudogene for human U4 RNA with a remarkable structure, *The EMBO Journal*, **1**, 737–739.

Hawley, R. G., Schulman, M. J., Murialdo, H., Gibson, D. M., and Hozumi, N. (1982) Mutant immunoglobulin genes have repetitive DNA elements inserted into their intervening sequences, *Proc. Natl. Acad. Sci. USA*, **79**, 7425–7429.

Hayward, W. S., Neel, B. G., and Astrin, S. M. (1981) Activation of a cellular *onc* gene by promoter insertion in ALV-induced lymphoid leukosis, *Nature*, **290**, 475–480.

Hickey, D. A. (1982) Selfish DNA: A sexually-transmitted nuclear parasite, *Genetics*, **101**, 519–531.

Jaenisch, R., Jähner, D., Nobis, P., Simon, I., Löhler, J., Harbers, K., and Grotkopp, D. (1981) Chromosomal position and activation of retroviral genomes inserted into the germ line of mice, *Cell*, **24**, 519–529.

Jain, H. K. (1980) Incidental DNA, *Nature*, **288**, 647–648.

Jenkins, N. A., Copeland, N. G., Taylor, B. A., and Lee, B. K. (1981) Dilute (*d*) coat colour mutation of DBA/2J mice is associated with the site of integration of an ecotropic MuLV genome, *Nature*, **293**, 370–374.

Kidwell, M. G. (1981) Hybrid dysgenesis in *Drosophila melanogaster:* The genetics of cytotype determination in a neutral strain, *Genetics*, **98**, 275–290.

Kidwell, M. G., Kidwell, J. R., and Sved, J. A. (1977) Hybrid dysgenesis in *Drosophila melanogaster:* A syndrome of aberrant traits including mutation, sterility and male recombination, *Genetics*, **86**, 813–833.

Kleckner, N. (1977) Translocatable elements in procaryotes, *Cell*, **11**, 11–23.

Krolewski, J. J., Bertelsen, A. H., Humayun, M. Z., and Rush, M. G. (1982) Members of the *Alu* family of interspersed, repetitive DNA sequences are in the small circular DNA population of monkey cells grown in culture, *J. Mol. Biol.*, **154**, 399–415.

Kruger, K., Grabowski, P. J., Zaug, A. J., Sands, J., Gottschling, D. E., and Cech, T. R. (1982) Self-splicing RNA: Autoexcision and autocyclization of the ribosomal RNA intervening sequence of tetrahymena, *Cell*, **31**, 147–157.

Kugimiya, W., Ikenaga, H., and Saigo, K. (1983) A close relation between avian leukosis-sarcoma virus and *Drosophila copia*-like movable genetic elements, *17.6-297*: One base to one base correspondence in the nucleotide sequences of long terminal repeats and their flanking regions, *Proc. Natl. Acad. Sci. USA*, **80**, 3193–3197.

Langley, C. H., Brookfield, J. F. Y., Kaplan, N. (1983) Transposable elements in Mendelian populations. I. A theory, *Genetics*, **104**, 457–471.

Laverty, T. R., and Lim, J. K. (1982) Site-specific instability in *Drosophila melanogaster*: Evidence for transposition of destabilizing element, *Genetics*, **101**, 461–476.

Lazowska, J., Jacq, C., and Slonimski, P. P. (1980) Sequence of introns and flanking exons in wild-type and *box 3* mutants of cytochrome b reveals an interlaced splicing protein coded by an intron, *Cell*, 22, 333–348.

Leicht, M., Long, G. L., Chandra, T., Kurachi, K., Kidd, V. J., Mace, M., Davie, E. W., and Woo, S. L. C. (1982) Sequence homology and structural comparison between the chromosomal human α_1-antitrypsin and chicken ovalbumin gene, *Nature*, **297**, 655–659.

Leuders, K. K., and Kuff, E. L. (1981) Sequences homologous to retrovirus-like genes of the mouse are present in multiple copies in the Syrian hamster genome, *Nucl. Acids Res.*, **9**, 5917–5930.

Levis, R., Collins, M., and Rubin, G. M. (1982) FB elements are the common basis for the instability of the w^{DZL} and w^{c} Drosophila mutations, *Cell*, **30**, 551–565.

Levis, R., Dunsmuir, P., and Rubin, G. M. (1980) Terminal repeats of the Drosophila transposable element *copia*: nucleotide sequence and genomic organization, *Cell*, **21**, 581–588.

Lewin, R. (1982) Can genes jump between eukaryotic species?, *Science*, **217**, 42–43.

Lewontin, R. C. (1974) *The Genetic Basis of Evolutionary Change*, Columbia University Press, New York.

Machida, Y., Machida, C., and Ohtsubo, E. (1982) A novel type of transposon generated by insertion element IS102 present in a pSC101 derivative, *Cell*, **30**, 29–36.

Martin, G., Wiernasz, D., Schedl, P. (1983) Evolution of *Drosophila* repetitive-dispersed DNA, *J. Mol. Evol.*, **19**, 203–213.

McKnight, G. L., Cardillo, T. S., and Sherman, F. (1981) An extensive deletion causing overproduction of yeast iso-2-cytochrome c, *Cell*, **25**, 409–419.

Modolell, J., Bender, W., and Meselson, M. (1983) D. melanogaster mutations suppressible by the suppressor of hairy-wing are insertions of a 7.3 kb mobile element, *Proc. Natl. Acad. Sci. USA*, **80**, 1678–1682.

Monson, J. M., Friedman, J., and McCarthy, B. J. (1982) DNA sequence analysis of a mouse proα1(I) procollagen gene: Evidence for a mouse B1 element within the gene, *Mol. Cell. Biol.*, **2**, 1362–1371.

Montgomery, E. A., Langley, C. H. (1983) Transposable elements in Mendelian populations. II. Distribution of three copia-like elements in a natural population of *Drosophila melanogaster*, *Genetics*, **104**, 473–483.

Mount, S. M. (1982) A catalogue of splice junction sequences, *Nucl. Acids Res.*, **10**, 459–472.

Nishioka, Y., Leder, A., and Leder, P. (1980) Unusual α-globin-like gene that has cleanly lost both globin intervening sequences, *Proc. Natl. Acad. Sci., USA*, **77**, 2806–2809.

O'Hare, K., and Rubin, G. M. (1983) Structures of P transposable elements of *Drosophila melanogaster* and their sites of insertion and excision, *Cell*, 34, 25–35.

Ohno, S. (1970) *Evolution by Gene Duplication*, Springer-Verlag.

Orgel, L. E., Crick, F. H. C., and Sapienza, C. (1980) Selfish DNA, *Nature*, **288**, 645–646.

Payne, G. S., Bishop, J. M., and Varmus, H. E. (1982) Multiple arrangements of viral DNA and an activated host oncogene in bursal lymphomas, *Nature*, **295**, 209–214.

Pfeiffer, P., and Hohn, T. (1983) Involvement of reverse transcription in the replication of cauliflower mosaic virus: a detailed model and test of some aspects, *Cell*, **33**, 781–789.

Phillips, S. J., Birkenmeier, E. H., Callahan, R., and Eicher, E. M. (1982) Male and female mouse DNAs can be discriminated using retroviral probes, *Nature*, **297**, 241–243.

Pierce, D. A., and Lucchesi, J. C. (1981) Analysis of a dispersed repetitive DNA sequence in isogenic lines of Drosophila, *Chromosoma*, **82**, 471–492.

Potter, S. D., Truett, M., Phillips, M., and Maher, A. (1980) Eukaryotic transposable genetic elements with inverted terminal repeats, *Cell*, **17**, 415–427.

Rose, M. R., and Doolittle, W. F. (1983) Molecular biological mechanisms of speciation, *Science*, **220**, 157–162.

Ross, D. G., Swan, J., and Kleckner, N. (1979) Physical structures of Tn10-promoted deletions and inversions: Role of 1400 bp inverted repetitions, *Cell*, **16**, 721–731.

Roughgarden, J. (1979) *Theory of Population Genetics and Evolutionary Ecology: An Introduction*, MacMillan, New York, Ch. 5.

Rubin, G. M. (1983) Dispersed repetitive DNAs in Drosophila, in Shapiro, J. A. (ed.) *Mobile Genetic Elements*, Academic Press, New York.

Rubin, G. M., Brorein, W. J., Jr, Dunsmuir, P., Flavell, A. J., Levis, R., Strobel, E., Toole, J. J., and Young, E. (1981) *Copia*-like transposable elements in the Drosophila genome, *Cold Spring Harbor Symp. Quant. Biol.*, **45**, 619–628.

Rubin, G. M., Kidwell, M. G., and Bingham, P. M. (1982) The molecular basis of P-M hybrid dysgenesis: The nature of induced mutations, *Cell*, **29**, 987–994.

Rubin, G. M., and Spradling, A. C. (1982) Genetic transformation of Drosophila with transposable element vectors, *Science*, **218**, 348–353.

Sakonju, S., Bogenhagen, D. F., and Brown, D. D. (1980) A control region in the center of the 5S RNA gene directs the specific initiation of transcription: I. The 5′ border of the region; II. The 3′ border of the region, *Cell*, **19**, 13–35.

Scherer, G., Tshudi, C., Perera, J., and Delius, H. (1982) *B104*, a new dispersed repeated gene family in *Drosophila melanogaster* and its analogies with retroviruses, *J. Mol. Biol.*, **157**, 435–451.

Schmid, C. W., and Jelinek, W. R. (1982) The Alu family of dispersed repetitive sequences, *Science*, **216**, 1065–1070.

Schwartz, H. E., Lockett, T. J., and Young, M. W. (1982) Analysis of transcripts from two families of nomadic DNA, *J. Mol. Biol.*, **157**, 49–68.

Scolnick, E. M., Vass, W. C., Howk, R. S., and Duesberg, P. H. (1979) Defective retrovirus-like 30S RNA species of rat and mouse cells are infectious if packaged by type C helper virus, *J. Virol.*, **29**, 964–972.

Searles, L. L., Jokerst, R. S., Bingham, P. M., Voelker, R. A., and Greenleaf, A. L. (1983) Molecular cloning of sequences from a *Drosophila* RNA polymerase locus by P element tagging, *Cell*, **31**, 585–592.

Shapiro, J. A., and Cordell, B. (1982) Eukaryotic mobile and repeated genetic elements, *Biol. Cell*, **43**, 31–54.

Shen-ong, G. L. C., and Cole, M. D. (1982) Differing populations of intracisternal A-particle genes in myeloma tumors and mouse subspecies, *J. Virol.*, **42**, 411–421.

Shiba, T., and Saigo, K. (1983) Retrovirus-like particles containing RNA homologous to the transposable element *copia* in *Drosophila melanogaster*, *Nature*, **302**, 119–124.

Shimotohno, K., and Temin, H. M. (1981) Evolution of retroviruses from cellular movable genetic elements, *Cold Spring Harbor Symp. Quant. Biol.*, **45**, 719–730.

Simmons, M. J., and Lim, J. K. (1980) Site specificity of mutations arising in dysgenic hybrids of *Drosophila melanogaster*, *Proc. Natl. Acad. Sci. USA*, **77**, 6042–6046.
Singer, M. F. (1982) Highly repeated sequences in mammalian genomes, *Int. Review of Cytology*, **76**, 67–112.
Spradling, A. C., and Rubin, G. M. (1982) Transposition of cloned P elements into Drosophila germ line chromosomes, *Science*, **218**, 341–347.
Stanfield, S., and Lengyel, J. A. (1980) Small circular deoxyribonucleic acid of *Drosophila melanogaster*: homologous transcripts in the nucleus and cytoplasm, *Biochemistry*, **19**, 3873–3877.
Stewart, C. L., Stuhlmann, H., Jähner, D., and Jaenisch, R. (1982) *De novo* methylation, expression, and infectivity of retroviral genomes introduced into embryonal carcinoma cells, *Proc. Natl. Acad. Sci. USA*, **79**, 4098–4102.
Streek, R. E. (1982) A multicopy insertion sequence in the bovine genome with structural homology to the long terminal repeats of retroviruses, *Nature*, **298**, 767–769.
Strobel, E., Dunsmuir, P., and Rubin, G. M. (1979) Polymorphism in the chromosomal locations of elements of the *412*, *copia* and *297* dispersed repeated gene families in *Drosophila*, *Cell*, **17**, 429–439.
Summers, J., and Mason, W. S. (1982) Replication of the genome of a hepatitis B-like virus by reverse transcription of an RNA intermediate, *Cell*, **29**, 403–415.
Suzuki, A., Kitasato, H., Kawakami, M., and Ono, M. (1982) Molecular cloning of retrovirus-like gene present in multiple copies in the Syrian hamster genome, *Nucl. Acids Res.*, **10**, 5733–5746.
Temin, H. M. (1976) The DNA provirus hypothesis, *Science*, **192**, 1075–1080.
Temin, H. M. (1980) Origin of retroviruses from cellular moveable genetic elements, *Cell*, **21**, 599–600.
Temin, H. M. (1982) Viruses, protoviruses, development, and evolution, *J. Cell. Biochem.*, **19**, 105–118.
Tereba, A., and Astrin, S. M. (1982) Chromosomal clustering of five defined endogenous retrovirus loci in white leghorn chickens, *J. Virol.*, **43**, 737–740.
Todaro, G. J. (1980) Interspecies transmission of mammalian retroviruses, in Stephenson, J. R. (ed.) *Molecular Biology of RNA Tumor Viruses*, Academic Press, New York, pp. 47–76.
Tooze, J. (ed.) (1973) *The Molecular Biology of Tumor Viruses*, Cold Spring Harbor Laboratory, Cold Spring Harbor, New York.
Truett, M. A., Jones, R. S., and Potter, S. S. (1981) Unusual structure of the FB family of transposable elements in *Drosophila*, *Cell*, **24**, 753–763.
Vanin, E. F., Goldberg, G. I., Tucker, P. W., and Smithies, O. (1980) A mouse α-globin-related pseudogene lacking intervening sequences, *Nature*, **286**, 222–226.
Weiss, R., Teich, N., Varmus, H. E., and Coffin, J. (eds.) (1982) *RNA Tumor Viruses*, Cold Spring Harbor Laboratory, Cold Spring Harbor, New York.
Wensink, P. C., Tabata, S., and Pachl, C. (1979) The clustered and scrambled arrangement of moderately repetitive elements in *Drosophila* DNA, *Cell*, **18**, 1231–1246.
Wilde, C. D., Crowther, C. E., and Cowan, N. J. (1982) Diverse mechanisms in the generation of human β-tubulin pseudogenes, *Science*, **217**, 549–552.
Young, M. W. (1979) Middle repetitive DNA: A fluid component of the *Drosophila* genome, *Proc. Natl. Acad. Sci. USA*, **76**, 6374–6378.
Young, M. W., and Schwartz, H. E. (1981) Nomadic gene families in *Drosophila*, *Cold Spring Harbor Symp. Quant. Biol.*, **45**, 629–641.
Zakut, R., Shani, M., Givol, D., Neuman, S., Yaffe, D., and Nudel, U. (1982) Nucleotide sequence of the rat skeletal muscle actin gene, *Nature*, **298**, 857–859.

Evolutionary Theory: Paths into the Future
Edited by J. W. Pollard

Chapter 8

Environmentally induced DNA changes

CHRISTOPHER A. CULLIS
John Innes Institute,
Colney Lane,
Norwich, Norfolk NR4 7UH,
UK

8.1. INTRODUCTION

The genomes of higher eukaryotes are very complex. In spite of this complexity and diversity of the DNA both within and between species a general understanding of the origins of this complexity is beginning to emerge. Much of the diversity can be explained by recognizing that the DNA of higher organisms is subject to a variety of sequence rearrangement events including amplification, deletion, mutation, and translocation both within and between chromosomes. These changes can spread through populations so that certain segments of a genome can change rapidly during evolution. This type of phenomenon has been substantially described in the Gramineae, in which the rearrangements have played a role in determining the structure and sequence composition of the complex genomes, and how the fixation of different arrangements has created substantial differences between the genomes of closely related species (Flavell, 1982). These species comparisons describe the end-products of processes of amplification, deletion, etc. rather than the series of events by which the observed arrangement was produced.

In order to study the events by which the rearrangements of the genome occur it is necessary to have these types of changes taking place in a small number of generations and, ideally, within a single generation. One situation where this can occur is when there is an interaction between the genome and the external environment in which that genome develops. The interdependence of the genotype and of the environment in which it develops to give the observed phenotype has long been recognized. However, in the absence of selection, the response to environmental factors in one generation is not, as a rule, transmitted to subsequent generations. There are exceptions and specific environmental factors have been found to affect subsequent generations (Durrant, 1962; Hill,

1965). In these exceptional cases most of the individuals subjected to a particular set of environmental conditions survive and produce offspring all of which showed similar heritable changes. This response is in contrast to the effect where the environment has exerted a selection pressure. When selection is operating a proportion of the original population does not survive and contribute to subsequent generations. A question which arises is how the genome is altered as a result of the selection? Is the change limited to a small specific DNA sequence, such as the amplification or deletion of a structural gene for a specific enzyme, or is any change in a specific sequence which confers the selective advantage only a fraction of a much more extensive variation? A further consideration of this latter possibility is, if the sequences conferring the selective advantage are only a fraction of those varying, then are there specific subsets of sequences which are involved in the responses to different selection pressures.

In the following discussion a number of different examples will be considered. These are (a) the selection of methotrexate resistant cell lines; (b) the environmental induction of heritable changes in flax and in particular the characterization of the DNA sequences shown to be susceptible to these changes; and (c) the apparent genetic variation induced during the cycle of tissue culture and subsequent regeneration of plants from culture.

The question which arises from all these examples is how much alteration of the genome has occurred? If a substantial fraction of the genome was susceptible to change, what, if any, were the characteristics of this fraction compared to the remainder of the genome which did not change? A corollary to this question is whether or not the genome is compartmentalized in some way such that variability can be tolerated in some fractions while others are held constant. A comparison will be made between the examples discussed, with particular reference to common mechanisms which may be operating in all the systems. Consideration will also be given to the possible extent to which the environment may play a role in the generation of the diversity on which selection can subsequently act.

8.2. EXAMPLES

8.2.1. Gene amplification and drug resistance

Methotrexate is an analogue of folic acid which binds to the enzyme dihydrofolate reductase (DHFR) preventing the conversion of dihydrofolate to tetrahydrofolate. The latter compound is required for the generation of key precursors of DNA and protein synthesis (Schimke, 1980). Animal cells can usually be killed by low concentrations of the drug, but cells resistant to high levels of methotrexate can be obtained. There are three ways in which this high level of resistance can be achieved: (a) by alterations in the transport of the drug so that the intracellular levels of the drug are minimal (Fischer, 1962); (b) by a mutation in the enzyme DHFR so that it has a lowered affinity for methotrexate; and (c) by an increase in the level of the enzyme in the cell (Hakala *et al.*, 1961). Only the last of the three mechanisms will be considered further here although it is possible that some form of amplification mechanism may result in a change in transport mechanisms (Schimke, 1983).

A stepwise selection procedure has been used to obtain cells in which the level of DHFR has increased (Schimke *et al.*, 1978b). This was done by growing the cells initially on a low concentration of methotrexate which killed almost all the cells. Occasionally cells arose (approximately one in 10^5 in certain systems (Schimke, 1980)) which were resistant to this low level of methotrexate. The resistance cells having more DHFR than the starting cells were then grown in a higher concentration of methotrexate. Once again most of the cells were killed but the survivors had a further increase in the level of DHFR. In this manner cells can be selected for much higher levels of methotrexate resistance and can have 400 times the level of DHFR than the original starting sensitive cells.

It has been shown that the number of copies of the gene for DHFR has been increased in the resistant cell line, as has the amount of mRNA for this gene (Schimke *et al.*, 1978a). In addition the degree of resistance to methotrexate was dependent on the number of copies of the gene present.

Two types of resistant cells with increased DHFR have been observed (Dolnick *et al.*, 1979; Kaufman *et al.*, 1979). In one case the resistance was retained irrespective of the presence or absence of methotrexate, and in the other case the resistance was unstable, so that in the absence of methotrexate the resistance was lost over a number of generations.

In the stably resistant lines the amplified genes have been integrated into the chromosome(s) and so segregate normally at each division. In the methotrexate-resistant murine lymphoblastoid cell line, the amplified DHFR genes have been located, by *in situ* hybridization, in the large homogeneously-staining region of chromosome 2 (Dolnick *et al.*, 1979). This segment probably consists of a tandem array of a repeating unit approximately 800 kilobases long. The DHFR gene, including the introns, spans some 40 kilobases (Kaufman *et al.*, 1978) so that the increase in the number of DHFR genes only accounts for about 5% of the total amplified DNA in this case.

In unstably resistant cell lines the amplified DHFR gene copies are present as extrachromosomal elements termed double-minute chromosomes. The double-minutes probably do not contain centromeric DNA and do not associate with the spindle apparatus at mitosis. This means that they segregate randomly and unequally so generating a heterogeneous population of cells with respect to the number of copies of the DHFR gene. Cells with a lower number of copies of the DHFR gene have a faster generation time with the consequence that in the absence of selection the reversion could be accounted for by the faster growth of cells with fewer double-minutes. It has also been reported that there can be association between the double minutes at mitosis with the subsequent formation of micronuclei which can be expelled (Levan and Levan, 1978). This process could account for the single-step loss of DHFR genes. As was the case of the homogeneously-staining regions, the extra number of copies of the DHFR genes cannot account for all the DNA in the double-minutes so that in this case as well the sequence providing the selectable advantage only contributes a small fraction of the amplified DNA.

Cell lines with amplified numbers of particular genes have been selected in two other cases. These are the increase in the number of genes for aspartate transcarbamylase (Wahl *et al.*, 1979) and for metallothionin-I (Beach and Palmiter, 1981). In both of these cases the size of the piece of DNA amplified was very much larger than the gene for the corresponding mRNA. Thus the amplification of the gene conferring the selectable advantage on the cells only accounted for a small percentage of the total DNA change.

8.2.2. Environmentally induced heritable changes in flax

Heritable changes can be found in some flax varieties after they have been grown in particular environments for a single generation (Durrant, 1962, 1971; Cullis, 1981a). The stable lines produced (termed genotrophs) could differ from each other and the original line from which they were derived (termed P1) in a number of characters, including the total nuclear DNA amount (as determined by Feulgen staining) (Evans *et al.*, 1966), the number of genes coding for the 18S and 25S ribosomal RNAs (rDNA) (Cullis, 1976, 1979), the number of genes coding for the 5S RNA (5SDNA) (Goldsbrough *et al.*, 1981) and in plant weight and height. The initial characterization was on the basis of plant weight so that the genotrophs were designated large (L) and small (S) in relation to this character.

There is no obvious adaptive significance in any of the characters known to vary during the environmental induction of heritable changes in flax. However, since only a limited number of phenotypic characters have been studied, it is possible that only a minor portion of the changes are adaptive. The remainder of the variation may be neutral in the particular inducing environments, and since this fraction may be the major portion any adaptive changes may appear insignificant among the wide range of variation observed.

8.2.2.1. Nuclear DNA Variation

The nuclear DNA amounts of a number of genotrophs have been determined by Feulgen cytophotometry (Evans *et al.*, 1966; Joarder *et al.*, 1975). The difference between the L and S genotrophs observed in these studies was highly significant and has been repeated by many workers. These measurements have shown that the difference can be up to 15% of the total nuclear DNA.

The genome of flax is small compared to most plants (1.5 pg/2C nucleus)

(Timmis and Ingle, 1973) and it has a diploid complement of 30 small chromosomes, all of which are approximately the same size. If all the extra DNA, by which L differed from S was located, or lost, from a single chromosome then there should be one or two noticeably different chromosomes when L and S were compared (corresponding to the homogeneously staining region or double minutes observed in the DHFR system discussed above). However, in both cases all the chromosomes appeared approximately the same size and could not even be distinguished in autotetraploids of L × S (Evans, 1968).

The analysis of the flax genome by renaturation kinetics has shown that the single copy and repeated sequences were arranged in a long period interspersion pattern, and that the genome appeared to be effectively tetraploid (Cullis, 1981b). The extent to which the genomes of the genotrophs differed after the environmental treatments has been determined by an analysis of the renaturation kinetics of the DNA from a number of genotrophs and two which showed large differences were further characterized for a number of sets of DNA sequences including the rDNA, 5SDNA and other cloned sequences (Cullis 1983a, 1983b).

8.2.2.1.1. Analysis of total nuclear DNA

The DNAs from twelve genotrophs were characterized by renaturation kinetics using the hydroxyapatite method (Britten *et al.*, 1974) with DNA of average single strand length of 300 base pairs, and the Cot curves obtained were analysed by a non-linear regression computer program (Cullis, 1983a). Wide variation for both the rate constants and the proportions of the components of the fitted curves were obtained when the DNAs from different genotrophs were analysed. These findings would be consistent with an effect on a wide range of sequences by the events leading to the formation of the genotrophs.

That a wide range of sequences could be affected was confirmed by a more detailed analysis of the DNAs of two genotrophs, one with a high nuclear DNA content (L^H) and the other with a low nuclear DNA content (L_6). These were used in renaturation experiments in which a small amount of the DNA (termed the tracer DNA) from one genotroph was renatured in the presence of a vast excess of DNA (termed the driver DNA) from either itself or from the other line. The reciprocal experiment was also performed. The resulting Cot curves are shown in Figure 1. It can be seen that when the tracer DNA was from the low DNA line a similar curve was obtained irrespective of the driver DNA (Figure 1a). However, when the tracer DNA was from the high DNA line then there were distinct differences between the two curves (Figure 1b). When the driver DNA was from the low DNA line the renaturation reaction was slower in both the highly repeated and intermediately repeated regions. However, in both sets of experiments the reactions were driven to the same extent so giving credence to the hypothesis that there is the same information content in the nuclear DNA in both

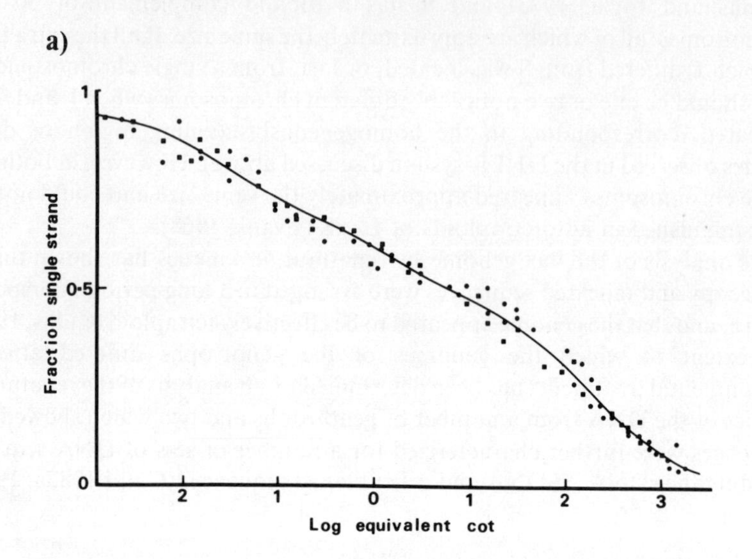

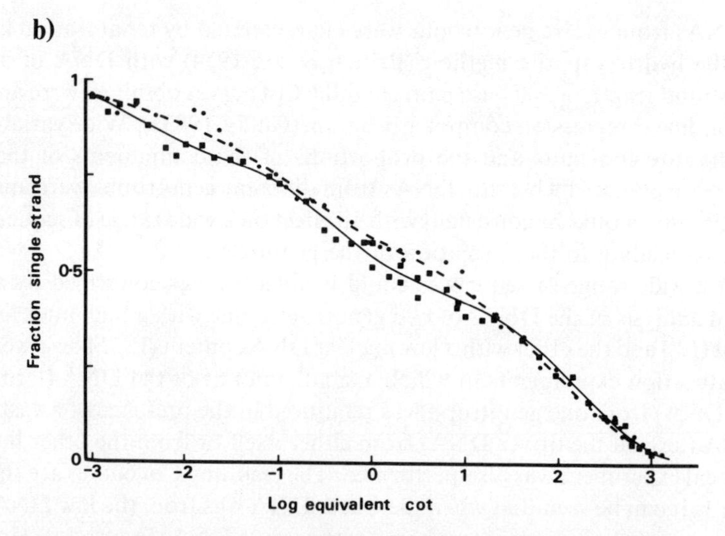

Figure 1a: Reassociation kinetics of L_6 tracer DNA driven by L_6 and L^H DNAs sheared to a single strand length of 300 nucleotides. L_6 driver ●——●; L^H driver ■——■.

b: Reassociation kinetics of L^H tracer DNA driven by L^H and L_6 DNAs sheared to a single strand length of 300 nucleotides. L_6 driver ●– –●; L^H driver ■——■

genotrophs and the differences are mainly due to differences in the number of copies of various sequences.

The measurements from renaturation kinetics are not of sufficient accuracy to determine small changes, especially those occurring in the unique or low copy number sequences. Thus if the differences observed in the expression of particular genes between the genotrophs (for example the isozyme band patterns for peroxidase) were due to the deletion of a single or low copy number sequence such a difference would not have been observed in these experiments. For the determination of this type of difference cloned probes for those specific genes would be needed.

That a wide range of sequences could be affected was confirmed by a more detailed analysis of the DNAs of two genotrophs, one with a high nuclear DNA content and the other with a low nuclear DNA content. It was shown that a substantial fraction of both the highly repetitive DNA and the intermediately repetitive DNA were present in greater abundance in the high line than in the low line (Cullis, 1983a,b). The difference between these two lines in these fractions was calculated to represent about 15% of the total nuclear DNA (Cullis, 1983a) which is consistent with the nuclear DNA differences as estimated by Feulgen staining between the high and low DNA lines. The proportions of chloroplast and mitochondrial DNAs in the preparations of DNA from both the high and low lines were monitored using cloned fragments of chloroplast and mitochondrial DNAs and were found not to differ in the preparations used. Thus the differences observed were not due to varying amounts of organelle DNA in the preparations.

8.2.2.2. Ribosomal DNA variation

The rDNA of flax consists of a homogeneous set of genes arranged in tandem arrays with a repeat length of 8.6 kilobases (Goldsbrough and Cullis, 1981). There was no detectable variation in the size of this repeat nor variation in the position of a number of restriction enzyme sites in the rDNAs from various genotrophs. However, the number of ribosomal genes per genome has been determined in a number of genotrophs and has been shown to vary nearly three-fold (Cullis, 1976). The rDNA gene number has also been determined during the growth of plants under specific conditions (Cullis and Charlton, 1981). It was shown that the rDNA amount in different parts of the plant varied during growth in these environments and that the changes observed were transmitted to the progeny. Thus the plants growing in these specific environments were chimeric for rDNA, having different amounts per cell at the base and the apex. As described above there was no detectable variation in the rDNA genes so that it was impossible to assay what size the units of variation might be, or whether or not any sub-fraction of the rDNA was involved in the amplification or deletion events.

8.2.2.3. 5S DNA variation

The 5S DNA of flax is arranged in tandem arrays of a 0.35 kilobase to 0.37 kilobase repeating sequence (Goldsbrough *et al.*, 1981). There is both length and sequence heterogeneity and the number of copies can vary by more than two-fold. In the genotrophs with the highest number of copies this sequence can comprise some 3% of the genome. It has been possible to differentiate a sub-population of the 5S genes which had been affected. A class of sequences which lacked a TaqI site had been reduced in two independent lines with low 5S gene numbers compared to the lines with high 5S gene numbers (Goldsbrough *et al.*, 1981).

8.2.2.4. Constant DNA Sequences

Attention has been focussed on the DNA sequences which differed, in some way, between the genotrophs. However, it must not be overlooked that the majority of the genome does not differ between the genotrophs. A number of clones representing sequences which are tandemly arrayed repeats within the genome have been isolated and shown not to vary between genotrophs. Among these is the satellite DNA which comprises some fifteen per cent of the total genome (Cullis, 1983b). This sequence is unlikely to be vital for viability at this level since numerous other *Linum* species do not have any significant amount of this sequence.

The question to be answered is in what way are the sequences, or even the sub-populations of the same set of repeating unit, different when those which are susceptible to change are compared to those which remain invariant? At present there is no answer to this question in the flax system but this question will be returned to when the possible mechanisms by which the changes occur are discussed.

8.2.3. Somoclonal variation

Historically plant tissue culture was essentially seen as a method for cloning a particular genotype. From this standpoint it was assumed that all the plants arising from a tissue culture should be exact copies of the parental plant. However, phenotypic variants have frequently been observed among regenerated plants. In addition variation in the sub-clones of particular perental cell lines has also occurred. Thus it would seem that the process of tissue culture *per se* may be a source of variability (Larkin and Scowcroft, 1981).

There are a number of possibilities for the mechanisms by which the culture variation can arise. These include karyotype changes, chromosome rearrangements and gene amplification and deletion events.

Karyotype changes have been frequently observed in tissue cultured plant cells (Skirvin, 1978). However, phenotypic variation has been observed in the absence of gross karyotype changes so that this type of mechanism cannot account for all of the tissue culture variability.

In a number of tissue culture systems chromosome rearrangements such as breakage/fusion events (Orton, 1980), reciprocal translocations, deletions, and inversions (Ahloowalia, 1978) have been observed. All of these events could have effects on the phenotype both by the loss of genetic material during the rearrangement process as well as by position effects due to the altered chromosomal locations of the rearranged material. Such position effects have been extensively documented in *Drosophila* species (Spofford, 1976).

Particular gene amplification and deletion events have not been extensively demonstrated at the DNA level in plant tissue cultures. One case where the number of the genes coding for the ribosomal RNA have been altered by changes in the culture medium has been described (Jackson and Lark, 1983). In addition, when plant cells have been subjected to stepwise selection in culture, in a manner analogous to that used to obtain methotrexate resistant cells, then variants resistant to higher levels of the toxic agent have been derived. This has been demonstrated for resistance to acetohydroxamate acquired by increases in urease in cultured tobacco cells (Yamaya and Filner, 1981). Many of the properties of the increased urease level appear to be analagous to the increases in DHFR, namely: (a) the level of urease increased in a stepwise fashion, initially to four times the starting level after 40 generations and could subsequently increase up to twelve-fold; (b) the initial increase in urease level is unstable in the absence of the acetohydroxamate, but after prolonged growth in the presence of this agent the urease level becomes independent of the presence or absence of aceto-hydroxamate. It has not been shown that the increase in urease level was caused by an amplification of the gene but the characteristics of the system make this probable and this is being further investigated.

8.3. DISCUSSION

The changes in the genome observed in response to particular components of the environment must be compared to the occurrence of variation in the absence of those particular components. How common is gene amplification? Based on the studies in *Salmonella* (Anderson and Roth, 1977) it was suggested that as many as five genes may be duplicated in each cell at any one time. In higher organisms this frequency would not appear to be nearly as high. However, given the increasing number of examples in which gene amplification occurs (Schimke, 1982) it is possible that it is relatively common. In gene duplications in both bacteria and higher organisms the amplified genes appear to be unstable initially so that many instances of amplification may have been missed due to its transient nature. In general the existence of an amplification event is most easily demonstrated when it has been stabilized.

Some form of gene duplication is a common theme in development. Perhaps the simplest forms of such duplications are the generation of polyploidy in certain cell types and polytenization of insect chromosomes in specific tissues (Nagl, 1978). More selective examples would be the DNA loss and amplification in the macronucleus of various protozoa (Lawn *et al.*, 1977; Yao *et al.*, 1978). The selective amplification of specific genes was first described for the ribosomal genes in *Xenopus* (Brown and Dawid, 1968) and has also been shown for the Chorion genes of *Drosophila* (Spradling, 1982) and the α-actin genes in chick embryo (Zimmer and Schwartz, 1982). Each of these examples is interesting in comparison to other organisms undertaking comparable development in which there has been no demonstration of amplification. Many other examples of developmental amplification may have gone undetected because of the particular organism studied or because of the transient nature of the phenomenon. Similarly the extent of the responses to environmental stimuli may have been underestimated due to the material used.

The mechanisms by which the duplications occur are not known in all cases. However, for the amplification of the rDNA in *Xenopus* it appears to take place by an extrachromosomal mechanism. By what mechanisms do the amplifications occur in response to the environmental stimuli? In the case of the amplification of the DHFR genes in response to selection on methotrexate three general mechanisms have been considered (Schimke, 1982). These are: (a) DNA uptake from killed cells; (b) unequal sister chromatid exchange; and (c) disproportionate replication. A model based on the third of these mechanisms has been proposed (Schimke, 1982) as it is the only one which is also compatible with the developmentally regulated amplifications.

The central feature of this disproportionate replication model is the finite probability that replication can be initiated at the same origin, or at an origin within the amplified DNA sequence, more than once in a given S phase of the cell cycle. A second feature is the concept that the DNA replication-elongation slows down and/or ceases at some site within the chromosome prior to joining with a replication fork proceeding from an adjacent origin of replication. The consequence of multiple rounds of replication is the presence of many strands of DNA which are unattached within the chromosome and can be released as linear or circular molecules. If such a structure could replicate then it would constitute a double minute chromosome. The generation of such duplicated segments could be at random and if there was no selection for their retention then they would be lost. In this case there would be no evidence of their existence. Chromosomally located sequences could be generated by the end-to-end ligation of the free DNA strands followed by their insertion into the chromosome either in the vicinity of the resident gene or elsewhere.

This model of disproportionate replication for the generation of extra-chromosomal sequences is compatible with an earlier model proposed to account for the envrionmentally induced changes in flax (Cullis, 1977). In the flax model it was proposed that the DNA changes were mediated via an extrachromosomal intermediate with subsequent over or under replication occurring in the extrachromosomal segments. The disproportionate replication model proposes a mechanism by which these extrachromosomal segments can be generated. The definitive demonstration of the applicability of the flax model system requires the demonstration of the extrachromosomal segments of DNA in flax and, more particularly, of those sequences known to vary between the genotrophs.

This model raises questions concerning the interaction of the cellular replicative process and the DNA structure in generating faithful DNA replication and the generation of the amplified sequences. What constitutes an origin of replication in higher organisms? How is replication controlled and how often is there more than one initiation of a DNA sequence within a single S phase? How may alterations in the cell metabolism and alterations in DNA structure imposed by a variety of environmental agents change the 'normal' replicative process of the cells?

If the multiple initiation of replication is responsible for the amplification events how can this be reconciled with the apparent separation of the sequences of the genome in those which appear to be variable and those which appear to be constant. One possibility is that the time in S when a sequence is replicated is related to the probability that multiple initiations can occur. Thus for sequences which are replicated early in S there is a higher probability that multiple initiations could occur and so these sequences should be more likely to be present in any amplified units. This prediction remains to be tested although the appropriate material may be available with the range of variable and constant sequences in the flax system.

The model described above gives a possible direct action to the environment in the generation of amplified sequences and not just a passive one in the selection of more fitted cells due to the 'normal' generation of random amplified segments. It is possible that the selection procedures used to select cells may themselves perturb the metabolism of the cells and so increase the rate of production of amplified segments. If the environment was having this effect it would mean that the environment was acting to generate the variability on which it was subsequently acting as the selective agent and not passively present as a screen through which a relatively constant genotype with rare variants had to pass.

The question as to how general a phenomenon the environmental induction of DNA changes in plants is has yet to be answered. It has been shown directly in the case of flax and perhaps indirectly in Nicotiana but has not been looked for directly in any other systems. It would seem unlikely that these two sets of plants are the only ones to respond in this way. In addition if the disproportionate replication model has any credibility then most plant species, when placed under the appropriate stress should have their DNA replication cycle interrupted with the concommitant change in DNA. The process may account for the wide variation seen in the copy numbers of various repetitive DNAs when closely related lines are compared. However, the appropriate experiments to determine whether or not these differences can occur in a single generation have not been performed. With the availability of a large number of genomic libraries in plants and so the presence of specific probes the question of the plasticity of the plant genome, in a general sense, should be answered in the near future.

ACKNOWLEDGEMENTS

This chapter was written while the author was in receipt of a Civil Service (Nuffield and Leverhulme) Travelling Fellowship.

8.4. REFERENCES

Ahloowalia, B. S. (1978) Novel ryegrass genotypes regenerated from embryo-callus culture, *Fourth Intl. Congr. Plant Tissue Culture*, **Abstract, 162**. Calgary, Canada.

Anderson, R. P., and Roth, J. R. (1977) Tandem genetic duplications in phage and bacteria, *Ann. Rev. Microbiol.*, **31**, 473–504.

Beach, L. R., and Palmiter, R. D. (1981) Amplification of the metallothionin-I gene in cadmium resistant mouse cells, *Proc. Natl. Acad. Sci. USA*, **78**, 2110–2114.

Britten, R. J., Graham, D. E., and Neufeld, B. R. (1974) Analysis of repeating DNA sequences by reassociation kinetics, in Grossman, L., and Moldave, K. (eds.) *Methods in Enzymology*, Academic Press, New York, Vol. 29, pp. 363–405.

Brown, D. D., and Dawid, I. D. (1968) Specific gene amplification in oocytes, *Science*, **160**, 272–280.

Cullis, C. A. (1976) Environmentally induced changes in ribosomal RNA cistron number in flax, *Heredity*, **36**, 73–79.

Cullis, C. A. (1977) Molecular aspects of the environmental induction of heritable changes in flax. *Heredity*, **38**, 129–154.

Cullis, C. A. (1979) Quantitative variation of ribosomal RNA genes in flax genotrophs, *Heredity*, **42**, 237–246.

Cullis, C. A. (1981a) Environmental induction of heritable changes in flax: Defined environments inducing changes in rDNA and peroxidase isozyme band pattern, *Heredity*, **47**, 87–94.

Cullis, C. A. (1981b) DNA sequence organization in the flax genome, *Biochim. et Biophys. Acta*, **652**, 1–15.

Cullis, C. A. (1983a) Environmentally induced DNA changes in plants, *CRC Critical Reviews in Plant Science*, **1**, 117–131.

Cullis, C. A. (1983b) Variable DNA sequences in flax, in Chater, K. F., Cullis, C. A., Hopwood, D. A., Johnston, A. W. B., and Wollhouse, H. W. (eds.) *Genetic Rearrangement. The Fifth John Innes Symposium*, Croon Helm, UK, pp. 253–264.

Cullis, C. A., and Charlton, L. (1981) The induction of ribosomal DNA changes in flax, *Plant Science Letters*, **20**, 213–217.

Dolnick, B. J., Berenson, R. J., Bertino, J. R., Kaufman, R. J., Nunberg, J. H., and Schimke, R. (1979) Correlation of dihydrofolate reductase elevation with gene amplification in a homogeneously staining chromosomal region in C5178Y cells, *J. Cell Biol.*, **83**, 394–402.

Durrant, A. (1962) The environmental induction of heritable changes in *Linum*, *Heredity*, **17**, 27–61.

Durrant, A. (1971) Induction and growth of flax genotrophs, *Heredity*, **27**, 277–298.

Evans, G. M. (1968) Nuclear changes in flax, *Heredity*, **23**, 25–38.

Evans, G. M., Durrant, A., and Rees, H. (1966) Associated nuclear changes in the induction of flax genotrophs, *Nature*, **212**, 697–699.

Fisher, G. A. (1962) Defective transport of amethopterin (methotrexate) as a mechanism of resistance to the antimetabolite in L5178Y leukemia cells, *Biochemical Pharmacology*, **11**, 1233–1234.

Flavell, R. B. (1982) Sequence amplification, deletion and rearrangement: major sources of variation during species divergence, in Flavell, R. B., and Dover, G. A. (eds.) *Genome Evolution*, Academic Press, New York, pp. 301–323.

Goldsbrough, P. B., and Cullis, C. A. (1981) Characterization of the genes for ribosomal RNA in flax, *Nucleic Acids Research*, **9**, 1301–1309.

Goldsbrough, P. B., Ellis, T. H. N., and Cullis, C. A. (1981) Organization of the 5S RNA genes in flax, *Nucleic Acids Research*, **9**, 5895–5904.

Hakala, M. T., Zakrzewshi, S. F., and Nichol, C. A. (1961) Relation of folic acid reductase to amethopterin resistance in cultured mammalian cells, *J. Biol. Chem.*, **236**, 952–958.

Hill, J. (1965) Environmental induction of heritable changes in Nicotiana rustica, *Nature*, **207**, 732.

Jackson, P. J., and Lark, K. G. (1983) Inherited changes in frequencies of different ribosomal RNA cistrons which occur in soybean (*Glycine max*) suspension cultures as a result of altered growth conditions, *Molec. gen. genet.* (submitted).

Joarder, I. O., Al-Saheal, Y., Begum, J., and Durrant, A. (1975) Environments inducing changes in the amount of DNA in flax, *Heredity*, **34**, 247–253.

Kaufman, R. J., Bertino, J. R. and Schimke, R. T. (1978) Quantitation of dihydrofolate reductase in individual parental and methotrexate-resistant murine cells, *J. Biol. Chem.*, **253**, 5852–5860.

Kaufman, R. J., Brown, P. C., and Schimke, R. T. (1979) Amplified dihydrofolate reductase genes in unstably methotrexate-resistance cells are associated with double minute chromosomes, *Proc. Natl. Acad. Sci. USA*, **76**, 5669–5673.

Larkin, P. J., and Scowcroft, W. R. (1981) Somoclonal variation—a novel source of variability from cell cultures for plant improvement, *Theor. Appl. Genet.*, **60**, 197–219.

Lawn, R. M., Heumann, J. M., Herrick, G., and Prescott, D. M. (1978) The gene sized molecules in *Oxytricha*, *Cold Spring Harbor Symposia Quant. Biol.*, **42**, 483–492.

Levan, A., and Levan, G. (1978) Have double minutes functioning centromeres? *Hereditas*, **88**, 81–92.

Nagl, W. (1978) *Endopolyploidy and Polyteny in Differentiation and Evolution*. Elsevier North-Holland Biomedical Press, Amsterdam.

Orton, T. J. (1980) Chromosomal variability in tissue cultures and regenerated plants of *Hordeum*, *Theoret. Appl. Genet.*, **56**, 101–109.

Schimke, R. T. (1980) Gene amplification and and drug resistance, *Scientific American*, **243(5)**, 50–59.

Schimke, R. T. (1982) *Gene Amplification*, Cold Spring Harbor Laboratory, New York.

Schimke, R. T. (1983) Gene amplification in mammalian somatic cells, in Chater, K. F., Cullis, C. A., Hopwood, D. A., Johnston, A. W. B., and Wollhouse, H. W. (eds.) *Genetic Rearrangement. The Fifth John Innes Symposium*, Croom Helm, UK, pp. 235–251.

Schimke, R. T., Alt, F. W., Kellems, R. F., Kaufman, R. J., and Bertino, J. R. (1978a) Amplification of dihydrofolate reductase genes in methotrexate-resistant cultured cells, *Cold Spring Harbor Symp. Quant. Biol.*, **42**, 649–657.

Schimke, R. T., Kaufman, R. J., Alt, F. W., and Kellems, R. F. (1978b) Gene amplification and drug resistance in cultured murine cells, *Science*, **202**, 1051–1055.

Skirvin, R. M. (1978) Natural and induced variation in tissue culture, *Euphytica*, **27**, 241–256.

Spofford, J. B. (1976) Position-effect variagation in *Drosophila*, in Ashburner, M., and Novitiski, E. (eds.) *The Genetics and Biology of Drosophila*. Academic Press, London, Vol. 1C, pp. 955–978.

Spradling, A. C. (1982) Chorion gene amplification during the development of *Drosophila* follicle cells, in Schimke, R. (ed.) *Gene Amplification*, Cold Spring Harbor, New York.

Timmis, J. N., and Ingle, J. (1973) Environmentally induced changes in rRNA gene redundancy, *Nature New Biology*, **244**, 235–236.

Wahl, G. A., Padgett, R. A., and Stark, G. R. (1979) Gene amplification causes the overproduction of the first three enzymes of UMP synthesis in N-(phosphono-acetyl)-L-aspartate-resistant hamster cells, *J. Biol. Chem.*, **254**, 8679–8689.

Yamaya, T., and Filner, P. (1981) Resistance to acetohydroxamate acquired by slow adaptive increases in urease in cultures tobacco cells, *Plant Physiol.*, **67**, 1133–1140.

Yao, M. C., Blackburn, E., and Gall, J. G. (1979) Amplification of the rRNA genes in *Tetrahymena*, *Cold Spring Harbor Symp. Quant. Biol.*, **42**, 1293–1296.

Zimmer, W. E., and Schwartz, R. J. (1982) Amplification of chicken actin genes during myogenesis, in Schimke, R. (ed.) *Gene amplification*, Cold Spring Harbor, New York.

Evolutionary Theory: Paths into the Future
Edited by J. W. Pollard

Chapter 9

The somatic selection of acquired characters

EDWARD J. STEELE
Department of Immunology,
The John Curtin School of Medical Research,
Australian National University,
Canberra, Australia 2601

REGINALD M. GORCZYNSKI
Ontario Cancer Institute,
500 Sherbourne Street,
Toronto, Ontario M4X 1K9,
Canada

and

JEFFREY W. POLLARD
MRC Human Genetic Diseases Research Group,
Department of Biochemistry,
Queen Elizabeth College,
University of London,
Campden Hill, London W8 7AH,
UK

9.1. INTRODUCTION

During the latter part of the nineteenth century, the German biologist August Weismann developed his theory of 'the continuity of the germ-plasm'. This theory was to become one of the cornerstones of twentieth century evolutionary biology and may be summarized by the following quotation:

> '...all permanent—i.e. hereditary—variations of the body proceed from primary modifications of the primary constituents of the germ, and that neither injuries,

> functional hypertrophy and atrophy, structural variations due to the effect of temperature or nutrition, nor any other influence of environment on the body, can be communicated to the germ cells, and so become transmissible' (Weismann, 1893, p. 395).

Weismann's theoretical and experimental work effectively delineated the germ-plasm (the DNA of the germline in modern terms; for an up-to-date review see Medvedev, 1981) as the only site of genetic variation relevant to heredity and therefore, for change upon which natural selection may act. Thus Weismann's barrier separating the germline genes from somatic influence provided the conceptual framework for the rise of Mendelian genetics and its evolutionary offshoot the synthetic or neo-Darwinian theory. This neo-Darwinian theory of chance mutation in the germline genes generating gradual changes in phenotype upon which natural selection acts allowing survival of the fittest as both *necessary* and *sufficient* explanation for the mechanism of evolutionary change (see, for example, Mayr, 1959), is under increasing challenge from a variety of biological disciplines (Ho and Saunders, 1984; and Chapter 5). Among these critics are many who feel that in order to explain phenomena such as homology, parallel evolution, the apparent rate of evolutionary change, and the perfect fit between environment and organism, known as adaptation, there must be some interaction between development, the environment and the evolutionary process over and above the selection against less fit individuals. Explanations for such a relationship range from within neo-Darwinian terms (Waddington, 1961) to epigenetic theories of inheritance (Ho and Saunders, 1979). Recently, however, in contravention to Weismann's doctrine, Steele (1979) has proposed a plausible genetic mechanism, known as the somatic selection hypothesis, whereby acquired somatic genetic modifications selected in response to environmental changes, may become inherited. It is the purpose of this chapter to examine the evidence for the permeability of Weismann's barrier and the inheritance of acquired characters and specifically to adjudge the data in relation to Steele's somatic selection hypothesis.

9.2. THE SOMATIC SELECTION HYPOTHESIS FOR THE INHERITANCE OF ACQUIRED CHARACTERS

The theory states that clonally selected somatic genetic variants will express these variations such that the information (probably mRNA) may be captured by endogenous retroviruses which, in turn, will transmit the information across the tissue barriers partitioning the soma from the germline and infect the germ cells. Once in the germline the somatic information carried as RNA will be reverse transcribed into DNA by the viral encoded RNA-dependent DNA polymerase (reverse transcriptase), allowing the integration of the captured genes into the appropriate position in the germline DNA (Figure 1). The above hypothesis is best illustrated by consideration of the immune system.

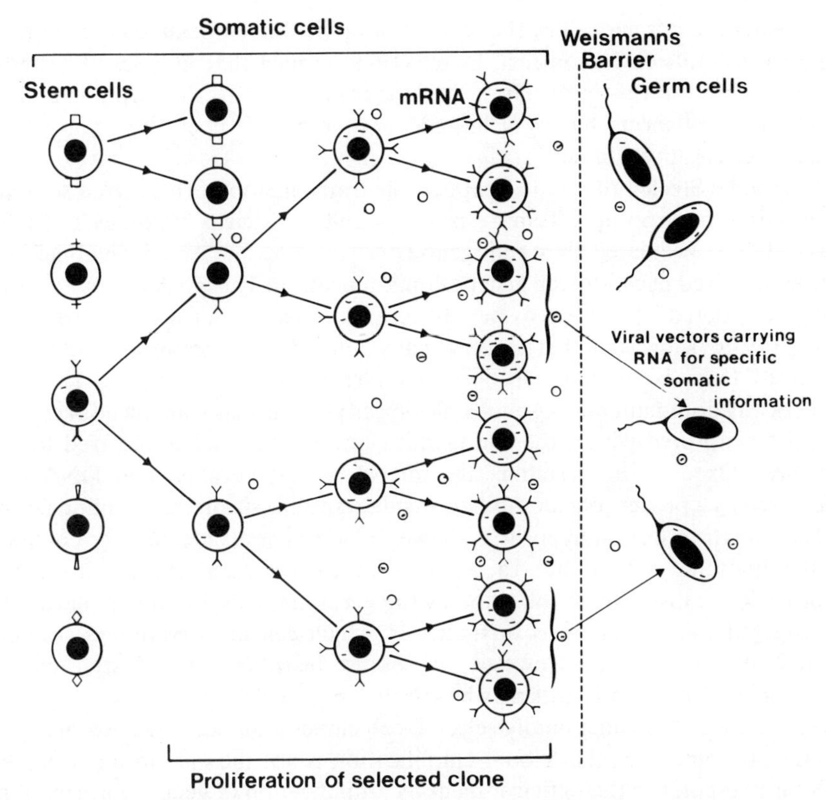

Figure 1. Schematic diagram of the somatic selection hypothesis.
A single cell displaying a unique character is caused by an environmental feature to selectively proliferate. During its proliferation, the cell produced mRNA (~) coding for the specific protein (Y) which is captured by endogenous retroviruses (○) and carried across Weismann's barrier to the germ cells

Every day stem cells produce large numbers of new lymphocytes, each displaying a unique antigenic specificity that is, in part, defined by germline encoded genes and, in part, by somatic mutation (Cohn *et al.*, 1980; Tonegawa, 1983). This process results in an enormous diversity of antigenic specificities such that it can be confidently predicted that any antigen found in nature, or synthesized by man, will have an antibody directed against it. Each specificity is normally borne on only a few precursor cells, but upon interaction with an antigen the clone displaying the appropriate specificity is stimulated to divide, until in cases of extreme antigen exposure, the somatic compartment may be dominated by that particular clonal specificity. This clonal selection of acquired immunity is Darwinism at a somatic level (Burnet, 1959). However, under conventional evolutionary theory, since these new genetic variants arise in the soma, they will have no impact on the antigenic specificity of the offspring, except to the extent that it may allow the parents to survive epidemics and contribute to the next generation. In contrast, Steele (1979) argued that in cases of extreme somatic selection, this information may be transmitted to the germline and so result in the offspring having a biased immunological repertoire upon which natural selection would act.

The model Steele proposed to explain the transmission of the altered somatic information incorporated Temin's provirus and protovirus hypotheses. Temin (1971, 1976) speculated that endogenous retroviruses may be normal cellular entities involved in cell-to-cell genetic communication. The RNA tumour viruses being considered, therefore, to be pathological offshoots of normal processes. Steele (1979) suggested that if genetically altered somatic information were amplified (e.g., by clonal selection), then representative copies of the now numerically dominant mRNA molecules specifying the new somatic information would be captured within the endogenous retroviruses and transferred to the germline. Once in the germline, the mRNA would be copied to DNA and integrated by a process of recombination into a specific site in the germline DNA.

The somatic selection hypothesis, however, was not considered to be confined to the immune system, but to be relevant to a general theory of genetic information transfer from the soma to the germline. It is, of course, harder to envisage similar mechanisms in tissues in which cell proliferation is severely limited, e.g. the brain. However, it should be noted that firstly, during development, specific variants may be selected, e.g. in the immune system specific B-cell variants may functionally select T-cell clones and vice versa. Secondly, it has now become clear that clonal amplification is not the only means whereby cells may respond to their environment by amplification of genetic information. There are many examples of functional gene amplification. Thus, in cells resistant to the anti-tumour drug, methotrexate, which inhibits the enzyme dihydrofolate reductase (DHFR), the gene for DHFR may be amplified many hundred-fold, such that DHFR can represent 5% of cell protein (Schimke *et al.*, 1978). Similarly, gene amplification can occur for the multifunctional CAD protein in cells resistant to *N*-(phosphonoacetyl)-L-aspartate, an inhibitor of aspartate transcarboxylase (Wahl *et al.*, 1979). Thus, it may be that gene amplification, first described for rRNA genes (Long and Dawid, 1980) is a general adaptive response of eukaryotic cells and may, of course, involve mutant genes (Lewis *et al.*, 1982) and in some circumstances, may be recalled by renewed environmental selection (Coderre *et al.*, 1983). Thirdly, Steele (1981a) has suggested that the concept of idiotypy (somatically variable sensor proteins) may not be restricted to the immune system, but be a general feature of recognition systems that respond to the environment in an adaptive fashion; the most obvious examples being the brain and liver. In a recent review, for example, Nerbert (1979) has drawn parallels between the liver drug detoxifying cytochrome P450 system, and the immune system. These parallels include the ability to respond to chemicals synthesized by man and thus not anticipated in nature, the presence of multiple cytochrome P450s with different specificities and the presence of a specific memory. The exact genetic basis of these parallels will have to await the molecular analysis of individually cloned cytochrome P450 genes. These genes however, do occur in multigene families and they contain hyper-variable regions (Atchison and Adesnik, 1983). Following on from these speculations, in the next two sections, we will consider the experimental evidence consistent with the somatic selection hypothesis.

9.3. EVIDENCE IN SUPPORT OF THE SOMATIC SELECTION HYPOTHESIS

Gorczynski and Steele (1980, 1981) using the classical experimental system of acquired neonatal tolerance to foreign histocompatibility antigens developed by Medawar and co-workers (Billingham *et al.*, 1956), provided evidence in support of the somatic selection hypothesis. The experimental design is shown in Figure 2. In brief, neonatal mice were injected intraperitoneally with a high dose ($\geq 5 \times 10^7$) of living bone marrow and spleen cells from F_1 mice displaying foreign histocompatibility antigens. Thereafter, all mice received biweekly injections of the same high dose of F_1 lymphoid cells to maintain chimerism and a profound state of tolerance. After eight weeks, male mice were mated to females of the same inbred strain which had not been exposed to the foreign antigen. In all breeding programmes, age-matched controls were bred alongside the test animals and breeding was carried out through the males to exclude interpretation involving direct maternal effects across the placenta or via the colostrum. Progeny were tested for their reactivity to foreign antigens by an *in vitro* five day cell-mediated lympholysis assay (CML). Other progeny were bred to give further generations, either by brother–sister matings, or by out-breeding to untreated females of the same strain.

In the first group of experiments (Gorczynski and Steele, 1980) CBA mice (H-$2K^kD^k$) were rendered tolerant of major histocompatibility antigens of A/J mice by neonatal injections of $(CBA \times A/J)F_1$ (H-$2K^kD^{k/d}$) cells. A significant proportion (50–60 %) of progeny of tolerant fathers failed to produce detectable anti-A/J cytotoxic responses as determined in an *in vitro* spleen cell CML assay. Responsiveness to unrelated third party antigens C57BL/6J and B10.A(2R) was normal, suggesting that the effect seen in the progeny was specific for the A/J determinant used to induce tolerance in the father (Figure 3). Second generation offspring also showed a high (20–40 %) proportion of mice with a diminished anti-A/J response, regardless of whether they were in- or out-bred.

In the second set of experiments, Gorczynski and Steele (1981) bred from congenic B10.SgN(H-$2K^bD^b$) mice tolerized to two different haplotypes, B10.D2(H-$2K^dD^d$) and B10.BR(H-$2K^kD^k$). It was shown that a specific tolerance state appears in the progeny at high frequency for each of the haplotypes independently and often simultaneously in both first and second generation mice

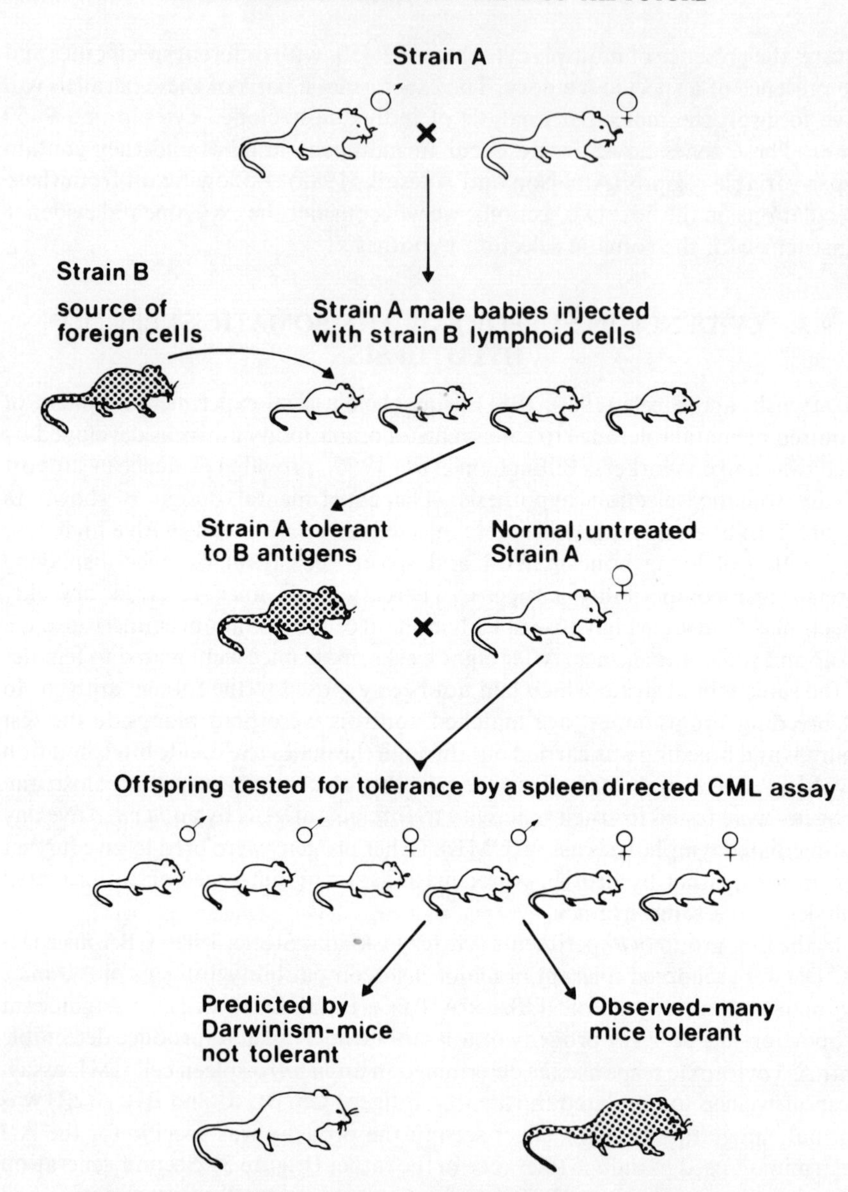

Figure 2. Simplified protocol of the Gorczynski–Steele experiments. Strain A male mice are made tolerant to strain B antigens by neo-natal injection of $F_1(A \times B)$ lymphoid cells. These mice are bred to normal strain A females and the progeny tested by a CML assay for tolerance to strain B antigens. By conventional hypothesis the progeny should not be tolerant, but experimentally many mice were observed to be tolerant

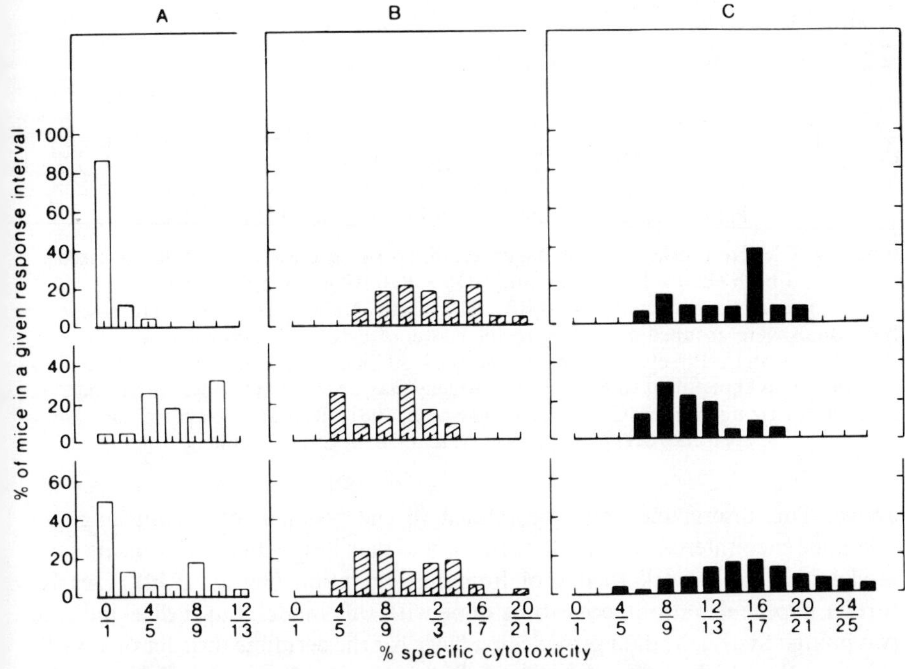

Figure 3. Cytotoxic responses produced by 35 normal (CBA × A/J)F_1 (top), 32 CBA mice (middle) and 83 first-generation tolerant progeny (bottom).

Cytotoxic responses were measured for A/J (A), B10.A (2R) (B), and CB7BL/6J (C) antigens. Both the F_1 and tolerant progeny displayed significantly lower responses to A/J compared to the CBA controls. Data from Gorczynski and Steele (1980)

raised in the absence of deliberate exposure to the tolerizing antigen (Figure 4). Animals tolerant to only one haplotype either maintained or lost that tolerance in following generations in the manner expected of a single dominant Mendelian allele.

In a further study, Steele and Gorczynski (1981) examined the ability of male CBA mice exposed since birth to repeated doses of sheep red blood cells (SRBC, an inert antigen), to transmit to offspring modifications in B-cell function. The spleen cells of the progeny were tested *in vitro* by measuring the number of haemolytic plaque forming cells (PFC) in response to the B-cell mitogen, LPS. In many progeny of male mice exposed repeatedly to SRBC (considered operationally 'tolerant'), low LPS induced PFC responses occurred in a pattern similar to the 'tolerant' parents.

This study also assayed the cytotoxic T-cell responsiveness in a CML assay, but in this case the major observation was a significant perturbation in the patterns of response, rather than the more restricted reproducible immunomodulation seen in the studies of tolerance to single determinants described

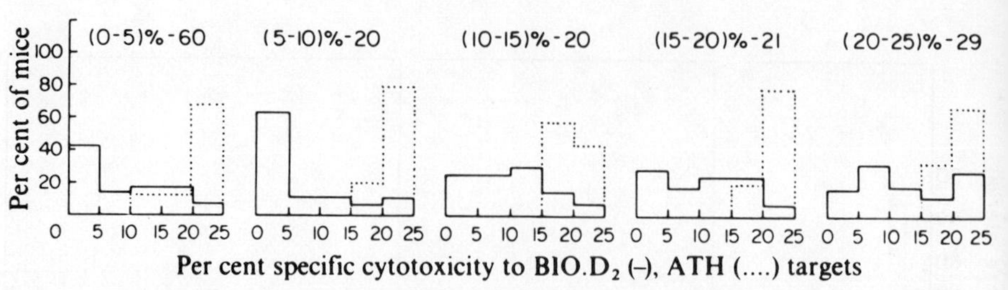

Figure 4. CML response profile of first generation progeny born to B10 males tolerized as neonates to both (B10 × B10.D_2)F_1 and (B10 × B10.BR)F_1 lymphoid cells.

Cytotoxic responses against B10.BR, B10.D_2 or ATH spleen cells were assessed. Individuals were grouped according to their level of cytotoxic response to one antigen (B10.BR; (0–5), (5–10), etc.) and the degree of cytotoxicity to the other antigen (B10.D_2) recorded. This experiment shows that the progeny may be tolerant to one, both or neither parental tolerizing antigen (Gorczynski and Steele, 1981). Reprinted by permission from Nature, **289**, 679, © 1981 Macmillan Journals Limited

above. This discrepancy may be related to the problem of establishing full tolerance encountered in this SRBC study, and may be rationalized by using such models as the network theory of immune regulation (Jerne, 1974). Clearly, further experimentation needs to be done with this model, but it does indicate two points: firstly, the data provides evidence for the germline transfer of a B-cell response (low LPS responses) to both first and second generations. Secondly, since an inert antigen was used, it excludes models which suggest that donor lymphocytes from the initial tolerizing lymphoid cells were transmitted in the seminal fluid, followed by establishment and replication in the developing embryo.

Several groups have attempted to reproduce the above experimental results, with both positive and negative outcomes (Brent *et al.*, 1981, 1982; Smith, 1981; Mullbacher *et al.*, 1983). In the most extensive of these, Brent and co-workers (1981, 1982) followed the general design of the CBA tolerant to A/J experiment testing the progeny for tolerance both in a spleen and peripheral blood cell mediated CML assay and also their ability to accept (CBA × A/J)F_1 skin grafts. In both groups of experiments, Brent and co-workers claimed to have failed to reproduce the data of Gorczynski and Steele (1980). Steele (1981b) however, in an independently confirmed analysis (Josovic, 1982; but see Brent, 1981), re-examined the skin graft data and suggested that there was a statistically significant delay in rejection time of (A/J × CBA)F_1 skin. In a further study, Brent *et al.* (1982) failed to find a significant delay in rejection time of F_1 skin grafts in the progeny of tolerant mice. More recently, however, Gorczynski *et al.* (1983) have repeated the original (CBA × A/J)F_1 tolerizing schedule and tested the progeny not only for their level of cytotoxicity against A/J antigens using a peripheral blood cell mediated CML assay, but also their ability to accept (CBA × A/J)F_1 skin. The data support the original results of Gorczynski and

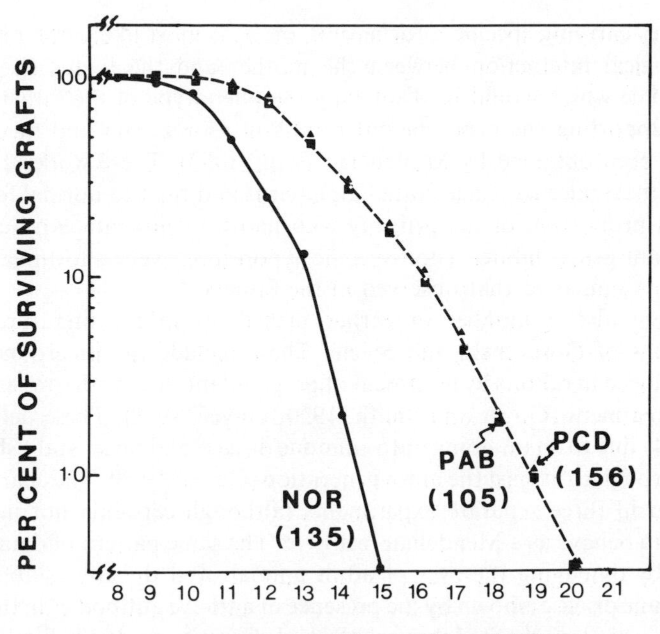

Figure 5. Skin graft survival curves for (CBA × A/J)F_1 female tail skin grafted to progeny mice from: NOR—normal untreated CBA mice (○); PAB—tolerant males bred with tolerant females (○); PCD—tolerant males bred with normal females (×).
The numbers in brackets represent the number of mice tested. Both the PAB and PCD curves show a significant delay in rejection of the graft when compared with the normal mice. Data from Gorczynski *et al.* (1983)

Steele (1981) and also showed a significant delay in the rejection time of A/J antigen bearing skin, although complete acceptance of the skin grafts was not observed (Figure 5).

The discrepancy in the data between these groups needs to be resolved. Gorczynski *et al.* (1983) have suggested that it may, in part, be due to the *in vitro* CML assay employed by Brent *et al.* (1981, 1982), since the very high level of lysis obtained makes interpretation difficult and also, since not all the assays are proportional to dilution, determination at a single target: effector ratio is liable to lead to false conclusions. In this study, Gorczynski *et al.* (1983) also bred to normal males the females that had borne at least three waves of progeny to tolerant males. Surprisingly, the offspring of this breeding programme also showed specific tolerance to the A/J antigens and may indicate a maternal component to the transmission of tolerance. This may be due to immunization of the females with the sperm or seminal fluid, infection of the oocytes with retroviruses carrying specific information, or, as is most likely, the result of an immunological interaction between the mother and the foetus sired by the tolerant male which would be displaying the phenotype of the father.

Data supporting the experimental results of Gorczynski and Steele (1980, 1981) has been obtained by Mullbacher *et al.* (1983). These workers tolerized neo-natal male mice to γ-inactivated alphavirus and bred to normal females. A significant proportion of the progeny who had no deliberate exposure to the tolerizing antigen, exhibited a non-specific hyporesponsiveness to alphavirus and alloantigens similar to that observed in the fathers.

There are also a number of earlier precedents in the literature for the observations of Gorczynski and Steele. These include the inheritance of eye defects induced in rabbits by treatment of pregnant mothers with fowl anti-rabbit eye lens treatment (Guyer and Smith, 1920; Guyer, 1928). These defects were presumably due to an on-going auto-immune disease and once established could be bred through to at least the ninth generation (Guyer, 1928). The character was established in three separate experiments (although certainly not in all) and appeared to behave as a Mendelian recessive. The same pattern of data was also obtained by damaging the eyes of adult animals and thereby establishing an auto-immune disease (shown by the presence of anti-eye antibodies in the blood). This defect was also inherited through the male line (Guyer, 1928). Similar results to those of Gorczynski and Steele showing the inheritance of acquired tolerance, were performed by Guttman and co-workers (Guttman and Aust, 1963; Guttman *et al.*, 1964) and by Kanazawa and Imai (1974). In the former, and more simply interpreted, C3H males were made tolerant to A strain histocompatibility antigens and mated to $(A \times C3H)F_1$ females. The backcross progeny, together with normal backcross offspring, were tested for growth of $(A \times C3H)F_1$ passaged tumour and for the ability to accept $(A \times C3H)F_1$ skin. Among the normal backcross progeny, about 8% allowed growth of the $(A \times C3H)F_1$ tumour, whereas of the backcross progeny of the A-tolerant C3H males, 90% allowed growth of the tumour. Similarly, normal progeny failed to accept F_1 skin (3% accepted), while 100% (11/11) of the tolerant progeny accepted the $(A \times C3H)F_1$ skin and in secondary testing all the animals rejected $(C57B1/1 \times C3H)F_1$ skins, indicating that the transmission of tolerance was specific. Unfortunately, this work was not carried on to the second generation and in attempts to repeat (Steinmuller, 1967; Brent *et al.*, 1982) positive data was not obtained. It should be pointed out, however, that Steinmuller (1967) looked for complete acceptance of the graft in progeny and failed to find it. Since he did not have a concurrent control breeding programme, any delay in rejection time would have been overlooked.

Steele (1979) in his original monograph, also reworked the data on the inheritance of rabbit idiotipy (Kelus and Gell, 1968; Oudin and Michell, 1969; Eichmann and Kindt, 1971; Winfield *et al.*, 1972). He concluded that only in experiments when the antigen was given before breeding was the idiotypic pattern inherited, i.e. only by creating an acquired immune state before breeding was there evidence of inheritance of the B-cell specificity.

The phenomena described above are all associated with the immune system, but to have general application to the theory of evolution, they must at least be extended to tissues other than the immune system, especially those able to mount an adaptive response to the environment. In fact, this is important in a more general sense, since neo-Darwinists have argued that 'even if the Gorczynski–Steele data were correct, the effect is unlikely to be general' and by corollary to be of no significance in evolution (Maynard-Smith, 1980; Ruse, 1982). Thus, although maternal inheritance is a well-documented phenomenon, it is considered to be unlikely to be genetic (see, for example, Skolnick *et al.*, 1980), despite the lack of knowledge of the information transmitted (Deneberg and Rozenberg, 1967). In considering other experiments in support of the somatic selection hypothesis, therefore, we will continue to restrict our discussion, to effects bred through males, and likely to be genetic.

Okamoto (1965) experimentally induced diabetes in rats, rabbits, and guinea-pigs using a variety of diabetogenic drugs. Progeny from matings of affected males with normal females displayed many diabetic characters, including a reduction in the β-cells of the islets of Langerhans. In later generations (F_7) spontaneous diabetes developed in the untreated progeny. In another well-controlled study carried out over several years, animals treated with the diabetogenic drug, alloxan, were shown to transmit the resultant sub-diabetic state (abnormal glucose tolerance) to their offspring (Goldner and Spergel, 1972). This apparent inheritance of an 'acquired diabetic syndrome' was as efficient through the males as through the females and could be bred to the third generation with a progressive deterioration in glucose clearance rate and a reduction in the size and number of β-, but not α-cells. In these studies it is not possible to rule out the direct action of alloxan on the germline. The specificity of the response, however, argues against this possibility (see Campbell, 1982 for possible models). Also, Okamoto (1965) prevented the transmission of the diabetic syndrome by reversing the father's diabetes with insulin. This strongly implies an effect via somatic modification.

In another group of studies (see review by Bakke *et al.*, 1975), thyroid function was impaired by treatment of neonatal rats with large dosages of thyroxine (T4). This treatment resulted in abnormal development, impaired thyroid and pituitary development, and lower plasma levels of both protein-bound and free T4. Although in the majority of cases maternal effects on offspring were studied, in one group of experiments paternal transmission of abnormalities was demonstrated. These abnormalities included increased pituitary, ovary and thyroid weight. They tended to be mirror images of the parental effect, although in the reduced responsiveness to thyroid release hormone the offspring showed similar responses to their fathers. Bakke *et al.* (1975) interpreted this data as indicating an augmented secretion of thyroid stimulating hormone owing to an alteration in the thyostat set point in the offspring. There are other reports in the literature, but there is little point in enumerating them, since they are broadly similar to those described above and may be found in Campbell, 1982; Steele, 1981a; Pollard, 1984).

There are also a number of observations that strictly fall outside the bounds of the somatic selection hypothesis, but which clearly show environmental influences of genotype. The best documented of these is the effect of environment on the genotype in flax, described in detail in Chapter 8 by Cullis. In brief, different genotypes may be produced in flax by growing them under different conditions which range from a large to a small type. These types may be discriminated morphologically and by biochemical markers such as ribosomal DNA cistron number and isozyme patterns for peroxidase and acid phosphatases (Cullis, 1977, 1979). Interestingly, once the altered isozyme patterns have been established they are inherited as Mendelian characters (Cullis, 1979). The variation does not seem to reflect any somatic modification, but instead to be an environmentally induced increased rate of genetic variation.

However, before dismissing this evidence in support of the somatic selection hypothesis, it should be appreciated that only a few characters have been studied and therefore it cannot be excluded that there are other characters that are adaptive. Nevertheless, to date the evidence indicates the observed variation in DNA sequences is due to a direct effect of the environment on the germline, causing disproportionate replication of gene segments (Chapter 8).

9.4. THE VECTOR FOR SOMATICALLY DERIVED GENETIC INFORMATION

A plausible model for the inheritance of somatically acquired characters demands a vector capable of transducing the somatic information. Steele (1979) proposed that this vector may be a class of retroviruses directed at the germline and able to carry as passengers, RNA molecules specifying representative genetic information from the soma. Once the germline is infected, the viral or germline encoded reverse transcriptase would replicate the passenger RNA molecules to DNA sequences that are capable of becoming inserted into the correct genomic site by recombination. This model was an extension of Temin's proto- and provirus hypothesis (1971, 1976), in which the RNA copies of these proviruses may play an important role as mediators of genetic exchange between somatic cells. The view that virus-like agents may be pathological offspring of normal cellular process is growing for other infectious agents, e.g., for viroids (Diener, 1981) and for agents involved in Kuru and Scrapie (R. Kimberlin, personal communication).

A substantial amount of information has been gathered over the last few years pertinent to both the provirus and the somatic selection hypotheses. It is our intention to limit discussion here to these points and not extend the argument to other vectors or to a general speculation of the flexible genome. Such speculations may be found elsewhere (see Chapter 7; Pollard, 1984).

Sequence analysis of retroviral genomes have shown striking similarities between the terminal repeats of several different retroviruses and the terminal repeats of cellular transposable elements (Shimotohno *et al.*, 1980; Majors *et al.*, 1980; Temin, 1980). Furthermore, one group of movable genetic elements, the Alu group of dispersed repetitive elements appear to transpose via an RNA intermediate (Schmid and Jelinek, 1982). These similarities between transposons and retroviruses comprise a compelling argument for an evolutionary relationship between the two (Shimotohno *et al.*, 1980; Temin, 1980; Flavell and Ish-Horowicz, 1981). This relationship could be either the result of divergence or convergence (resemblance as a result of similar function), but Temin (1980) has argued strongly against the latter. Movable genetic elements have been demonstrated in bacteria, higher plants, yeasts, insects and vertebrates (Calos and Miller, 1980; see Chapter 7). Since the above organisms represent a reasonable span of the biological world, it seems that these elements will be ubiquitous. It is attractive to argue that they have a function in controlling normal cellular differentiation, e.g. control of yeast mating strains (Nasmyth and Tatchell, 1980; Pays *et al.*, 1983), both within cells and between cells. A role in development in higher organisms, however, has yet to be demonstrated (Calos and Miller, 1980) and it may be that these elements replicate parasitically (Orgel *et al.*, 1980). Retroviruses are also known to recombine with cellular genes (Baxt and Meinkoth, 1978) and this is thought to be the method of acquiring oncogenic potential (Weinberg, 1980) through picking up cellular oncogenes and transposing them to a new site under different regulatory controls. Since all mammalian cells appear to contain endogenous retroviruses and these are inherited, it is clearly possible that similar mechanisms for genes other than oncogenes may exist, and in fact this may be a normal cellular process involved in inter-cellular communication.

Retroviral particles also contain a variety of RNA molecules unrelated to the viral genome (Faras *et al.*, 1973; Ikawa *et al.*, 1974). These RNAs include a variety of small RNAs and mRNAs and appear representative of the RNA population of the cytoplasm through which the virus matures (Sawyer and Dahlberg, 1973; Ikawa *et al.*, 1974). Retrovirus-like particles have also been observed in *Drosophila* that contain RNA homologous to the transposable element, copia (Shiba and Saigo, 1983) and it may be that this is one method of transposition of these sequences (Flavell and Ish-Horowicz, 1983). Thus, in contrast to the covalent linkage of the genes to retroviral sequences, Steele (1979, 1981a) in his model focussed on the transport of non-covalently linked mRNA molecules within the viral particle. This was because covalent linkage would result in integration of the mutant somatic information next to the viral genome and therefore exclude the site-specific recombination at the normal genomic site and consequent gene correction demanded by his model. Furthermore, covalent linkage would lead to these 'passenger' genes being flanked by, and under the control of, viral long terminal repeat with the concomitant disruption of normal developmental controls. Steele, therefore, argued that in somatically selected cells the new genetic information could be expressed at high levels in the form of mRNA and thus may be captured by an endogenous retrovirus (Figure 2). These viruses will carry the mRNA to the germline where it will be reverse transcribed either by the viral or germline encoded enzyme into DNA which would integrate into the germline DNA by recombination. Although, as yet, there is no direct evidence for such a process, it is noteworthy that pseudo-genes (i.e., genes similar to functional genes but containing information that prevents functional expression) are thought to be derived from reverse transcription of mRNA molecules (Lemischka and Sharp, 1982; Reilly *et al.*, 1982; Hollis *et al.*, 1982). What is generally overlooked, however, is that the processed pseudo-gene is found in genomic DNA present in every cell, thus at some point these genes must have been introduced into the germline by a process involving reverse transcription. Since immunoglobulin and adult globin genes are not thought to be expressed in germ cells, this process may have involved a transduction of mRNA from the soma, possibly via a retroviral vector. Pseudo-genes, by definition, are functionally dead, but we would conjecture that they represent the error level of a normal cellular process that enables the establishment of somatic information in the germline. These pseudo-genes also presumably provide sites for recombination with functional genes, as well as sequences through which new genes may evolve. Recently, some authors (Jeffries, 1982; Busslinger *et al.*, 1982) have even argued for horizontal gene transfer between species which presumably involves a viral vector, perhaps a retrovirus. If this is the case, this obviously represents a genetic penetrance of Weismann's barrier and a possible extension of Steele's hypothesis.

The above discussion has restricted itself to the transfer and copying of mRNA sequences. It is pertinent to ask what type of information may be transferred, since in all cases of directed somatic transfer discussed above, the systems appear to involve complex cellular regulatory phenomena, e.g., tolerance to specific major histocompatibility antigens, or to complex endocrine control systems. Some of these systems also at the first sight are of the instructionalist type, specifically excluded by Steele's selectionist hypothesis, or by Weismann's tail-cutting experiment (1893). However, on closer examination, each is a regulatory system involving complex interactions between cells and thus it may be that the genetic information carried by the retroviruses are controlling elements rather than mRNAs for structural genes. It is clear that there can be little speculation about this point since little is known about the genetic basis of complex regulatory phenomena occurring during development. Good candidates may be small RNAs, whose complexity changes dramatically during development (Davidson and Britten, 1979), a sub-group of which are known to be involved in RNA splicing. It is interesting that several snRNA pseudogenes are present in the germline owing to reverse transcription of nuclear RNA (Van Arsdell *et al.*,

1981), but it would be unwise to speculate further until the molecular genetic basis of development is more fully understood.

9.5. FUTURE PATHS OF EXPERIMENTATION

Two paths of experimentation offer themselves in this area; the first is to track specific gene sequences using recombinant DNA technology and the second is to construct new animal strains by breeding somatically induced variations through the male line. One of the major criticisms of the Gorczynski–Steele experiments is that neonatal tolerance is poorly understood at the cellular level, let alone at the genetic level. In fact, some workers claim that somatic genetic variation of the type found in immunoglobulin production has no role to play in the establishment of neonatal tolerance. That being as it may, the definitive experiment to show the inheritance of acquired characters is to sequence a somatically altered gene and show that it is transmitted to the germline. This experiment is now technically feasible since foreign genes may be introduced with surprising ease into somatic tissues, embryos and oocytes by DNA transformation. Genes introduced into fertilized eggs or early stage developing embryos become integrated into the DNA of putative germ-cells and thus become established in the germline and become stably inherited (Jaenisch *et al.*, 1981; Wagner *et al.*, 1981; Constantini and Lacy, 1981) and in some cases expressed in the progeny (Wagner *et al.*, 1981b). Similarly, genes have been introduced into stem cells of the bone marrow of mice *in vitro* and reintroduced into an adult mouse (Cline *et al.*, 1980; Bar-Eli *et al.*, 1983). Thus, a tenable experiment to test the inheritance of acquired characters is to breed from such reconstituted male mice and test for the presence of the artificially introduced gene by sequence or restriction endonuclease analysis of progeny.

An alternative approach has been taken by E. J. Steele (unpublished observations and Steele, 1984) following the work of Biozzi and co-workers (1979). These investigators established high and low antibody responder mice by selection within a *randomly* bred strain of mice. Thus Steele immunized or 'tolerized' (see Figure 6) inbred CBA/H male mice with rat erythrocytes and mated them to normal females of the same strain. This is an example of the antigen before breeding strategy, and if the somatic selection hypothesis is correct, populations similar to those described by Biozzi *et al.* (1979) should appear within one generation even in inbred mouse strains. Biozzi *et al.* (1979) of course considered they were selecting variants already present in the randomly bred mouse population. E. J. Steele (unpublished observations) has observed a significant downward biphasic shift in the magnitude of antibody response in the progeny of male immune CBA/H mice compared to control mice from untreated parents (Figure 6). No systematic shift occurred in the progeny of tolerized male mice. While the mechanism(s) of this transmission phenomenon is currently under investigation, it is nevertheless clear that a population shift in the magnitude of the primary antibody response of inbred mice has occurred within

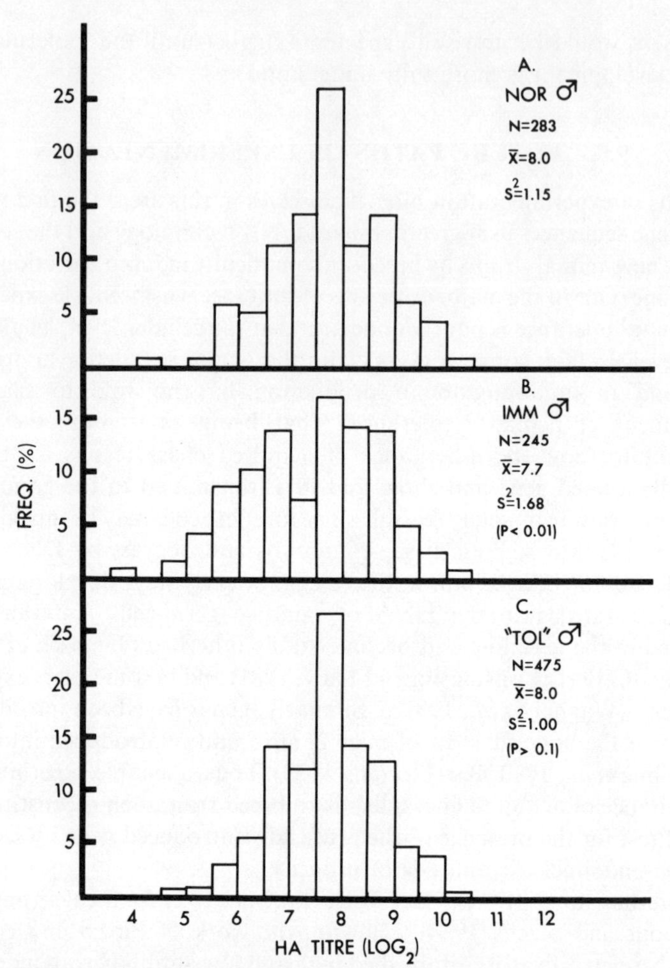

Figure 6. Frequency distributions of anti-rat erythrocyte haemagglutination (HA) titres in the sera of first generation offspring one week after intraperitoneal (i.p.) immunization with 1×10^8 rat erythrocytes.

Progeny mice of each type were coded and housed together after weaning (three weeks of age) in the same holding boxes and were challenged with antigen between one and a half and two months of age. All sera were titrated and HA titres recorded double blind. Each parental breeding cage contained two normal CBA/H females mated to either one normal CBA/H male (panel A; ten cages), one CBA/H male immunized in adult life with 1×10^8 rat erythrocytes i.p. (panel B; nine cages), or one CBA/H male exposed since birth to repeated injections (16–17 in total) of 5×10^8 rat erythrocytes (panel C; 14 cages). There was no significant difference between the responses of male and female offspring one generation by immunization of the male parent with antigen. The shift is of a similar magnitude to those observed by Biozzi *et al.* (1979) for random bred Swiss mice. Breeding of these low responders needs to be carried out to conclude the nature of the phenomenon, since it appears to be of the non-directional group of acquired phenotypes described above and may be related to a non-specific low responsive B-cell populations found in offspring of fathers tolerized to SRBC (see above). Nevertheless, these types of experiments will define the constraints on the inheritance of acquired characters and may facilitate genetic analysis of such phenomena.

9.6. CONCLUSION

There is a substantial amount of data in support of the somatic selection hypothesis including molecular evidence for the type of vector proposed and physiological evidence for the transmission of acquired phenotypes. A definitive molecular experiment showing that a somatically altered genetic sequence is transmitted to the germline, however, has yet to be performed. Such experiments, although difficult, are now technically feasible.

From an evolutionary standpoint evidence indicating that Weismann's barrier is permeable to exogenous genetic information is contrary to the neo-Darwinism theory and thus such proposals inevitably become the centre of controversy. However, they do offer new perspectives to age old problems in conventional evolutionary hypotheses. Firstly they provide a mechanism for directed change such that there is a feedback between the environment (via selection of the somatic information) and the germline coupled to a larger potential for genetic variability randomly generated in the soma during ontogeny and normal cellular regeneration. Secondly, since many of the examples of the inheritance of acquired characters described above involve complex control systems within disparate cellular populations, it provides a means of parallel evolution perhaps by co-selection of one somatic variant by another, or by common control elements that act intercellularly. Such mechanisms of inheritance of acquired characters provides a rationale for the genetic assimilation of altered developmental plans so eloquently described by Waddington (1975).

ACKNOWLEDGEMENTS

We thank Drs J. H. Campbell and R. V. Blanden for many helpful discussions and Dr H. Temin for critical comments on the manuscript.

9.7. REFERENCES

Atchison, M., and Adesnik, M. (1983) A cytochrome P-450 multigene family. Characterization of a gene activated by phenobarbital administration, *J. Biol. Chem.*, **258**, 11285–11295.

Bakke, J. L., Lawrence, N. L., Bennett, J., and Robinson, S. (1975) Endocrine syndrome produced by neonatal hyperthyroidism, hypothyroidism, or altered nutrition and effects seen in untreated progeny, in Fisher, D. A., and Burrow, G. H. (eds.) *Perinatal Thyroid Physiology and Disease*, Raven Press, New York, pp. 79–116.

Bar-Eli, M., Stang, H. D., Mercola, K. E., and Cline, M. J. (1983) Expression of methotrexate-resistant dihydrofolate reductase gene by transformed hematopoietic cells of mice, *Somatic Cell Genetics*, **9**, 55–67.

Baxt, W. G., and Meinkoth, J. L. (1978) Transfer of duck cell DNA sequences to the nucleus of 3T3 cells by Rous sarcoma virus, *Proc. Natl. Acad. Sci. USA*, **75**, 4252–4256.

Billingham, R. E., Brent, L., and Medawar, P. B. (1956) Quantitative studies on tissue transplantation immunity III: Actively acquired tolerance, *Phil. Trans. R. Soc. Lond. Ser. B*, **239**, 357–414.

Biozzi, G., Mouton, D., Sant'Anna, O. A., Possos, H. C., Gennasi, M., Reis, M. H., Ferreira, V. C. A., Heumann, A. M., Bouthillier, Y., Ibancz, O. M., Stiffel, C., and Jiqueira, M. (1979) Genetics of immunoresponsiveness to natural antigens in the mouse, *Current Topics in Microbiology and Immunology*, **85**, 31–98.

Brent, L. (1981) Lamarck and Immunity: the tables unturned, *New Scientist*, **90**, 493.

Brent, L., Chandler, P., Fiertz, W., Medawar, P. B. N., Rayfield, L. S., and Simpson, E. (1982) Further studies on supposed Lamarckian inheritance of immunological tolerance, *Nature*, **295**, 242–244.

Brent, L., Rayfield, L. S., Chandler, P., Fiertz, W., Medawar, P. B., and Simpson, E. (1981) Supposed Lamarckian inheritance of immunological tolerance, *Nature*, **290**, 508–512.

Burnet, F. M. (1959) *The Clonal Selection Theory of Acquired Immunity*, Cambridge University Press, London and New York.

Busslinger, M., Rusconi, S., and Birnsteil, M. L. (1982) An unusual evolutionary behaviour of a sea urchin histone gene cluster, *EMBO J.*, **1**, 27–33.

Calos, M. P., and Miller, J. H. (1980) Transposable elements, *Cell*, **20**, 579–595.

Campbell, J. H. (1982) Autoevolution, in Milkman, R. (ed.) *Perspectives on Evolution*, Sinauer Press, New York, pp. 190–201.

Cline, M. J., Stang, H. D., Mercola, K. E., Morse, L., Ruprecht, R., Browne, J., and Salser, W. (1980) Gene transfer in intact animals, *Nature*, **284**, 422–425.

Cohn, M., Langman, R., and Gleckeler, W. (1980) Diversity 1980, in Fougereau, M., and Dausset, J. (eds.) *Immunology 1980*, Academic Press, London and New York, pp. 153–201.

Coderre, J. A., Beverley, S. M., Schimke, R. T., and Santi, D. V. (1983) Overproduction of a bifunctional thymidylate synthetase-dihydrofolate reductase in methotrexate-resistant *Leishmania tropica*, *Proc. Natl. Acad. Sci. USA*, **80**, 2132–2136.

Constantini, F., and Lacy, E. (1981) Introduction of a rabbit β-globin gene into the mouse germline, *Nature*, **294**, 92–94.

Cullis, C. A. (1977) Molecular aspects of the environmental induction of heritable changes in flax, *Heredity*, **38**, 129–154.

Cullis, C. A. (1979) Segregation of the isozyme of flax genotrophs, *Biochemical-Genetics*, **17**, 391–401.

Davidson, E. H., and Britten, R. J. (1979) Regulation of gene expression: possible role of repetitive sequences, *Science*, **204**, 1052–1059.

Denenberg, V. H., and Rozenberg, K. M. (1967) Non-genetic transmission of information, *Nature*, **216**, 549–550.

Diener, T. O. (1981) Are viroids escaped introns?, *Proc. Natl. Acad. Sci. USA*, **78**, 5014–5015.

Eichmann, K., and Kindt, T. J. (1971) The inheritance of individual antigenic specificities of rabbit antibodies to streptococcal carbohydrates, *J. Exp. Med.*, **134**, 532–552.

Faras, A. J., Garapin, A. C., Levinson, W. E., Bishop, J. M., and Goodman, H. M. (1973) Characterization of the low molecular weight RNAs associated with the 70S RNA of Rous sarcoma virus, *J. Virol.*, **12**, 334–342.

Flavell, A. J., and Ish-Horowicz, D. E. (1981) Extrachromosomal circular copies of the eukaryotic transposable elements copia in cultured *Drosophila* cells, *Nature*, **292**, 591–595.

Flavell, A. J., and Ish-Horowicz, D. E. (1983) The origin of extrachromosomal circular copia elements, *Cell*, **34**, 415–419.

Goldner, M. G., and Spergel, G. (1972) On the transmission of alloxan diabetes and other diabetogenic influences, *Ad. Metabolic Diseases*, **6**, 57–72.

Gorczynski, R. M., Kennedy, M., Macrae, S., and Ciampi, A. (1983) A possible maternal effect in the abnormal hyporesponsiveness of specific alloantigens in offspring born to neonatally tolerant fathers, *J. Immunol.*, **131**, 1115–1120.

Gorczynski, R. M., and Steele, E. J. (1980) Inheritance of acquired immunological tolerance to foreign histocompatibility antigens in mice, *Proc. Natl. Acad. Sci. USA*, **77**, 2871–2875.

Gorczynski, R. M., and Steele, E. J. (1981) Simultaneous yet independent inheritance of somatically acquired tolerance to two distinct H-2 antigenic haplotype determinants in mice, *Nature*, **289**, 678–681.

Guttman, R. D., and Aust, J. B. (1963) A germplasm transmitted alteration of histocompatibility in the progeny of homograft tolerant mice, *Nature*, **197**, 1220–1221.

Guttman, R. D., Vosika, G. J., and Aust. J. B. (1964) Acceptance of allogenic tumour and skin grafts in progeny of a homograft tolerant male, *J. Natl. Cancer Inst.*, **33**, 1–5.

Guyer, M. F. (1928) *Being Well Born: An Introduction to Heredity and Eugenics*, Constable and Company, London.

Guyer, M. F., and Smith, E. A. (1920) Studies on cytolysins: II. Transmission of induced eye-defects, *J. Exp. Zool.*, **30**, 171–216.

Ho, M. W., and Saunders, P. T. (1979) Beyond Neo-Darwinism—An epigenetic approach to evolution, *J. Theor. Biol.*, **78**, 573–591.

Ho, M. W., and Saunders, P. T. (1984) (eds.) *Beyond Neo-Darwinism*, Academic Press, London (in press).

Hollis, G. F., Hieter, P. A., McBride, D., Swan, D., and Leder, P. (1982) Processed genes: A dispersed human immunoglobulin gene bearing evidence of RNA-type processing, *Nature*, **296**, 321–325.

Ikawa, Y., Ross, J., and Leder, P. (1974) An association between globin messenger RNA and 60S RNA derived from Friend leukaemia virus, *Proc. Natl. Acad. Sci. USA*, **71**, 1154–1158.

Jaenisch, R., Jahner, D., Nobis, P., Simon, I., Löhler, J., Harkers, K., and Grotkopp, D. (1981) Chromosomal position and activation of retroviral genomes inserted into the germline of mice, *Cell*, **24**, 519–529.

Jeffries, A. J. (1982) Evolution of globin genes, in Dover, G. A., and Flavell, R. B. (eds.) *Genome Evolution*, Academic Press, London, pp. 157–175.

Jerne, N. R. (1974) Towards a network theory of the immune system, *Ann. Immunol.* (*Inst. Pasteur*), **125c**, 373–389.

Josovic, J. (1982) *Lamarckian Inheritance—The Statistical Arguments*, M.Sc. Dissertation, City of London Polytechnic.

Kanazawa, K., and Imai, A. (1974) Parasexual–sexual hybridization. Heritable transformation of germ cells in chimeric mice, *Japan J. Exp. Med.*, **44**, 227–234.

Kelus, A. S., and Gell, P. G. H. (1968) Immunological analysis of rabbit anti-body systems, *J. Exp. Med.*, **127**, 215–234.

Lemischka, J., and Sharp, P. A. (1982) The sequences of an expressed rat alpha-tubulin gene and a pseudogene with an inserted repetitive element, *Nature*, **300**, 330–335.

Lewis, J. W., Davide, J. P., and Melera, P. W. (1982) Selective amplification of polymorphic dihydrofolate reductase gene loci in Chinese hamster lung cells, *Proc. Natl. Acad. Sci. USA*, **79**, 6961–6965.

Long, E. O., and Dawid, I. B. (1980) Repeated genes in eukaryotes, *Ann. Rev. Biochem.*, **49**, 727–764.

Majors, J. E., Swanstrom, R., Delorbe, W. J., Payne, G. S., Hughes, S. H., Ortiz, S., Quintrel, N., Bishop, J. M., and Varmus, H. E. (1980) DNA intermediates in the replication of retroviruses are structurally (and perhaps functionally) related to transposable elements, *Cold Spring Harbor Symp. Quant. Biol.*, **45**, 731–738.

Maynard-Smith, J. (1980) Regenerating Lamarck, *Times Literary Supplement*, October 18, p. 1185.

Mayr, E. (1959) Where are we?, *Cold Spring Harbor Symp. Quant. Biol.*, **24**, 1–14.

Medvedev, Z. A. (1981) On the immortality of the germline: Genetic and biochemical mechanisms. A review, *Mech. Ageing and Develop.*, **17**, 331–359.

Mullbacher, A., Ashman, R. B., and Blanden, R. V. (1983) Induction of T-cell hyporesponsiveness to Bebaru in mice, and abnormalities in the immune responses in progeny of hyporesponsive males, *Aust. J. Exp. Biol. Med. Sci.*, **61**, 187–191.

Nasmyth, K. A., and Tatchell, K. (1980) The structure of transposable yeast mating types, *Cell*, **19**, 753–764.

Nerbert, D. W. (1979) Multiple forms of inducible drug-metabolizing enzymes: A reasonable mechanism by which any organism can cope with adversity, *Mol. Cell. Biol.*, **27**, 27–46.

Okamoto, K. (1965) Apparent transmittance of factors to offspring by animals with experimental diabetes, in Leibel, B. S., and Wrenshall, G. A. (eds.) *On the Nature and Treatment of Diabetes*, Excerpta Medica Foundation, Amsterdam, pp. 627–637.

Orgel, L. E., Crick, F. H. C., and Sapienza, C. (1980) Selfish DNA, *Nature*, **288**, 645–646.

Oudin, J., and Michell, M. (1969) Idiotypy of rabbit antibodies I. Comparison of idiotypy of antibodies against Salmonella typhi with that of antibodies against other bacteria in the same rabbits, or of antibodies against Salmonella typhi in various rabbits, *J. Exp. Med.*, **130**, 595–617.

Pays, E., Assel, S. V., Laurent, M., Dorville, M., Vervoort, T., Van Meirvenne, N. V., and Steinert, M. (1983) Gene conversion as a mechanism for antigenic variation in Trypanosomes, *Cell*, **34**, 371–381.

Pollard, J. W. (1984) Is Weismann's Barrier Absolute? in Ho, M. W., and Saunders, P. (eds.) *Beyond Neo-Darwinism*, Academic Press, London and New York, pp. 291–314.

Reilly, J. G., Ogden, R., and Rossi, J. J. (1982) Isolation of a mouse pseudo tRNA gene encoding CCA—a possible example of reverse flow of genetic information, *Nature*, **300**, 287–289.

Ruse, M. (1982) *Darwinism Defended*, Reading, Addison-Wesley, New York.

Sawyer, R. C., and Dahlberg, J. E. (1973) Small RNAs of Rous sarcoma virus: Characterization by two-dimensional polyacrylamide gel electrophoresis and finger-print analysis, *J. Virol.*, **12**, 1226–1237.

Schimke, R. T., Kaufmann, R. J., Alt, F. W., and Kellems, R. F. (1978) Gene amplification and drug resistance in cultured murine cells, *Science*, **202**, 1051–1055.

Schmid, C. W., and Jelinek, W. R. (1982) The Alu family of dispersed repetitive sequences, *Science*, **216**, 1065–1070.

Shiba, T., and Saigo, K. (1983) Retrovirus-like particles containing RNA homologous to the transposable element copia in *Drosophila melanogaster*, *Nature*, **302**, 119–124.

Shimotohno, K., Mizutani, S., and Temin, H. M. (1980) Sequence of retrovirus provirus resembles that of bacterial transposable elements, *Nature*, **285**, 550–554.

Skolnick, N. J., Ackermann, S. H., Hofer, M. A., and Weiner, H. (1980) Vertical transmission of acquired ulcer susceptibility in the rat, *Science*, **208**, 1161–1163.

Smith, R. N. (1981) Inability of tolerant males to sire tolerant progeny, *Nature*, **292**, 767–768.
Steele, E. J. (1979) *Somatic Selection and Adaptive Evolution: On the Inheritance of Acquired Characters*, 1st Edn., Williams-Wallace Productions International Inc., Toronto.
Steele, E. J. (1981a) *Somatic Selection and Adaptive Evolution: On the Inheritance of Acquired Characters*, 2nd Edn., University of Chicago Press, Chicago.
Steele, E. J. (1981b) Lamarck and immunity: a conflict resolved, *New Scientist*, **89**, 360–361.
Steele, E. J. (1984) Acquired paternal influence in mice. II. Altered serum antibody response in the progeny population of immunized CBA/H males, *Aust. H. Exp. Biol. Med. Sci.* (In press).
Steele, E. J., and Gorczynski, R. M. (1981) Inheritance of acquired somatic modification of the immune system, in Hraba, T., and Hasek, M. (eds.), *Cellular and Molecular Mechanisms of Immunologic Tolerance*, Marcel Dekker Inc., New York, pp. 381–397.
Steinmuller, D. (1967) Behaviour of skin allografts in backcross progeny in tolerant males, *J. Natl. Cancer. Inst.*, **39**, 1247–1251.
Temin, H. M. (1971) The protovirus hypothesis: Speculations on the significance of RNA-directed DNA synthesis for normal development and for carcinogenesis, *J. Natl. Cancer. Inst.*, **46**, III–VII.
Temin, H. M. (1976) The DNA provirus hypothesis, *Science*, **192**, 1075–1080.
Temin, H. M. (1980) Origin of retroviruses from cellular moveable genetic elements, *Cell*, **21**, 599–600.
Tonegawa, S. (1983) Somatic generation of antibody diversity, *Nature*, **302**, 575–581.
Van Arsdell, S. W., Denizon, R. A., Bernstein, L. B., Weiner, A. M., Manser, T., and Gesteland, R. F. (1981) Direct repeats flank three small nuclear RNA pseudogenes in the human genome, *Cell*, **26**, 11–17.
Waddington, C. H. (1961) *The Nature of Life*, George Allen and Unwin, London.
Waddington, C. H. (1975) *Evolution of an Evolutionist*, Edinburgh University Press, Edinburgh.
Wagner, E. F., Stewart, T. A., and Mintz, B. (1981a) The human beta-globin gene and a functional viral thymidine kinase gene in developing mice, *Proc. Natl. Acad. Sci. USA*, **78**, 5016–5020.
Wagner, T. E., Hoppe, P. C., Jollick, J. D., Schill, D. R., Hodinka, R. L., and Gault, J. B. (1981b) Microinjection of a rabbit beta-globin gene into zygotes and its subsequent expression in adult mice and their offspring, *Proc. Natl. Acad. Sci. USA*, **78**, 6376–6380.
Wahl, G. M., Padgett, R. A., and Stark, G. R. (1979) Gene amplification causes overproduction of the first three enzymes of UMP synthesis in *N*-(phosphonoacetyl)-L-aspartate-resistant Hamster cells, *J. Biol. Chem.*, **254**, 8679–8689.
Weinberg, R. A. (1980) Origins and roles of endogenous retroviruses, *Cell*, **22**, 643–644.
Weismann, A. (1893) *The Germ Plasm: A Theory of Heredity*, Scott Publishing Co. Ltd, London, UK.
Winfield, J. B., Pincus, J. H., and Mage, R. G. (1972) Persistence and characterization of the idiotypes in pedigreed rabbits producing antibodies of restricted heterogeneity to Pneumococcal polysaccharides, *J. Immunol.*, **108**, 1278–1287.

Evolutionary Theory: Paths into the Future
Edited by J. W. Pollard

Chapter 10

Evolutionary epistemology*

KARL R. POPPER
London School of Economics & Political Science
University of London,
Houghton Street,
London WC2 2AE,
UK

10.1. INTRODUCTION

Epistemology is the English term for the theory of knowledge, especially of scientific knowledge. It is a theory that tries to explain the status of science and the growth of science. Donald Campbell called my epistemology 'evolutionary' because I look upon human language, upon human knowledge, and upon human science, as a product of biological evolution, especially of Darwinian evolution through natural selection.

I look upon the following as the main problems of evolutionary epistemology: the evolution of human language and the part it has played (and still plays) in the growth of human knowledge; the ideas of truth and falsity; the description of states of affairs; and the way states of affairs are picked out by language from the complexes of facts that constitute the world; that is, 'reality'.

To put this at once briefly and simply in the form of two theses:

First thesis: The specifically human ability to know, and also the ability to produce scientific knowledge, are the results of natural selection. They are closely connected with the evolution of a specifically human language. This first thesis is almost trivial. My second thesis is perhaps slightly less trivial.

Second thesis: The evolution of scientific knowledge is, in the main, the evolution of better and better theories. This is, again, a Darwinian process. The theories become better adapted through natural selection: they give us better and better information about reality. (They get nearer and nearer to the truth.) All organisms are problem solvers: problems arise together with life.

* This chapter is based on a public lecture given after the conference on *Open Questions on Quantum Physics*, in Bari, Italy, on May 7 1983.

We are always faced with practical problems; and out of these grow sometimes theoretical problems; for we try to solve some of our problems by proposing theories. In science these theories are highly competitive. We discuss them critically; we test them and we *eliminate* those theories which we judge to be less good in solving the problems which we wish to solve: so only the best theories, those which are most fit, survive in the struggle. This is the way science grows.

However, even the best theories are always our own inventions. They are beset by errors. What we do when we test our theories is this: we try to detect the errors which may be hidden in our theories. That is to say: we try to find the weak spots of our theories, the place where they break down. This is the critical method.

There is often much ingenuity needed in this critical testing process.

We can sum up the evolution of theories by the following diagram:

$$P_1 \rightarrow TT \rightarrow EE \rightarrow P_2$$

A problem (P_1) gives rise to attempts to solve it by tentative theories (TT). These are submitted to a critical process of error elimination (EE). The errors which we detect give rise to new problems (P_2). The distance between the old and the new problem is often very great: it indicates the progress made.

It is clear that this view of the progress of science is very similar to Darwin's view of natural selection by way of the elimination of the unfit: of the errors in the evolution of life, the errors in the attempts at *adaptation*, which is a trial and error process. Analogously, science works by trial (theory making) and by the elimination of the errors.

One may say: from the amoeba to Einstein there is only one step. Both work with the method of tentative trials (TT) and of error elimination (EE). *Where is the difference?*

The main difference between the amoeba and Einstein is not in the power of producing TT, tentative theories, but in EE, in the way of error elimination.

The amoeba is not aware of the process of EE. The main errors of the amoeba are eliminated by eliminating the amoeba: this is just natural selection.

As opposed to the amoeba, Einstein was aware of the need for EE: he criticized and tested his theories severely. (Einstein said that he produced and eliminated a theory every few minutes.)

What was it that enabled Einstein to go beyond the amoeba? The main thesis of this chapter will be the answer to this question.

Third thesis: What enables a human scientist like Einstein to go beyond the amoeba is the possession of what I call the *specifically human language*.

While the theories of the amoeba are part of the organism of the amoeba, Einstein could formulate his theories in language; if needed, in written language. He could in this way put his theories outside his organism. This enabled him to look upon a theory *as an object*; to look at it *critically*; to ask himself whether it can solve his problem, and whether it could possibly be true; and to eliminate it if he found that it could not stand up to this criticism.

It is *only* the specifically human language which can be used for this kind of purpose.

My three theses together give an outline of my evolutionary epistemology.

10.2. THE TRADITIONAL THEORY OF KNOWLEDGE

What is the usual approach to the theory of knowledge, to epistemology? It is totally different from my evolutionary approach which I have sketched in Section 10.1. The usual approach demands *justification* by *observation*. I reject both parts of this position.

It usually starts from a question like: 'how do we know?', and it usually takes this question to mean the same as: 'what kind of perception or observation is the justification of our assertions?' In other words, it is concerned with the justification of our assertions (in my way of putting it, of our theories); and it looks for this justification to our perceptions or to our observations. This epistemological approach may be called *observationism.*

Observationism assumes that our senses or our sense organs are the sources of our knowledge; that we are 'given' some so-called 'sense data' (a sense datum is something that is given to us by our senses) or some perceptions and that our 'knowledge' is the result of, or the digest of, these sense data, or of our perceptions, or of the information received. The place where the sense data are digested is, of course, the head, shown in Figure 1.

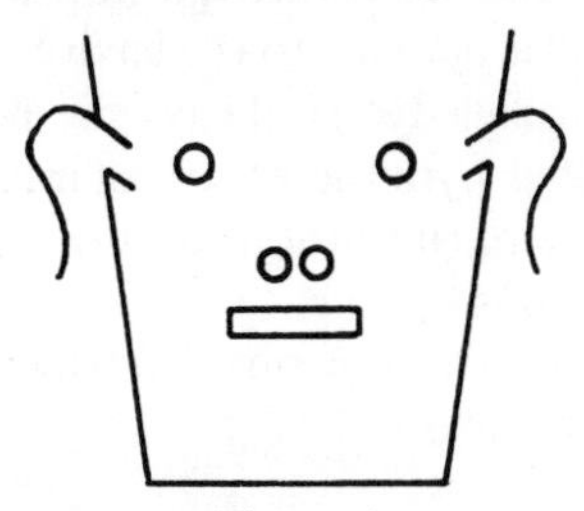

Figure 1

That is why I like to call observationism 'the bucket theory of the mind'.

The theory can also be put as follows. The sense data enter the bucket through the well-known seven holes, two eyes, two ears, one nose with two holes, the mouth; and also through our skin, the sense of touch. In the bucket, they are digested; or more especially, they are connected, associated, and classified. And then, from those which recur repeatedly, we obtain—by repetition, association, and by generalization and induction—our scientific theories.

The bucket theory, or observationism, is the standard theory of knowledge from Aristotle to some of my own contemporaries, for example, Bertrand Russell; or the great evolutionist J. B. S. Haldane or Rudolf Carnap.

But it is also the theory held by the man in the street.

The man in the street can put it very briefly: 'how do I know? Because I have had my eyes open, I have seen, and I have heard'. Carnap also identifies the question 'how do I know?' with 'what perceptions or observations are the sources of my knowledge?'.

These straightforward questions and answers by the man in the street give, of course, a reasonably true picture of the situation as he sees it. But it is not a view that may be elevated and transformed into a theory of knowledge that can be taken seriously.

Before proceeding to criticize the bucket theory of the human mind, I wish to mention that the opposition to it goes back to Greek times (Heraclitus, Xenophanes, Parmenides). Kant saw the problem fairly clearly; he stressed the difference between non-observational or *a priori* knowledge and observational or *a posteriori* knowledge. The idea that we may have *a priori* knowledge shocked many people; and the great ethologist and evolutionary epistemologist Konrad Lorenz proposed that the Kantian knowledge *a priori* may be knowledge that was, some thousands or millions of years ago, first acquired *a posteriori* (Lorenz, 1941), and that it became later genetically fixed by natural selection. However, in a book written between 1930 and 1932, and still available only in German (Popper, 1979; it is the book to which Donald Campbell referred when he described my theory of knowledge as 'evolutionary') I proposed that the *a priori* knowledge had never been *a posteriori*, and that, from the historical and genetical point of view, *all* our knowledge was the *invention* of animals and therefore *a priori* in its conception (although not, of course, *a priori* valid, in the Kantian sense). It becomes *adapted* to the environment by natural selection: the apparent *a posteriori* knowledge is always the result of the *elimination* of badly fitting *a priori* invented hypotheses, or adaptations. In other words, all knowledge is the result of *trial* (invention) and of the *elimination of error*, of badly fitting *a priori* inventions.

The trial and error method is the one by which we are actively seeking information from our environment.

10.3. CRITICISM

My *fourth thesis* (which I have been teaching and preaching for more than 60 years) is:

Every single aspect of the justificationist and observationist philosophy of knowledge is mistaken:

1. Sense data or similar experiences do not exist.
2. Associations do not exist.
3. Induction by repetition or generalization does not exist.
4. Our perceptions may mislead us.
5. Observationism or the bucket theory is a theory which says that knowledge

can stream from outside into the bucket through our sense organs. In fact, we, the organisms, are most active in our acquisition of knowledge—perhaps more active than in our acquisition of food. Information does not stream into us from the environment. Rather, it is we who explore the environment and suck information from it actively, like food. And humans are not only active but sometimes critical.

A famous experiment which refutes the bucket theory and especially the sense datum theory is due to Held and Hein (1963); it is reported in a book which Sir John Eccles and I wrote together (Popper and Eccles, 1977). It is the experiment of the active and the passive kitten. These two kittens are so linked that the active kitten causes the passive kitten to be moved in a kind of perambulator, passively, through exactly the same environment in which the normal kitten moves actively. So the passive kitten gets with great approximation the same perceptions as the active kitten. But tests of the kittens show that, while the active kitten learns a lot, the kitten kept passive has learned nothing.

The defenders of the observationist theory of knowledge could argue against this criticism that there is also a kinaesthetic sense, the sense of our movements, and that the absence of the kinaesthetic sense data in the sensory input of the passive kitten can explain, within the observationist theory, why the passive kitten learns nothing. What our experiment shows, the observationist might say, is nothing more than this: unless the kinaesthetic sense data are associated with the optical and acoustic sense data, the latter will not be useful.

In order to make my rejection of observationism or the bucket theory or sense datum theory independent of any such defence, I shall now introduce an argument which I regard as decisive. It is an argument that is typical for my evolutionary theory of knowledge.

The argument may be put as follows. The idea that theories are the digest of sense data or perceptions or observations *cannot be true* for the following reasons.

Theories are (and in fact all knowledge is) from the evolutionary point of view, part of our tentative *adaptations* to our environment. They are like expectations or anticipations. This is indeed their function: the biological function of all knowledge is to try to anticipate what will happen in our environment. Now our sense *organs*, for example our eyes, are also such adaptations. They are, seen in this light, theories: animal organisms have invented eyes, and evolved them in great detail, like an expectation or theory that light in the visible wavelength will be useful for extracting information from the environment; for sucking from the environment information that can be interpreted as indicating the *state* of the environment, both the long-term state and the short-term state.

Now our sense *organs*, obviously, are logically prior to our alleged sense *data* (even though there would be a feedback if sense data did exist; just as there may be a feedback from our perceptions upon the sense organs).

It is therefore impossible that all theories or theory-like entities are reached by induction or generalization from the alleged 'data', the allegedly 'given' influx of information, from perception or observation. For the sense organs that suck information from the environment are genetically, as well as logically, prior to this information. The camera, and its structure, comes before the photograph, and the organism, and its structure, comes before any information.

I think that this argument is decisive and that it leads to a new view of life.

10.4. LIFE AND THE ACQUISITION OF KNOWLEDGE

Life is usually characterized by the following powers or functions which largely depend on one another:

1. Procreation and heredity.
2. Growth.
3. Absorption and assimilation of food.
4. Sensitivity to stimuli.
   I suggest that this might be replaced by:
   a) Problem solving (problems which may emerge from the environmental or the internal situation of the organism). *All organisms are problem solvers.*
   b) Active exploration of the environment, often helped by random trial movements. (Even plants explore their environment.)
5. Construction of theories about the environment, in the form of physical organs or other changes of the anatomy or in the form of new behavioural repertoires, or changes to the existing behavioural repertoire.

All these functions originate in the organism. This is very important. They are all actions. They are not reactions to the environment.

This can also be put as follows. It is the organism, and the state in which it happens to be, that determines or chooses or selects what kind of changes in the environment may be significant for it, so that it may 'react' to them, as 'stimuli'.

One usually speaks of the stimulus which releases a reaction; and one usually implies that the environmental stimulus comes first and that it causes the release of the reaction of the organism. This leads to the (mistaken) interpretation that the stimulus is a bit of information, flowing into the organism from the outside, and, altogether, that the stimulus comes first; it is the cause, which comes before the response, the effect.

I think this is fundamentally mistaken.

It is a mistake due to a traditional model of physical causation that does not work with organisms, and that does not even work with motor cars or wireless sets; indeed, not with things that have some access to a source of energy which they can expend in *different* ways, and in different amounts.

Even a motor car or a wireless set *selects* according to its internal state the stimuli to which it responds. The motor car may not react properly to pressure on the accelerator if it is held still by its handbrake. And the wireless set will not be tempted by the most beautiful symphony unless it is tuned to the proper frequency.

The same holds for organisms; even more strongly, insofar as they have to be tuned in or programmed by themselves. They are tuned in, for example, by their gene structure, or by some hormone, or by the lack of food, or by curiosity, or by the hope of learning something interesting.

All this speaks very strongly against the bucket theory of the mind, which was often formulated by the following phrase: 'Nothing is in the intellect that was not before in the senses'—or in Latin: '*Nihil est in intellectu quid non antea fuerat in sensu*'. This is the slogan of observationism, of the bucket theory. Few people know of its prehistory. It stems from a contemptuous remark of the anti-observationist Parmenides who said something like this: 'There is just nothing at all in the erring intellects (*plakton noon* should be *plankton noon* in Parmenides, (Diels and Kranz, 1960)) of these people except what was previously in their much-erring (*polyplanktos*) sense organs'. (See my *Conjectures and Refutations*, pp. 410–413, of the Routledge & Kegan Paul editions from 1968 on.) I suggest that it may have been Protagoras who countered Parmenides by turning this into a proud slogan of observationism.

10.5. LANGUAGE

From these considerations we see the significance of the active and explorative behaviour of animals and men. And this insight in its turn is of great importance for the theory of evolution in general, not only for evolutionary epistemology. I must, however, come now to a central point of evolutionary epistemology—the evolutionary theory of the human language.

The most important contribution to the evolutionary theory of language known to me lies buried in a little paper published in 1918 by my former teacher Karl Bühler (1918). In this paper, too little regarded by modern students of linguistics, Bühler distinguishes three stages in the evolution of language (Table 1, to which I have added a fourth function). In each of these stages, language has a task, a certain biological function.

The lowest stage is the one where the only biological function of language is to *express* outwardly the inward state of an organism; perhaps by certain noises, or by certain gestures.

It is probably only for a comparatively short period that this is the *only* function of language. Very soon some other animals (either of the same species or of other species) will take notice of the *expressions* and *adapt* themselves to them: they discover how to suck information from them, how to include them into the stimuli of their environment to which they may respond with advantage. More especially, they may use the expression as a warning of some impending danger. For example, the roar of a lion which is a *self-expression* of the lion's inner state may be used, in this way, as a warning by the lion's possible prey. Or a cry of a

Table 1

		Functions	Values	
		(4) Argumentative function	Validity/ Invalidity	
		(3) Descriptive function	Falsity Truth	man
	Perhaps bees	(2) Signal function	Efficiency/ Inefficiency	
Animals, plants		(1) Expressive function	Revealing/ Not revealing	

goose expressing fear may be interpreted by other geese as a warning against a hawk; and a different cry as a warning against a fox. So the *expressions* may *release* in the receiving or responding animal a typical pre-formed reaction. The responding animal takes the *expression* as a *signal*, as a *sign* that releases a definite response. It thus comes in *communication* with the expressing animal.

At this stage, the original expressive function has changed. And what was originally an outward sign or symptom, although the expression of an inner state, has acquired a signal function or release function. It may now even be used by the expressor as a signal and may thus change its biological function from expressing to signalling, even to conscious signalling.

So we have now two evolutionary levels: *first*, the *pure expression*, and *second*, the expression which tends to become a *signal* since there will be receiving animals that will respond: or react to it as a *signal*: so we have *communication*.

Bühler's *third* evolutionary level is that of the human language. According to Bühler, human language and human language alone, introduces a revolutionary novelty: it can *describe*; that is, describe a state of affairs, a situation. It may be a state of affairs that is present at the instant of time at which the state of affairs is being described, such as: 'our friends are coming now'; or a state of affairs that has no present relevance, such as: 'my brother-in-law died 13 years ago', or a state of affairs that may have never taken place and may never take place, such as: 'far behind this mountain there is another mountain that is made of solid gold'.

Bühler calls the power of human language to describe possible or real states of affairs the '*descriptive function*' of human language. And he rightly stresses its immense importance.

Bühler shows that language never loses its expressive function. Even a highly unemotional description will retain some of it. Nor does language ever lose its signalling or communicative function. Even an uninteresting (and false) mathematical equation, such as '$10^5 = 1{,}000{,}000$' may perhaps provoke a correction and thus a reaction, even an angry emotional reaction, from a mathematician.

But neither the expressiveness nor the sign-character—the signal that produces the reaction—is characteristic of the human language; nor is it characteristic of it that it serves a family of organisms to communicate. The *descriptive character* is characteristic of the human species. And it is something new, and something truly revolutionary: *human language can convey information about a state of affairs, about a situation that may or may not be present or biologically relevant. It may not even exist.*

Bühler's simple and most important contribution has been neglected by almost all linguists. They still talk as if the essence of human language would be self-expression; or as if words like 'communication' or 'sign language', or 'symbol language' would sufficiently characterize the human language. (But signs or symbols are also used by other animals.)

Bühler of course never asserted that there are no further functions of human language: language may be used to beg, or to implore, or to admonish. It may be used for commanding, or for advising. It may be used to offend people or to hurt people, or to frighten them. And it may be used to comfort people or to make people feel loved and at ease. But on the human level, the descriptive language can be basic to all these uses.

10.6. HOW DID THE DESCRIPTIVE FUNCTION EVOLVE?

It is easily seen how the signalling function of language evolved once the expressive function had appeared. But it is very difficult to see how the descriptive function could have developed out of the signalling function. It must be admitted that the signalling function can come close to the descriptive function. A characteristic warning cry of a goose may mean 'hawk!' and another one may mean 'fox!'. And this is in many ways very close to a descriptive statement. 'A hawk is coming! hide!' or 'take to the air! a fox is coming!'. But there are great differences between such descriptive warning cries and human descriptive languages. These differences make it difficult to believe that the human descriptive languages evolved from warning cries, and other signals, such as war cries.

Admittedly, the dancing language of bees comes close to the human descriptive use of language. Bees can convey, by their dance, information about the direction and distance from the beehive concerning a place where food can be found and concerning the character of the food.

But there is a most important difference between the biological situations of the bee language and the human language: the descriptive information conveyed by the dancing bee is part of a signal for the other bees; and it is its main function to incite these other bees to an action that is useful here and now; the information conveyed is closely related to the present biological situation.

As opposed to this, the information conveyed by human language may not be immediately useful. It may not be useful at all, or useful only years later and in a totally different situation.

There is also a possible *playfulness* in the use of human descriptive language which makes it so different from warning cries or mating cries or from the bee language, which all serve in very serious biological situations. Natural selection might explain that the system of warning cries becomes richer, more differentiated. But one should expect that if this happens, it should also become more rigid. Human language, however, must have evolved by a process that combined a great increase in differentiation with a still greater increase in the *degrees of freedom* (the term 'degrees of freedom' may be taken either in the ordinary sense or in its mathematical or in its physical sense).

All this becomes clear if we look at one of the oldest ways of using human language: its use in story telling and in the invention of religious myths. Both these uses no doubt have functions that are biologically serious. But these functions are fairly remote from the situational urgency and the rigidity of warning cries.

It is, more especially, the rigidity of these biological signals (as we may call them) which creates the difficulty: it is difficult to conceive that the evolution of the biological signals may lead to the human language with its chatter, the variety of its uses, and its playfulness, on the one hand, and, on the other hand, its most serious biological functions, such as *its function in the acquisition of new knowledge*, such as the discovery of the use of fire.

However, there may be some ways out of this impasse, even though these ways out are purely speculative hypotheses. What I have to say are conjectures; but they may indicate how things may have happened in the evolution of language.

The *playfulness of young animals*, especially mammals, to which I wish to draw special attention, and its biological functions, raises tremendous problems, and some excellent books deal with this great subject. (See, for example, Baldwin (1895), Eigen and Winkler (1975), Groos (1896), Hochkeppel (1973), Lorenz (1973, 1977), and Morgan (1908).) The subject is too vast and important to go into here in any detail. But I conjecture that it may be the key to the problem of the evolution of freedom and of human language. I shall only refer to some recent discoveries which show the creativeness of the playfulness of young animals and its significance for new inventions. We can read about Japanese monkeys in Menzel (1965); for example, where he writes:

> 'It is usually juveniles rather than adults who are the originators of group adaptation processes and "pro-cultural" changes in relatively complex behaviours such as coming into a newly established feeding area, acquiring new food habits, or adopting new methods of collecting foods . . .' .

(See also, Frisch (1959), Itani (1958), Kawamura (1959), and Miyadi (1964).)

I suggest that the main phonetic apparatus of human language arises not from the closed system of warning calls or war cries and similar signals (systems that need to be rigid and may be genetically fixed), but rather from playful babbling and chatter of mothers with babies, and of gangs of children; and that the descriptive function of the human language—its use for describing states of affairs in the environment—may arise from make-believe plays, so-called 'representation plays' or 'imitation plays'; and especially from the play-acting of children trying to imitate playfully the behaviour of adults.

These imitation plays are widely established among many mammals: there are mock fights, mock war cries, also mock cries for help: and mock commands, impersonating certain adults. (This may lead to giving them names, possibly names intended to be descriptive.)

Play-acting may go with babble and chatter; and it may create the *need* for something like a descriptive or explanatory commentary. In this way a *need* for storytelling may develop, together with a situation in which the descriptive character of the story is clear from the beginning. Thus the human language, the descriptive language, may have been first invented by children playing or play-acting, perhaps as a secret gang language (they still invent such things sometimes). It may then have been taken over by the mothers (like the inventions of Japanese monkey children, see above), and only later, with modifications, by the adult males. (There are still some languages that contain grammatical forms indicating the sex of the speaker.) And out of the storytelling—or as part of it—and out of the description of a state of affairs may have developed *the explanatory story; the myth; and later the linguistically formulated explanatory theory.*

The *need* for the descriptive story, perhaps also for the prophecy, with its immense biological significance, may in time become genetically fixed. The tremendous superiority, especially in warfare, given by the possession of a descriptive language, creates a new selection pressure; and this, perhaps, explains the astonishingly rapid growth of the human brain.

It is unfortunate that it is hardly possible that a speculative conjecture like the foregoing could ever become *testable*. (It would not even be a test if we were to succeed in inducing the Japanese monkey children to do all I suggested above.) Yet even so, it has the advantage of telling us an explanatory story of how things *may* have happened—how a flexible and descriptive human language *may* have arisen: a descriptive language which from the start would have been open, capable of almost infinite development, stimulating the imagination, and leading to fairy tales, to myths, and to explanatory theories: to 'culture'.

I feel that I should draw attention here to the story of Helen Keller (see Popper and Eccles, 1977): it is one of the most interesting cases to show the inborn need of a child for actively acquiring the human language, and for its humanizing influence. We may conjecture that the need is encoded on the DNA, together with various other dispositions.

10.7. FROM THE AMOEBA TO EINSTEIN

Animals, and even plants, acquire new knowledge by the method of trial and error; or, more precisely, by the method of trying out certain active movements, certain *a priori* inventions, and the elimination of those which do not 'fit', which are not well-adapted. This holds for the amoeba (see H. S. Jennings, 1906); and it also holds for Einstein. Wherein lies the main difference between these two?

I think the elimination of errors works in different ways. In the case of the amoeba, any gross error may be eliminated by eliminating the amoeba. Clearly this does not hold for Einstein: he knows he will make mistakes and he is actively looking out for them. But it is not surprising to find that most men have inherited from the amoeba a strong aversion to making mistakes, and also to admitting that they have made them! However, there are exceptions: some people do not mind making mistakes as long as there is a chance of discovering them, and trying again if one is discovered. Einstein was one of these; and so are most creative scientists: in contradistinction to other organisms, human beings use the method of trial and error *consciously* (unless it has become 'second nature' to them). We have, it seems, two types of people: those who are under the spell of an inherited aversion to mistakes, and who therefore fear them and fear to admit them; and those who also wish to avoid mistakes, but know that we make mistakes more often than not, who have learned (by trial and error) that they may counter this by *actively searching for their own mistakes*. The first type of people are *thinking dogmatically*; the second are those *who have learned to think critically*. (By saying 'learned' I wish to convey my conjecture that the difference between the two types is not based on inheritance but on education.) I now come to my

Fifth thesis: in the evolution of man, the descriptive function of the human language has been the prerequisite for critical thinking: it is the descriptive function that makes critical thinking possible.

This important thesis can be established in various ways. Only a descriptive language of the kind described in the foregoing Section 6 raises *the problem of truth and falsity*: the problem whether or not some description corresponds to the facts. It is clear that the problem of truth precedes the evolution of critical thinking. Another argument is this. Prior to human descriptive language, all theories could be said to be part of the structure of organisms which were their carriers. They were either inherited organs, or inherited or acquired dispositions to behave, or inherited or acquired unconscious expectations. So they were part and parcel of their carrier.

But in order to make it possible to criticize a theory, the organism must be able *to regard the theory as an object*. The only way known to us to achieve this is to formulate the theory in a descriptive language and, preferably, in a written language.

In this way, our theories, our conjectures, the trials of our trial-and-error attempts, can become objects; objects like dead or living physical structures. They may become objects of a critical investigation. And they may be killed by us without killing their carriers. (Strangely enough, even critical thinkers frequently develop hostile feelings towards the carriers of theories they criticize.)

I may perhaps insert here a brief note on what I regard as a very minor problem:

whether or not the two types of people—the dogmatic thinkers and the critical thinkers—are hereditary types. As hinted above, I conjecture that they are not. My reason is, simply, that these 'types' are an invention. One may perhaps classify actual people according to this invented classification. But there is no reason to think that the classification is based on DNA—no more than there is reason to believe that liking or disliking golf is so based. (Or that what is called IQ measures 'intelligence': as pointed out by Peter Medawar, no competent agriculturist would dream of measuring the fertility of the soil by a measure depending on *one variable*. But some psychologists seem to believe that 'intelligence', which involves creativity, could be so measured.)

10.8. THREE WORLDS

Human language is, I suggest, the product of human inventiveness. It is a product of the human mind, of our mental experiences and dispositions. And the human mind, in its turn, is the product of its products: its dispositions are due to a feedback effect. One particularly important feedback effect, mentioned above, would be the disposition to invent arguments, to *give reasons* for accepting a story as true, or for rejecting it as false. Another very important feedback effect is the invention of the sequence of natural numbers.

First comes the dual and the plural: one, two, many. Then come the numbers up to 5; then come the numbers up to 10 and to 20. And then comes the invention of the principle that we can extend any series of numbers by adding one: the successor principle, the principle of forming a successor to every numeral.

Each such step is a *linguistic* innovation, an invention. The innovation is linguistic, and it is totally different from counting (for example, a shepherd carving one mark on a stick for each sheep that goes past). Each such step changes our mind—our mental picture of the world, our consciousness.

Thus there is feedback, or interaction, between our language and our mind. And with the growth of our language and of our mind, we can see more of our world. Language works like a searchlight: just as a searchlight picks up a plane, language may bring into focus certain aspects, certain states of affairs which it describes from the continuum of facts. Thus language not only interacts with our mind, it helps us to see things and possibilities, which we would never have seen without it. I suggest that early inventions such as the igniting and controlling of fire and, very much later, the invention of the wheel (unknown to certain highly cultured peoples) were made with the help of language: they were made possible (in the case of fire) by identifying very dissimilar situations. Without a language, only biological situations to which we *react* in the same way (food, dangers, etc.) can be identified.

There is at least one good argument in favour of the conjecture that descriptive language is much older than the control of fire: if deprived of language, children are scarcely human. Deprivation of language has even a physical effect on them, possibly worse than the deprivation of some vitamin, to say nothing of the devastating mental effect. Children deprived of language are mentally abnormal. But in a mild climate, nobody is dehumanized by being deprived of fire.

In fact, learning a language and learning to walk upright seem to be the only skills whose acquisitions are vital to us and, no doubt, are genetically based; and both of these are eagerly acquired by small children, largely on their own initiative, in almost any social setting. Learning a language is also a tremendous intellectual achievement. And it is one that all normal children master; probably because it is deeply needed by them (a fact that may be used as an argument against the doctrine that there are physically normal children of very low innate intelligence).

About twenty years ago I introduced a theory that divides the universe into three sub-universes which I called World 1, World 2, and World 3.

World 1 is the world of all physical bodies and forces, and fields of forces; also of organisms, of our own bodies and their parts, our brains, and all physical, chemical, and biological processes within living bodies.

World 2 I call the world of our mind: of conscious experiences of our thoughts, of our feelings of elation or depression, of our aims, of our plans of action.

World 3 I call the world of the products of the human mind, and especially the world of our human language; of our stories, our myths, our explanatory theories; the world of our mathematical and physical theories, and of our technologies and of our biological and medical theories. But beyond this, also the world of human creation in art, in architecture and in music—the world of all those products of our minds, which, I suggest, could never have arisen without human language.

World 3 may be called the world of culture. But my theory, which is highly conjectural, stresses the central role played by descriptive language in human culture. World 3 comprises all books, all libraries, all theories, including, of course, false theories, and even inconsistent theories. And it attributes a central role to the ideas of truth and of falsity.

As indicated before, the human mind lives and grows in interaction with its products. It is greatly influenced by the feedback from the objects or inmates of World 3. And World 3, in its turn, consists largely of physical objects, such as books and buildings and sculptures.

Books, buildings, and sculptures, which are products of the human mind, are, of course, not only inmates of World 3 but also inmates of World 1. But in World 3 there are also symphonies, and mathematical proofs, and theories. And symphonies, proofs, and theories are strangely abstract objects: Beethoven's Ninth is not identical with his manuscript (which may get burned without the Ninth getting burned) or with any or all of the printed copies, or records, or with any or all of its performances. Nor is it identical with human experiences or memories. The situation is analogous to Euclid's proof of the prime number theorem and to Newton's theory of gravitation.

The objects that constitute World 3 are highly diverse. There are marble sculptures like those of Michelangelo. These are not only material, physical bodies, but unique physical bodies. The status of painting, of architectural works of art, and of manuscripts of music is somewhat similar; and so is even the status of rare copies of printed books. But as a rule, the status of a book as a World 3 object is utterly different. If I ask a physics student whether he knows Newton's theory of gravity, then I do not refer to a material book and certainly not to a unique physical body, but to the objective *content* of Newton's thought or, rather, to the objective content of his writings. And I do not refer to Newton's actual thought processes which of course belong to World 2, but to something far more abstract; to something that belongs to World 3, and which was developed by Newton by a critical process of improving upon it, again and again, at different periods of his life.

It is difficult to make this quite clear, but it is very important. The main problem is the status of a statement: and the logical relations between statements, or more precisely between the logical *contents* of statements.

Now all the purely logical relations between statements, such as contradictoriness, compatibility, deducibility (the relation of logical consequence) are World 3 relations. They are relations holding *only* between World 3 objects. They are, decidedly, *not* psychological World 2 relations: they hold independently of whether anybody has ever thought about them, or has ever believed that they hold. On the other hand, they can very easily be 'grasped': they can be easily understood; we can, mentally, think it all out, in World 2, and we may experience that a deducibility relation holds and that it is trivially convincing; and this is a World 2 experience. Of course, with difficult theories, such as mathematical or physical theories, it may even happen that we grasp them, understand them, without at the same time being convinced that they are true.

Our minds, belonging to World 2, are thus capable of standing in close contact with World 3 objects. Yet World 2 objects—our subjective experiences—should be clearly distinguished from the objective World 3 statements, theories, conjectures, and also open problems.

I have spoken before about the interaction between World 2 and World 3, and I will illustrate this by another arithmetical example. The sequence of natural numbers, 1, 2, 3 . . . is a human invention. As I emphasized before, it is a *linguistic invention*, as opposed to the invention of counting. Spoken and perhaps written languages co-operated in inventing and perfecting the system of natural numbers. But we do not invent the distinction between odd numbers and even numbers: we *discover* this within the World 3 object—the sequence of natural numbers—which we have produced, or invented. Similarly we discover that there are divisible numbers and prime numbers. And we discover that the prime numbers are at first very frequent (up to 7, even the majority) 2, 3, 5, 7, 11, 13 . . .; but after 13 they slowly become rarer and rarer. These are facts which we have not made, but which are unintended and unforeseeable and inescapable consequences of the invention of the sequence of natural numbers. They are objective facts of World 3. That they are unforeseeable may become clear when I point out that there are open problems here. For example, we have found that primes sometimes come in pairs, such as 11 and 13, or 17 and 19, or 29 and 31. These are called twin primes, and they become very rare when we proceed to large numbers.

But in spite of much research, we do not know yet whether twin primes fizzle out completely or whether they turn up again and again; or in other words, we do not know yet whether or not there exists a greatest pair of twin primes. (The so-called twin prime conjecture proposes that no such greatest pair exists; or in other words, that the sequence of twin primes is infinite.)

Thus there are open problems in World 3; and we work on discovering these open problems, and on attempts to solve them. This shows very clearly the objectivity of World 3, and the way World 2 and World 3 interact: not only can World 2 work on discovering and solving World 3 problems, but World 3 can act upon World 2 (and through World 2 on World 1).

We can distinguish between the almost always conjectural knowledge in the World 3 sense—knowledge in the objective sense—and the World 2 knowledge that is, information carried by us in our heads: knowledge in the subjective sense. The distinction between knowledge in the subjective sense (or World 2 sense) and knowledge in the objective sense (World 3 knowledge: formulated knowledge, for example, in books, or stored in computers, and possibly unknown to any person) is of the greatest importance. What we call 'science', and what we try to advance, is, in the first instance, true *knowledge in the objective sense*. But it is most important, of course, that knowledge in the subjective sense should also spread—together with the knowledge how little we know.

The incredible thing about the human mind, about life, evolution, and mental growth is the interaction, the feedback, the give and take, between World 2 and World 3; between our mental growth and the growth of the objective World 3 which is the result of our endeavours, and our talents, our gifts, and which helps us to transcend ourselves.

It is this self-transcendence which appears to me to be the most important fact of all life and of all evolution: in our interaction with World 3 we can learn, and thanks to our invention of language, our fallible human minds can grow into lights that illuminate the universe.

10.9. REFERENCES

Baldwin, J. M. (1895) *Mental Development in the Child and in the Race*, MacMillan and Co., New York.

Bühler, K. (1918) Kritische Musterung der neueren Theorien des Satzes, *Indogermanisches Jahrbuch*, **6**, 1–20.

Diels, H., and Kranz, W. (1964) *Fragmente der Vorsokratiker*, Weidmann, Dublin and Zürich.

Eigen, M., and Winkler, R. (1975) *Das Spiel*, R. Piper and Co., Verlag, München.
Frisch, J. E. (1959) Research on Primate Behavior in Japan, *American Anthropologist*, **61**, 584–596.
Groos, K. (1896) *Die Spiele der Thiere*, Verlag von Gustav Fischer, Jena.
Held, R., and Hein, A. (1963) Movement produced stimulation in the development of visually guided behaviour, *Journal of Comparative Physiological Psychology*, **56**, 872–876.
Hochkeppel, W. (1973) *Denken als Spiel*, Deutscher Taschenbuch Verlag, München.
Itani, J. (1958) On the acquisition and propagation of a new food habit in the natural group of the Japanese monkey at Takasakiyama, *Primates*, **1**, 84–98.
Jennings, H. E. (1906) *The Behaviour of the Lower Organisms*, Columbia University Press, New York.
Kawamura, S. (1959) The process of sub-culture propagation among Japanese macaques, *Primates*, **2**, 43–60.
Lorenz, K. Z. (1941) Kants Lehre vom Apriorischen im Lichte gegenwärtiger Biologie, *Blätter f. Dt. Philos.*, **15**, 1941. New impression in *Das Wirkungsgefüge und das Schicksal des Menschen*, Serie Piper 309 (1983).
Lorenz, K. Z. (1973) *Die Rückseite des Spiegels*, Piper, München.
Lorenz, K. Z. (1977) *Behind the Mirror*, Methuen, London.
Lorenz, K. Z. (1978) *Vergleichende Verhaltungsforschung, Grundlagen der Ethologie*, Springer Verlag, Wien/New York.
Menzel, E. W. (1966) Responsiveness to objects in free-ranging Japanese monkeys, *Behaviour*, **26**, 130–150.
Miyadi, D. (1964) Social life of Japanese monkeys, *Science*, **143**, 783–786.
Morgan, C. (1908) *Animal Behaviour*, Edward Arnold, London.
Popper, K. R. (1934) *Logik der Forschung*, Julius Springer, Vienna; 8th edition (1984) J. C. B. Mohr (Paul Siebeck), Tübingen; also (1983) *The Logic of Scientific Discovery*, 11th impression, Hutchinson, London.
Popper, K. R. (1963, 1981) *Conjectures and Refutations*, Routledge and Kegan Paul, London.
Popper, K. R. (1979, written 1930–32) *Die beiden Grundprobleme der Erkenntnistheorie*, J. C. B. Mohr (Paul Siebeck), Tübingen.
Popper, K. R., and Eccles, J. C. (1977) *The Self and Its Brain*, Springer International, Berlin, Heidelberg, London, New York, pp. 404–405. Now also as a paperback at Routledge and Kegan Paul, London (1984).

Author Index

Subject Index